Zoosporic Plant Pathogens

Zoosporic Plant Pathogens

A Modern Perspective

Edited by

S. T. BUCZACKI

National Vegetable Research Station
Wellesbourne, Warwick, UK

1983

ACADEMIC PRESS

A Subsidiary of Harcourt Brace Jovanovich, Publishers

London   New York
Paris   San Diego   San Francisco   São Paulo
Sydney   Tokyo   Toronto

ACADEMIC PRESS INC. (LONDON) LTD.
24/28 Oval Road
London NW1 7DX

United States Edition published by
ACADEMIC PRESS INC.
111 Fifth Avenue
New York, New York 10003

British Library Cataloguing in Publication Data
Zoosporic plant pathogens.
1. Spores (Botany)
I. Buczacki, S. T.
581.4'6   QK662
ISBN 0-12-139180-9
LCCCN 82-72596

Typeset by Eta Services (Typesetters) Ltd., Beccles, Suffolk
and printed in Great Britain by Galliard Printers Ltd., Great Yarmouth, Norfolk

Contributors

D. J. S. Barr Biosystematics Research Institute, Agriculture Canada, Central Experimental Farm, Ottawa, Ontario KIA OC6, Canada.

G. C. A. Bruin Department of Environmental Biology, University of Guelph, Guelph, Ontario, Canada.

S. T. Buczacki National Vegetable Research Station, Wellesbourne, Warwick CV35 9EF, UK.

W. D. Campbell Department of Plant Pathology and Plant Genetics, University of Georgia, Athens, Georgia 30601, USA.

M. W. Dick Department of Botany, University of Reading, Reading, Berks RG6 2AH, UK.

D. P. Dylewski Department of Botany, Ohio University, Athens, Ohio 45701, USA.

L. V. Edgington Department of Environmental Biology, University of Guelph, Guelph, Ontario, Canada.

F. F. Hendrix Department of Plant Pathology and Plant Genetics, University of Georgia, Athens, Georgia 30601, USA.

Lene Lange Danish Government Institute of Seed Pathology for Developing Countries, 2900 Hellerup, Copenhagen, Denmark.

C. E. Miller Department of Botany, Ohio University, Athens, Ohio 45701, USA.

L. W. Olson Institute of Genetics, University of Copenhagen, 1353 Copenhagen A, Denmark.

D. S. Shaw School of Plant Biology, University of North Wales, Bangor, Gwynedd LL57 2UW, UK.

D. S. Teakle Department of Microbiology, University of Queensland, St. Lucia, Brisbane, Queensland 4067, Australia.

Preface

A zoospore is a unicellular propagating or disseminating body possessing one or two flagella and having limited powers of mobility. It is a feature of what are commonly considered relatively primitive organisms—the "lower" fungi and algae, as well as slime moulds and similar groups and some protozoans. One reason for the conventional association of the zoospore with primitiveness is its inevitable requirement for free water in which to move, for it is generally considered that organisms less advanced in the evolutionary sense are aquatic rather than terrestrial. Zoospores have fascinated biologists for over 150 years but the interest all began with Benedict Prévost, at first sight an improbable individual, more usually associated with his classic treatise on bunt in wheat. Nonetheless, whilst we shall never know who first set eyes upon a fungal zoospore, it was Prévost who first recorded and published his observation (Prévost, 1807). Satisfyingly, the spores he observed were almost certainly those of a plant pathogen, *Albugo portulacae* D.C. (Kunze) (Ainsworth, 1976), although the actual term spore was not widely accepted at that time and in Prévost's account, his bodies, even if he recognised their possible functions, were merely "globules". Forty-five years elapsed before fungal zoospores were recorded again and this time, almost inevitably, the observer was Anton de Bary (de Bary, 1852). Subsequently, the nineteenth century mycological literature became studded with contributions on zoospores and zoosporic fungi by the likes of de Bary himself, Cornu, Dangeard, Woronin and Zopf. In the present century a number of further milestones stand out: Coker (1923) monographed the Saprolegniaceae; Mathews (1931), the Pythiaceae; Karling (1942), the Plasmodiophorales and Sparrow (1943), in one of a number of eminent contributions on the subject, all those "aquatic phycomycetes" not covered in earlier texts. Whilst Mathews and Karling in particular were concerned in fairly large measure with plant pathogenic organisms, there has hitherto been no overall attempt to view zoosporic plant pathogens as a group and to examine the role and importance of the zoospore itself as a common feature. This book is an attempt to fill that void. In no sense, however, is it intended or should it be viewed as either monographic or definitive. It presents a series of essays by internationally recognised authorities on various aspects of the taxonomy, structure, biology, pathogenicity and control of zoosporic plant pathogens. The contributors, all of whom are still actively concerned with the problems that their subjects present, were invited to stand back from their research and, in effect, to set down on paper, not only their results and observations, but also their thoughts and ideas. They were invited to speculate as freely as they felt appropriate. Some did so more than others with the result that the book presents a blend of essays ranging from the reporting of novel results through the considered review of particular fields, to speculation that, if not quite heretical, certainly puts forward views counter to most modern orthodoxy. If the book stimulates discussion or even argument, it will have fulfilled one purpose. If it stimulates further interest and research into a group of organisms of quite dazzling variety and quite stunning importance and intractability, it will have an even more useful, practical role.

Although, inevitably, the book is concerned primarily with fungi, its title was chosen quite deliberately to enable zoosporic plant pathogenic algae, protozoa and slime moulds to be included. A few non-zoosporic, plant pathogenic genera, such as *Bremia*, which belong to predominantly zoosporic groups are referred to when appropriate as, inevitably, are certain saprophytic members of predominantly pathogenic groupings. Nonetheless, in as far as it is meaningful, it is the importance of the zoospore in relation to plant diseases with which this work is concerned.

The classification of zoosporic pathogens is considered by Barr in Chapter 2, and throughout the book, his preferred scheme, outlined in Appendix 3 is generally followed. One name may be unfamiliar; Hyphochytriomycetes is used instead of Hyphochytridiomycetes and a validation for this change appears as Appendix 1, together with an explanation for it.

January 1983   S.T.B.

Contents

List of contributors   v

Preface   vii

1   The Fungal Zoospore—Its Structure and Biological Significance
     Lene Lange and L. W. Olson   1

2   The Zoosporic Grouping of Plant Pathogens—Entity or Non-Entity?
     D. J. S. Barr   43

3   The Peronosporales—A Fungal Geneticist's Nightmare
     D. S. Shaw   85

4   Some Pythiaceous Fungi—New Roles for Old Organisms
     F. F. Hendrix Jr. and W. D. Campbell   123

5   *Plasmodiophora*—An Inter-Relationship Between Biological and Practical Problems
     S. T. Buczacki   161

6   The Chemical Control of Diseases Caused by Zoosporic Fungi—A Many-Sided Problem
     G. C. A. Bruin and L. V. Edgington   193

7   Zoosporic Fungi and Viruses—Double Trouble
     D. S. Teakle   233

8   Zoosporic Fungal Pathogens of Lower Plants—What Can be Learned From the Likes of *Woronina*?
     C. E. Miller and D. P. Dylewski   249

Appendix I   Validation of the Class Name Hyphochytriomycetes
     M. W. Dick   285

Appendix II   Some Methods for Studying Zoosporic Plant Pathogenic Fungi
     Lene Lange and L. W. Olson   287

Appendix III   The Genera of Phytopathogenic Zoosporic Fungi
        D. J. S. Barr   293

References   303

Index   331

1

The Fungal Zoospore

Its Structure and Biological Significance

LENE LANGE and LAURITZ W. OLSON

Institute of Seed Pathology for Developing Countries, Hellerup, Denmark and Institute of Genetics, University of Copenhagen, Copenhagen, Denmark

I. Introduction	2
II. Zoospore Ultrastructure	3
A. Technical Difficulties in Working with Fungal Parasites.	3
B. Zoospore Morphology in Relation to Its Ultrastructure.	5
C. Location and Shape of the Zoospore Nucleus	9
D. Ribosome Arrangement	13
E. Type and Organisation of the Mitochondria	15
F. Lipid Bodies, Microbodies and the Microbody–Lipid Body Complex	16
G. Zoospore Microtubule Systems	17
H. Rhizoplasts, Rootlets and Other Kinetosome-Associated Characters	20
I. Vesicles and Vacuoles	21
J. Ultrastructural Zoospore Types	22
III. The Use of Ultrastructural Studies	25
A. What Can We Learn from Electron Microscope Studies of Zoosporic Fungi?	25
B. Future, Applied Perspectives of Ultrastructural Studies of Zoosporic Fungi	27
IV. The Biological Significance of the Fungal Zoospore	28
A. Requirements for Free Water	28
B. Spread of Zoosporic Plant Pathogens	30
C. Rapidity of Zoosporogenesis and the Length of the Motile Period	31
D. Epidemiological Patterns	31
E. Tactic Response in Zoosporic Fungi	34
F. Zoospores as Vectors for Plant Viruses	37
V. Grouping of Zoosporic Plant Pathogens According to Zoospore Behaviour	38
VI. Exclusion of Zoosporic Fungi from Plant Pathological Tests	41

ZOOSPORIC PLANT PATHOGENS
ISBN 0 12 139180 9

I. Introduction

Just what is it that makes the trained mycologist recognise a fungal zoospore among all the other free-swimming, unicellular organisms in a biological community, just as one recognises a face in a crowd? Is it the way in which the zoospore moves—the characteristic long-wave movement of the bi-flagellates, the energetic swimming motion of the chytrids, which rapidly change direction every second or so with a sudden jerking action, or the more steamboat-like motion of the blastocladian spore? Does this characteristic type of swimming motion set the motile spore off from the rather undetermined, unsteady, swimming pattern found in so many protozoan and other unicellular animals and algae? Or is it the somewhat opalescent, uninuclear fungal cytoplasm, the flagellation of the fungal spore, or the characteristic refractile droplet found in many uniflagellate spores?

In spite of some similarities it is unfortunately not possible to give a single, complete description of the appearance of a fungal zoospore, for this the group is simply too diverse. This diversity, as observed in the light microscope (which is most striking when a living, well-aereated mount is studied in phase contrast), is seen in differences in the ultrastructural organisation among the groups. Significantly different patterns in the cytoplasmic organisation have been observed, not only among the major groupings of the fungal zoospores (i.e. the anteriorly uniflagellate, the posteriorly uniflagellate, and the biflagellate fungi (Fuller, 1966)), but different types of cytoplasmic organisation have also been observed within the individual groups (Lange and Olson, 1979). In recent years, much emphasis has been put on recognition of these differences; and in the future even more emphasis, in the evolutionary, phylogenetic, and taxonomic fields, may come to be focused on them, for the structural organisation of the zoospore is the most stable characteristic of these fungi.

The essential features of the fungal zoospore are as follows: the zoospore is formed inside a sporangium (with the final cleavage taking place either inside or outside it); the zoospore is discharged from the sporangium to become a free-swimming stage; after a motile period, the length of which is influenced by environmental conditions, the zoospore encysts on a suitable substrate or host. In this way the fungal zoospore is the propagule for the spread of the organism. But the key position of the zoospore in the life cycle of the plant parasitic species is more striking when one considers that it is the specific or non-specific movement of these zoospores that enables the plant parasite to locate a new host plant. And, for a plant parasitic fungus, encystment must be coupled with a highly sophisticated penetration process by which the parasite finally reaches its goal—the cytoplasm of the host plant cell.

The term zoospore has been used above to cover the unicellular, motile portion of the fungal life cycle. However, to be more specific and more correct, the term must be defined further and differentiated from other, related terms: (1) true zoospores are formed in a zoosporangium, by mitotic nuclear divisions, and directly give rise to a vegetative thallus; (2) meiospores are formed as the result of a meiotic nuclear division in a meiosporangium which often represents the resting stage of the fungus (e.g. in *Physoderma maydis* Shaw (Lange and Olson, 1980b)); (3) gametes are formed in gametangia and must undergo cell fusion before being able to grow; (4) zygotes are usually short-lived and result from fusion of gametes. Among these four different types of motile fungal spores (which later in the chapter will be referred to jointly as zoospores), the true zoospores, meiospores and zygotes may all initiate infection in the host plant whereas the gametes are not directly infectious (although there was a suggestion (Ingram and Tommerup, 1972) that secondary infection by *Plasmodiophora brassicae* Woronin might be initiated by gametes). Moreover, while gametes and meiospores are haploid, and zygotes are always diploid, zoospores may be either haploid or diploid. In the plant parasitic species of the Peronosporales the infectious motile spore stage will be diploid if the site of meiosis is assumed to be the same as that observed in other similar but non-parasitic organisms. Much less is known about ploidy in the uniflagellate fungal plant parasites. In spite of many thorough studies, neither the ploidy level nor the correct designation of the infectious motile spore are known for the genera *Olpidium* and *Synchytrium*; no conclusive evidence has been presented for the fusion of gametes or for the site of meiosis in the life cycle of these organisms. However, more is known for *Physoderma*, and in *P. maydis* the most important of the flagellated spore stages is the haploid meiospore which is formed in the thick-walled resting sporangium (Lange and Olson, 1980b).

II. Zoospore Ultrastructure

A. TECHNICAL DIFFICULTIES IN WORKING WITH FUNGAL PARASITES

The zoospores of each group of the aquatic fungi possess a unique combination of ultrastructural characteristics (Fuller, 1966; Colhoun, 1966: Aist and Williams, 1971; Lange and Olson, 1979) with the flagellation of the zoospore (posteriorly uniflagellate, anteriorly uniflagellate, heterokont biflagellate or isokont biflagellate) defining the major groups. The flagellum itself with the $9 + 2$ arrangement of microtubules in the axoneme is probably the most conservative structure in the biological kingdom in the phylogenetic sense.

At the present time our knowledge of the ultrastructural organisation of the obligate plant parasitic fungi is at a very primitive level, while our knowledge of the saprophytic and facultative parasitic fungi has reached a fairly advanced state. This is, however, quite understandable when the following practical difficulties are taken into account.

For ultrastructural studies certain technical demands are placed on the material to be examined. The amount of material available at a given time must be sufficient to sustain the inevitable losses which occur during the preparation of the material for electron microscopy. The material must be relatively free of other organic material; ultra-thin sectioning becomes almost unbearable if it is necessary to search, solely to locate the cell of interest. The material must also be completely free of inorganic particles which may not only make sectioning very difficult, but may ruin the glass or diamond knife.

The fulfilment of the above requirements is obviously more easily achieved when working with fungi which can be grown in the laboratory in pure culture on a well defined artificial medium. And the task is even easier if it is possible to manipulate the organism so that zoospore production can be synchronised.

When working with obligate plant parasites one is left with an almost hopeless set of problems. Where is the fungus located in its host? How do you find the fungus in the host? And how do you relate the ultrastructure of the fungus and its host to what one sees in the light microscope?

For obligate plant parasites the formation of zoospores, and the encystment and host penetration process must be studied in situ in host tissue. To examine the ultrastructure of the zoospore it may be necessary to examine zoosporangia in the host itself after spore cleavage but prior to release. When trying to determine the optimal fixation procedure for such material one must tolerate an undesirable compromise. A fixation procedure which is strong enough to reach the fungus in the plant cells will almost inevitably be too strong a treatment for the delicate zoosporic cytoplasm. The high aesthetic standards of electron microscopists have not encouraged their willingness to work with the obligate plant parasites.

But why take the time and effort to tame obligate parasites to "behave" in the laboratory so that electron microscope studies can be made? Could not close, saprophytic, relatives of these fungi be used and the experience from such studies extrapolated to cover their parasitic kin? On the basis of our ultrastructural knowledge of the zoosporic fungi the answer would have to be no. It appears to be a general phenomenon (at least in the uniflagellate group), that the ultrastructural criteria used to characterise and group these fungi simply do not function for the parasitic species; the obligate parasites represent the exception to the rule (Barr, 1980a,b; Lange and Olson, 1979). A possible explanation for this could be that the evolution of an obligately parasitic life cycle, infecting higher plants, had led to such specialisation that the organism has diverged structurally from its non-parasitic relatives.

B. ZOOSPORE MORPHOLOGY IN RELATION TO ITS ULTRASTRUCTURE

Ultrastructural aspects of the fungal zoospore influence also its gross morphology. The conspicuous refractile globule typical of the chytrid zoospore and shown by ultrastructural and cytochemical studies to be a storage body of lipoid nature, is an obvious example of this phenomenon. Even more striking is the relationship between zoospore shape, as observed in the light microscope, and the cytoskeletal organisation of the cytoplasmic microtubules: among the orders included in the Chytridiomycetes, the most definite zoospore shape (here triangular in section or pear-shaped) is found in the Blastocladiales while a small, ovoid to spherical, zoospore is typical for the Chytridiales. This observation is correlated with the fact that a basket-like arrangement of 9-triplet microtubules is typical of the spores of the Blastocladiales, while in general a rather disorganised system of cytoplasmic microtubules is found in the zoospores of the Chytridiales (Lange and Olson, 1979). A similar comparison can be made between the primary zoospore and the secondary zoospore of the orders of the biflagellate Phycomycetes. The secondary biflagellate zoospore characteristically has a lateral groove, in the middle of which the heterokont flagella are attached; this lateral groove is found to have a very elaborate cytoskeletal structure composed of up to eight systems of cytoplasmic microtubules (Holloway and Heath, 1977a,b; and here Fig. 11).

Further details about the importance of the organisation of the cytoplasmic microtubules in the fungal zoospore will be dealt with in Section II, p. 17; but in this context it should be noted that it appears from recent studies that the cytoplasmic microtubules also play an important, governing role, in the cleavage of the zoospores during zoosporogenesis (Heath, 1976; Olson *et al.*, 1981).

In spite of the increased knowledge achieved during the last decades of research into zoospore ultrastructure, it still holds true that the structure 'par excellence' for separation of the groups of aquatic fungi is the type of flagellation (Sparrow, 1960). The division of the aquatic fungi into the posteriorly uniflagellates, the anteriorly uniflagellates, the heterokont biflagellates, and biflagellates having two flagella, both of the whiplash type (i.e. isokont), is still the most basic and most valid means of separating the zoosporic fungi. However, the attempt to recognise another flagellar character, its length, as a taxonomic criterion (Powell and Koch, 1977) is apparently not justifiable (Lange and Olson, 1979). Estimates of flagellar length seem to vary, often because of different methods of measurement, but it is remarkable that in the great majority of zoosporic groups (uniflagellate as well as biflagellate) the flagellar length is close to 20 μm.

Another example of a morphological character which cannot be used for

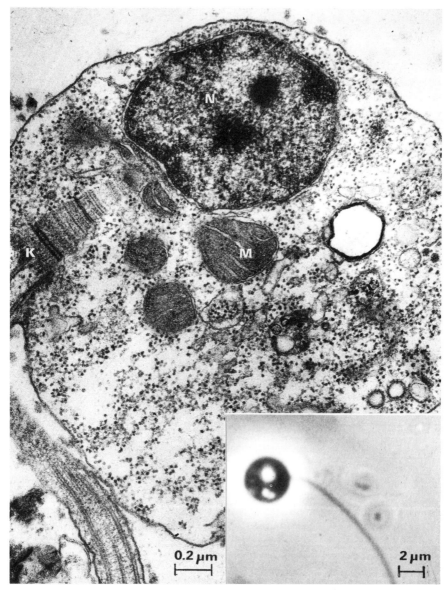

Fig. 1. Longitudinal section of a zoospore of *Olpidium brassicae*, showing the general arrangement of the organelles. Abbreviations: N, nucleus; M, mitochondrion; K, kinetosomes. Inset: phase-contrast view of a zoospore of *O. brassicae*.

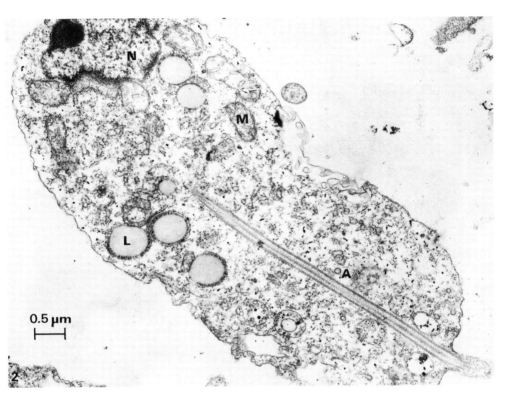

Fig. 2. Longitudinal section of a zoospore of *Olpidium radicale*, showing the general arrangement of the organelles. Abbreviations: N, nucleus; M, mitochondrion; L, lipid body; A, axoneme.

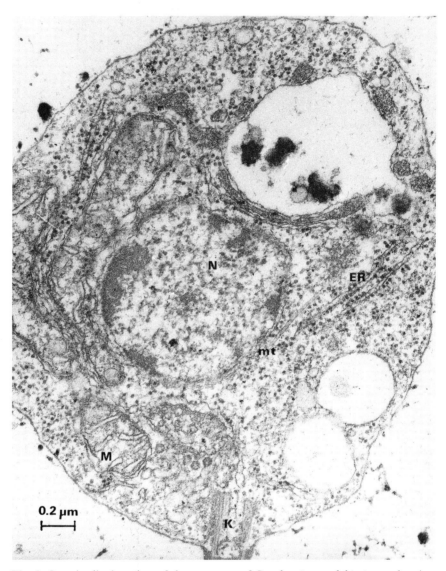

Fig. 3. Longitudinal section of the zoospore of *Synchytrium endobioticum*, showing the general arrangement of the organelles. Abbreviations: N, nucleus; M, mitochondrion; K, kinetosome; ER, endoplasmic reticulum; mt, cytoplasmic microtubules.

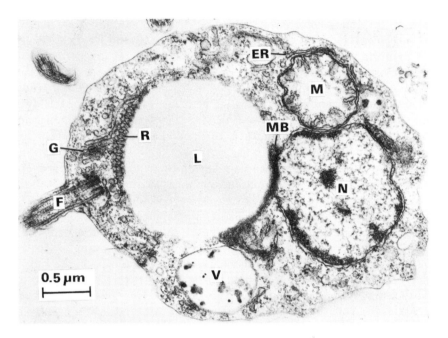

Fig. 4. Longitudinal median section of a zoospore of *Synchytrium macrosporum.*
Courtesy of C. M. Montecillo, C. E. Bracker and M. J. Powell. Abbreviations: N,
nucleus; M, mitochondrion; L, lipid body; MB, microbody; R, rumposome; G,
dictyosome; ER, endoplasmic reticulum; V, vacuole.

the taxonomic or biological grouping of the aquatic fungi is the size of the
zoospore. The diameter of a fungal zoospore is found to vary with its
physiological condition, to differ between rather closely related species, and
to be similar in species which are only very distantly related, if related at all.
There is, however, a tendency to find the larger sized zoospores among the
biflagellates (e.g. *Pythium* 6–16 µm, *Phytophthora* 9–13 µm, *Sclerophthora*
9–12 µm) and the smallest zoospores among the uniflagellates, especially
among the Chytridiales (e.g. *Synchytricum endobioticum* (Schilberszky)
Percival *c.* 3 µm, see Fig. 3; *Olpidium brassicae* (Woronin) Dang. *c.* 4 µm, see
Fig. 1): an exception among the Chytridiales is the giant zoospore of
Olpidium radicale, Schwartz and Cook measuring *c.* 7–8 µm in length, see
Fig. 2).

C. LOCATION AND SHAPE OF THE ZOOSPORE NUCLEUS

For certain groups of zoosporic fungi, the shape and location of the nucleus
is one of the most conspicuous ultrastructural features. In the primary,
biflagellate zoospore the nucleus is pyriform with the beaked portion

adjacent to the kinetosomes; and it is located anteriorly in the cell (*Saprolegnia*, Heath and Greenwood, 1971; Holloway and Heath, 1977a,b). The pyriform shape and the close association between the nuclear beak and the kinetosomes are also found in the secondary kidney-shaped biflagellate zoospore but here the nucleus is located in the middle of the lateral groove (*Pythium*: Lunney and Bland, 1976; and here, Fig. 6; *Saprolegnia*: Holloway and Heath, 1977a,b; and here, Fig. 11; *Phytophthora*: Bimpong and Hickman, 1975; Grove and Bracker, 1978; and here, Fig. 7; *Aphanomyces*: Hoch and Mitchell, 1972).

The close association between flagellar attachment, zoospore shape, and the shape and location of the nucleus is emphasised by the fact that in the kidney-shaped primary zoospore of *Lagenidium* (Bland and Amerson, 1973) and *Lagenisma* (Schnepf *et al.*, 1978a,b)—which, in spite of the shape of the zoospore, is interpreted to be of the primary zoospore type (Schnepf *et al.*, 1978a,b)—the shape and location of the nucleus are similar to the organisation found in the true secondary biflagellate zoospore (see Figs. 6 and 7). The shape of the nucleus in the zoospores of species belonging to the Blastocladiales is exceptional and very characteristic of the order: the nucleus is triangular in longitudinal section and surrounded by a basket-like arrangement of microtubules. It is located posteriorly, i.e. with the pointed

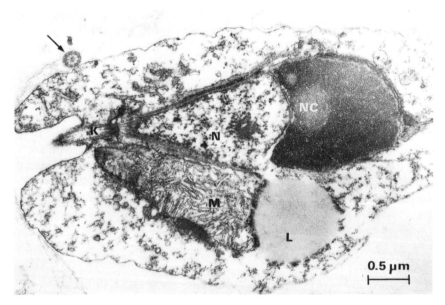

Fig. 5. Longitudinal section of a zoospore of *Physoderma maydis*, showing the general arrangement of the organelles. Abbreviations: N, nucleus; NC, nuclear cap; M, mitochondrion; L, lipid body; K, kinetosome; arrow, flagellum in cross section.

end of the nucleus adjacent to the kinetosomes (e.g. *Physoderma*, Olson and Lange, 1978; Lange and Olson, 1979; and here, Fig. 5).

In the Chytridiales, the Monoblepharidales and the Hyphochytriales the shape and location of the nucleus is less distinct. The nucleus is spherical to spheroid and more or less centrally located in the zoospore body but strict anterior or posterior placement of the nucleus has also been described.

A connection between the nucleus and the kinetosomes appears to be a common feature of almost all zoospores described. This character is obvious in the ultrastructural organisation found in the primary and the secondary biflagellate zoospore and the blastocladian zoospore, but such an association may also exist for spores of the Chytridiales, Monoblepharidales and the

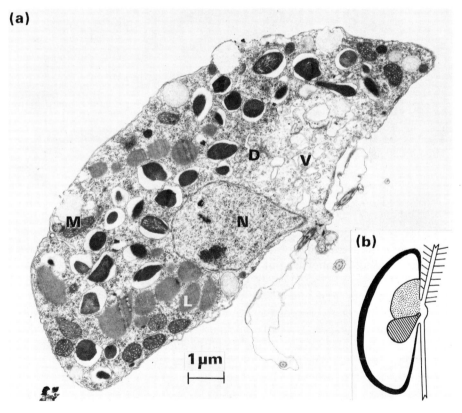

(a)

(b)

Fig. 6. (a) Longitudinal section through the groove region of the biflagellate zoospore of *Pythium aphanidermatum*. By courtesy of S. N. Grove and C. E. Bracker. Abbreviations: N, nucleus; M, mitochondrion; L, lipid body; V, water expulsion vacuole. (b) Schematic drawing of the compartmentalisation of the cytoplasm of the zoospore of *P. aphanidermatum*. By courtesy of S. N. Grove and C. E. Bracker.

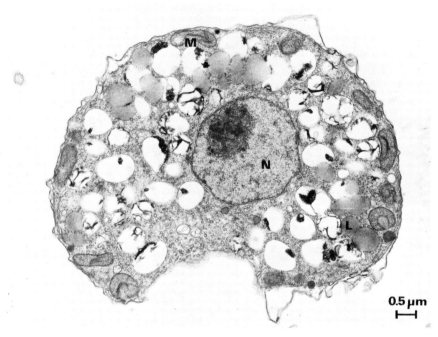

Fig. 7. Transverse section through the groove region of the zoospore of *Phytophthora infestans*. Courtesy of L. Chang and M. D. Coffey. Abbreviations: N, nucleus; M, mitochondrion; L, lipid body.

Hyphochytriales. In these three orders the association of the nucleus with the kinetosomes, if it is not located adjacent to them, is brought about by a system of cytoplasmic microtubules which originate from the vicinity of the functional kinetosome, flare out from it and terminate in the vicinity of the nucleus (Lange and Olson, 1979). Other, more conspicuous, ways in which the nucleus and kinetosome are structurally associated have also been described; the most impressive structural association is found in the extensive, cone-shaped striated rhizoplast in the zoospore of *Olpidium brassicae* (Lange and Olson, 1976a,b) and in *Rhizophlyctis rosea* de Bary & Woronin (Barr and Hartmann, 1977); other structural associations have been reported in *O. radicale* (Lange and Olson, 1978a), where the axoneme traverses the elongated zoospore body and in *Phlyctochytrium arcticum* Barr where a solid rhizoplast is found to be composed of three electron dense strands (Barr, 1970a,b).

From this pattern of observations it is tempting to make a general conclusion on the necessity of a connection between the functional kinetosome(s) and the nucleus. However, there are exceptions. In the

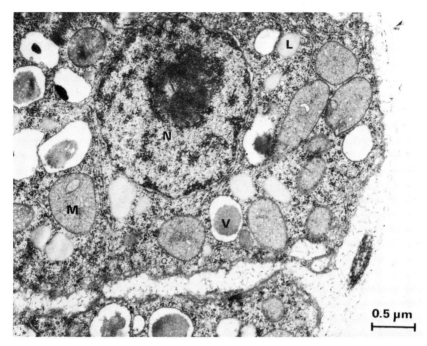

Fig. 8. Transverse section of the zoospore of *Sclerospora graminicola* while still enclosed in the zoosporangium. Abbreviations: N, nucleus; M, mitochondrion; L, lipid body; V, finger-print vacuole. Below, portion of the neighbouring zoospore.

zoospore of *Synchytrium macrosporum* Karling (see Fig. 4) a single prominent lipid body is present between the nucleus and the kinetosomes (Montecillo *et al.*, 1980) and in *Rozella allomycis* Foust one large spheroidal mitochondrion is located in a characteristic spatial arrangement between the nucleus and the striated rhizoplast (Held, 1975). Regarding the necessity for a structural association between the nucleus and the kinetosome(s) therefore, our knowledge is still insufficient.

D. RIBOSOME ARRANGEMENT

In the uniflagellate fungal zoospore four major types of ribosomal arrangement have been described (Lange and Olson, 1979). The lowest degree of organisation, where the ribosomes are evenly distributed in the zoospore body, is found in *Olpidium* spp. and *Rozella allomycis* (Olpidiaceae), in *Synchytrium endobioticum* (Synchytriaceae), and in the species of the Phlyctidiaceae which Barr (1980a,b) proposed should be grouped in the new family Spizellomycetaceae.

In the anteriorly uniflagellate zoospore the ribosomes are loosely aggregated, being confined to the central area of the zoospore body. The ribosome containing area is not delimited by a rigid organelle arrangement or by a bounding membrane.

A slightly higher level of ribosomal arrangement, where the ribosomes are aggregated in a so-called nuclear cap area, is found in zoospores of *Chytridiomyces* spp., *Chytridium* spp., *Rhizophydium* spp. and a few species belonging to the genus *Phlyctochytrium* (i.e. all species included in the revised Chytridiaceae *sensu* Barr, 1980a), in the polycentric genera *Cladochytrium* and *Nowakowskiella* (Lucarotti, personal communication), and in species belonging to the Monoblepharidales. In zoospores having a nuclear cap area (*sensu* Lange and Olson, 1979) the aggregation of ribosomes is partially delimited by membranes and/or organelles.

The presence of a true nuclear cap, in which the ribosome aggregation is completely bounded by a nuclear cap envelope, is an important and very specific characteristic of all members of the Blastocladiales. This order includes several obligate and facultative parasites of which the plant pathogenic species are included in the genus *Physoderma* (Sparrow, 1960; Lange and Olson, 1980a).

No structure which could be compared to the nuclear cap area or the true nuclear cap has been described for any of the biflagellate zoospores studied so far. There is, however, a tendency for the compartmentalisation of ribosomes in the cytoplasm which may resemble the ribosome arrangement found in the anteriorly uniflagellate zoospore. In the biflagellate spore this compartmentalisation is most clearly described from the zoospores of *Pythium aphanidermatum* (see Fig. 6 from Grove and Bracker, 1978). In the peripheral zone of the zoospore and the groove region few or no ribosomes are observed whereas they are evenly dispersed in the rest of the zoospore body. A similar arrangement of the ribosomes is found in other secondary, biflagellate zoospores of *Pythium proliferum* (Lunney and Bland, 1976), *Phytophthora nicotianae* Van Breda Haan var. *parasitica* (Dastur) Waterhouse (Reichle, 1969), *Phytophthora palmivora* (Bimpong and Hickman, 1975) and *Aphanomyces euteiches* (Hoch and Mitchell, 1972). This ribosomal arrangement appears to be less distinct in the reniform, but primary zoospores of *Lagenidium* (Bland and Amerson, 1973) and *Lagenisma* (Schnepf *et al.*, 1978a,b).

When dealing with the ribosomal arrangement found in fungal zoospores, the results of Murphy and Lovett (1966) in their work with *Blastocladiella* should be noted. They observed that the packaging of the ribosomes into the nuclear cap coincided with cessation of protein synthesis. And the presence of a nuclear cap or a nuclear cap area seems to be strictly associated with the motile zoospore stage. It is one of the last structures to be formed during zoosporogenesis in *Physoderma* (Lange and Olson, 1980b) and it is lost upon encystment (Olson, 1973; Lovett, 1975).

The ratio of the cytoplasmic ribosomes that are found to be associated with the endoplasmic reticulum (in rough endoplasmic reticulum (RER)) to those that have free ribosomes varies considerably. However, at least one or a few configurations of RER appear to be constantly present in most motile fungal spores.

E. TYPE AND ORGANISATION OF THE MITOCHONDRIA

The typical mitochondrion of a posteriorly uniflagellate zoospore has few, long wing-like cristae in an electron opaque matrix; while the typical mitochondrion found in the anteriorly uniflagellate zoospore and in all biflagellate zoospores (primary as well as secondary) is characterised by having closely packed swollen and undulating tubular cristae. The latter type of mitochondria may have an almost reticulate appearance in cross section (see Figs. 6–8 and 17).

The various spatial organisations of the mitochondria of the uniflagellate zoospore are most important and conspicuous ultrastructural characters of this group. Lange and Olson (1979) described five different levels of mitochondrial disposition ranging from numerous evenly dispersed mito-chondria, which occur in zoospores with a low level of ultrastructural organisation, to the single mitochondrion of the blastocladian zoospore which forms part of a very complex organellar arrangement called the side body complex.

The mitochondria found in the biflagellate zoospores are characteristically many in number, somewhat irregular in shape and apparently random in distribution within the zoospore body. As such, the mitochondrial organisa-tion in the biflagellate zoospore may be considered to be at a very low level. However, other less conspicuous structural associations may be present; features which cannot be distinguished in random ultra-thin sections. Here, however, it should be noted that the apparently random distribution of several mitochondria in the zoospore of *Olpidium brassicae*, was shown upon serial section reconstructions to represent one or two elaborately branched mitochondria (Lange and Olson, 1976b).

Astonishingly little is known about the formation of the mitochondria during zoosporogenesis. Does each spore primordium regularly receive a single mitochondrion initial which later either divides or remains as a single entity or are the mitochondria formed by a coalescence of numerous smaller mitochondria?

Among the various groups of posteriorly uniflagellate zoospores a tendency for packing the mitochondria close to the kinetosomal region is obvious. Heath (1976) made a tentative interpretation of this phenomenon from a functional point of view saying that this organisation pattern ensures that the energy generating system is placed close to the main source of energy utilisation during motility, *viz.* the flagellum.

F. LIPID BODIES, MICROBODIES AND THE MICROBODY–LIPID BODY
COMPLEX

The zoospore (*sensu lato*) is a propagative structure and in order to fulfil its function—active spread and initiation of the new generation—an energy resource, in the form of storage material, is obligatory. So far no zoosporic fungus has been described which does not contain lipid or lipidiaceous storage material in its zoospore.

The composition (i.e. saturated or unsaturated) and the packing of the lipid in the zoospore varies considerably. The refractile globule, classically considered to be an important taxonomic character in the uniflagellate chytrid zoospore, may, by ultrastructural studies, prove to be composed of several closely packed lipid bodies (Lange and Olson, 1979). The biflagellate spore has, in general, numerous lipid bodies, more or less peripherally located in the zoospore body. The intimate association between the single mitochondrion and the lipid body or bodies found in what is believed to be a highly advanced type of zoospore organisation in the Blastocladiales, appears functionally to be a very logical association. The main energy source is thus placed adjacent to the respiratory centre (Heath, 1976).

In the last decade, microbodies and catalase and glyoxylate enzymic activities associated with these organelles have attracted considerable interest (Vigil, 1973; Maxwell *et al.*, 1975). However, many of the descriptions of the occurrence of microbodies in the zoosporic fungi are based exclusively on the ultrastructural appearance of these organelles and enzymic studies have not been made to demonstrate conclusively that the observed structures are associated with the essential enzymic activity. Moreover, the ultrastructural appearance of what is termed a "microbody" varies greatly; that found among the biflagellate spores is most striking in that a prominent inclusion body is found in its centre, while the microbody typical of the uniflagellate zoospore has a homogeneous granular appearance.

The significance of the structural association of the microbody with a lipid body forming the so-called microbody–lipid body complex (MLC) was first described in a systematised way by Powell (1978). She described it to be a unique and intricate association of organelles which is confined to zoospores of the posteriorly uniflagellate fungi. As such it could be used as a distinctive character to separate this group of fungi from all other groups of Phycomycetes. She proposed a system for categorising the MLC into four major types which corresponded to the orders Chytridiales, Harpochytriales, Monoblepharidales and the Blastocladiales. These types were further divided into nine subgroups.

A very interesting phylogenetic approach can be applied to the various MLC types. The following sequence of structural associations has a falling level of complexity: the side body complex, the rumposomal structure, and the membrane-covered lipid body located adjacent to the plasmalemma. These structural associations may all be regarded as deriving from the same original structure, and serving the same function in the motile cell. This original structure may further be related to the eye-spot bodies found in various algal groups (Moestrup, 1978; Moestrup and Thomsen, 1974).

The next logical question to ask would be: how is the supposed functional relationship of a microbody–lipid body complex carried out in zoospores such as those of the anteriorly uniflagellate fungi and the biflagellate fungi which do not possess an MLC? There is no valid answer to this at present, but it may be that the presence of individual microbodies, lipid bodies and the endomembrane system may together serve the same purpose as an MLC.

G. ZOOSPORE MICROTUBULE SYSTEMS

The arrangement of the microtubules (cytoplasmic, kinetosomal and axonemal) in the zoospore has attracted great interest. This interest has not only been of a descriptive nature, but has also embraced morphogenetic, functional and phylogenetic aspects.

The most "advanced" arrangement of microtubules observed in the uniflagellate zoospore is found in those of the Blastocladiales (the plant parasitic genus *Physoderma* has recently been transferred to this order (Lange and Olson, 1980a)), where the microtubules which radiate from the functional kinetosome are arranged in groups (usually nine triplets) and form a basket-like structure which holds (and shapes?) the triangular nucleus.

A system of cytoplasmic microtubules with a lower level of organisation is found in the zoospores of most of the genera usually referred to the orders Chytridiales and Monoblepharidales (the most important plant pathogen of this type is *Synchytrium endobioticum* (Lange, 1978; Lange and Olson, 1978b)). Here, the microtubules which flare out from the functional kinetosome occur as singlets, and they do not have a definite structural relation to the nucleus.

The cytoplasmic microtubules found in the few anteriorly uniflagellate species examined to date also flare out from the functional kinetosome as singlets and these few and wavy microtubules lie almost parallel to the plasmalemma.

A "bridge-structure" formed by microtubules connecting the functional kinetosome to the channels or tubules of the rumposomal structure in a tangential manner is found in some members of the Chytridiales although this arrangement is not known to occur in any plant pathogenic species (see Lange and Olson, 1979). It is tempting to suggest that this "bridge-structure" is (or has been) capable of transmitting a response from the supposedly light-sensitive rumposomal structure on the surface of a lipid body to the functional kinetosome, and in this way mediates a motile response by

18   LENE LANGE AND LAURITZ W. OLSON

flagellar movement to a light stimulus. Although some circumstantial
evidence favours this hypothesis (the bridge is found to be more extensively
developed in a phototactic organism (Kazama, 1972; Lange and Olson,
1977)), the idea still has to be looked upon as "armchair philosophy".

There are only a few cases where the zoospore does not contain an
organised system of cytoplasmic microtubules; among these few are two of
plant pathological interest, *Olpidium brassicae* and *O. radicale*.

In the biflagellate (heterokont) zoospore the cytoplasmic microtubules are
basically arranged in two patterns. These two models more or less
correspond with the division into primary and secondary zoospore types.

The microtubular arrangement found in the primary zoospore is best
described from that of *Saprolegnia ferax* (Gruith.) Thuret (Heath and
Greenwood, 1971; Holloway and Heath, 1977a,b). It is said to consist of two
microtubule sets, the perinuclear cone of microtubules and the outer cone of
microtubules. The microtubules in the perinuclear cone run adjacent to the
envelope of the nuclear beak whereas the microtubules of the outer cone are
more divergent and pass through the cytoplasm close to the plasmalemma.
All the microtubules in these two cones originate in the electron-opaque
material found proximally to surround the kinetosomes (Heath and
Greenwood, 1971; Holloway and Heath, 1977a,b).

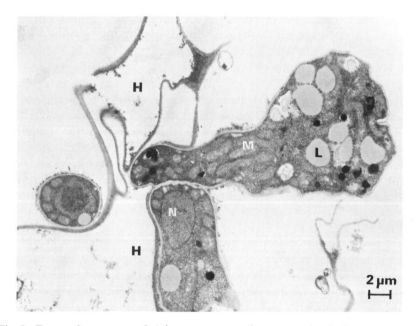

Fig. 9. Encysted zoospores of *Aphanomyces euteiches* penetrating its host, a pea root
(*Pisum sativum*). Abbreviations: N, nucleus; M, mitochondria; L, lipid body; H,
host cells.

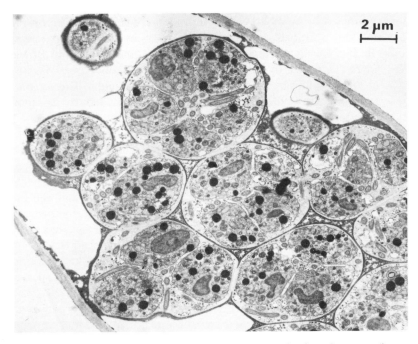

Fig. 10. A cluster of secondary zoosporangia of *Plasmodiophora brassicae* (in a root hair of *Brassica campestris* var. *chinensis*) after zoospore cleavage is completed. By courtesy of J. Falk. Numerous flagella are seen in both longitudinal section and in transverse section.

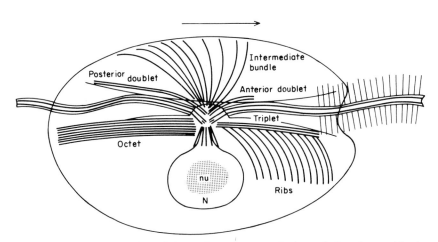

Fig. 11. A schematic drawing of the various systems of microtubules observed in the secondary heterokont biflagellate zoospore of *Saprolegnia ferax*. By courtesy of S. A. Holloway and I. B. Heath. Abbreviations: N, nucleus; nu, nucleolus. The arrow denotes the direction of swimming.

Many important examples of the secondary biflagellate zoospore type have been studied ultrastructurally in both pathogenic and non-pathogenic species: in *Pythium* by Lunney and Bland (1976) and Grove and Bracker, (1978); in *Phytophthora* by Reichle (1969) and by Bimpong and Hickman (1975); in *Aphanomyces* by Hoch and Mitchell (1972); and in *Saprolegnia* by Holloway and Heath (1977a,b). In the studies of some of these zoospores, special attention has been given to the arrangement of the microtubules—see for example the studies on *Saprolegnia ferax* (Holloway and Heath, 1977a,b); *Phytophthora nicotianae* var *parasitica* (Dastur) Waterhouse (Reichle, 1969) and *Aphanomyces euteiches* Drechsler (Hoch and Mitchell, 1972). It appears that the arrangement of the cytoplasmic microtubules in the secondary zoospores of these fungi belongs to the same model, and that the more thorough the ultrastructural study, the more complicated the arrangement of these microtubules appears. Or, to put it another way, the fact that a set of microtubules has not been reported does not necessarily mean that it is not present in the zoospore in question. Cytoplasmic microtubules are extremely sensitive to even slightly adverse conditions during fixation and specimen preparation for electron microscopy.

A schematic drawing (previously published by Holloway and Heath, 1977a) of the microtubular root system of the reniform secondary zoospore of *Saprolegnia ferax* is seen in Fig. 11. The terms suggested for the individual microtubule systems appear in the drawing (except for the perinuclear cone which shapes the nuclear beak). The complexity of this system is impressive. And even more so when it is noted that the different types of microtubules are seen to be formed and to vanish during zoosporogenesis, excystment and encystment, not synchronously, but with a specific timing.

For both the primary and the secondary zoospore types it is believed that microtubule systems function as an exoskeleton for the nuclear beak, anchoring the kinetosome to the nucleus and being partly responsible for the definite shape of the spore, especially the secondary zoospore.

It is now more widely accepted that cytoplasmic microtubules are involved also in ensuring the proper cleavage of the content of the phycomycete zoosporangium by mechanically delineating the zoospore primordia. It is tempting to suggest that some of the microtubules found in the mature zoospore (especially those found in the anteriorly uniflagellate zoospore) are actually remnants of microtubules which were active during sporogenesis; however, careful developmental studies must be carried out in order to confirm this hypothesis.

H. RHIZOPLASTS, ROOTLETS AND OTHER KINETOSOME-ASSOCIATED CHARACTERS

The rhizoplast and rootlet structures found in the uniflagellate fungal zoospore are diverse in their ultrastructural appearance and may serve different purposes in the cell. Lange and Olson (1979) attempted to group these variable structures using the following terms: (1) rootlets such as those found in the blastocladian spore which are striated or banded and rather extensively developed extending on both sides of the functional kinetosome; (2) a striated disc, a structure of multiple fibrillar elements radiating from the functional kinetosome and apparently unique for the Monoblepharidales; (3) a striated rhizoplast typical of *Olpidium brassicae*, *Rozella allomycis* and *Rhizophlyctis rosea*; (4) a solid rhizoplast (so far only described for *Phlyctochytrium arcticum* (Chong and Barr, 1973)); like the striated rhizoplast, the solid rhizoplast also connects the functional kinetosome with the nucleus; (5) simple electron-dense bars associated with the functional kinetosome; these structures are widely distributed among members of the Chytridiales. The structure observed to interconnect the kinetosomes in most of the biflagellate zoospores so far studied has several characteristics in common with the banded rootlet found in the blastocladian spore (see above). In the biflagellate spore it is generally called a banded or striated fibre (Reichle, 1969; Holloway and Heath, 1977a,b; Lunney and Bland, 1976).

A description of the phycomycete kinetosomes and their transition zone, including an evaluation of the significance of the structural variations observed in these structures is given by Barr in Chapter 2. Accordingly only a few comments are given here on this subject.

The angle between the two (functional) kinetosomes in the biflagellate spore (usually described to be *c*. 130°) and between the functional and the vestigial kinetosome in the uniflagellate spore has at times been emphasised as an ultrastructural characteristic of major importance (Powell, 1978). The observations, from zoospores of species belonging to the Phlyctidiaceae presented in our monograph on the uniflagellate zoospore (Lange and Olson, 1979) indicate strongly that this character may be highly variable even within one species.

At several points in the discussion of zoospore ultrastructure it has been emphasised that the anteriorly uniflagellate zoospore has characteristics in common with the biflagellate fungi. And this is also the case with the kinetosomal region where the location and appearance of the axonemal terminal plate, observed in these two groups of zoosporic fungi, is strikingly similar (compare, for example, the appearance of the terminal plate in *Phytophthora parasitica* (Reichle, 1969) with that in *Hyphochytrium catenoides* Karling (Lange and Olson, 1979)).

I. VESICLES AND VACUOLES

The present concept of the types of vesicles and vacuoles found in the uniflagellate zoospore is rather confused—in spite of the relatively large number of species and genera studied (Lange and Olson, 1979). Much of the confusion is due to the apparently unlimited variation which can be found in the content and shape. This variation has made a systematic grouping of these vesicles and vacuoles impossible. By contrast, in the biflagellate zoospore it has been possible to make a tentative grouping of the various structures observed (excellently presented by Lunney and Bland (1976)) while interesting observations on the time of occurrence of the various vacuolar structures have been presented by Grove and Bracker (1978).

The "peripheral vesicles" (*sensu* Lunney and Bland, 1976) are of special interest when dealing with the plant parasitic species as it is claimed that this type of vesicle is instrumental in bringing about spore adhesion (see the study of *Phytophthora palmivora*, by Singh and Bartnicki–Garcia (1975)). From the very thorough study by Grove and Bracker (1978) of the zoospore encystment in *Pythium aphanidermatum* it appears that glycoprotein vesicles are of major importance in forming an adhesive cyst coat. However, more studies are needed to understand the mechanism of zoospore adhesion and the subsequent penetration of the fungus into the plant or animal host. Supplementary experience can be drawn from saprophytic organisms but the primary observations must be obtained by careful ultrastructural and cytochemical studies of a wide range of zoosporic parasites.

Most of the zoosporic fungi studied ultrastructurally, and among them, all the plant pathogenic organisms, are fresh water inhabitants, the source of the water being for example, the soil or rain. One would expect therefore to find in the zoospore of these organisms a structure responsible for maintaining the osmotic pressure of the naked protoplast. However, a true contractile vacuole (or water expulsion vesicle) has only been described from the biflagellate zoospore (Grove and Bracker, 1978; Lunney and Bland, 1976; Hoch and Mitchell, 1972) and never from the uniflagellate spore. The observation of a non-specialised system of vacuoles and vesicles placed adjacent to the plasmalemma in several uniflagellate species led us to suggest that a contractile vacuole-like structure does also exist in the uniflagellate zoospore (Lange and Olson, 1979), but that this structure is not highly organised. Recent studies have shed new light on this hypothesis (Olson and Lange, 1983).

J. ULTRASTRUCTURAL ZOOSPORE TYPES

The evaluation of the various ultrastructural characteristics presented above gives rise to the recognition of a number of zoospore types.

Five such major types can be recognised (see Figs. 12 and 13); for a full description of them, see Lange and Olson (1979).

Type 1 (Fig. 12a) corresponds to the major zoospore types found in Spizellomycetales (*sensu* Barr, 1980a). Type 2 (Fig. 12b) corresponds to the major zoospore type found in the revised Chytridiales (*sensu* Barr, 1980a).

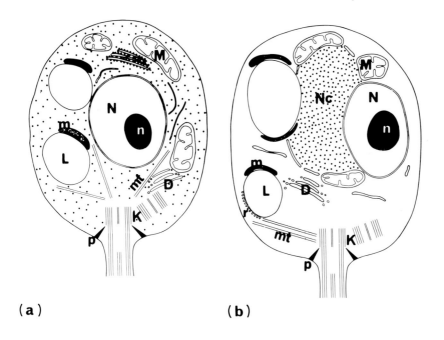

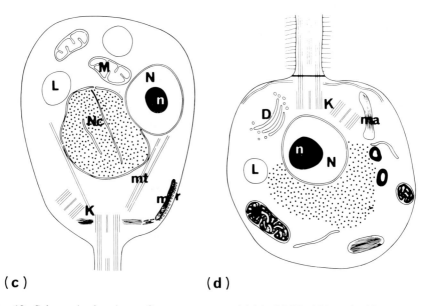

Fig. 12. Schematic drawings of zoospore types 1A(a), 2A(b), 3(c) and 5(d), *sensu* Lange & Olson (1979). Note, type 5 is anteriorly uniflagellate. Abbreviations: N, nucleus; n, nucleolus; M, mitochondrion; L, lipid body; m, microbody; K, kinetosomes; D, dictyosome; ER, endoplasmic reticulum; mt, microtubules; r, rumposomal structure; R, striated rhizoplast; p, props; Nc, nuclear cap area.

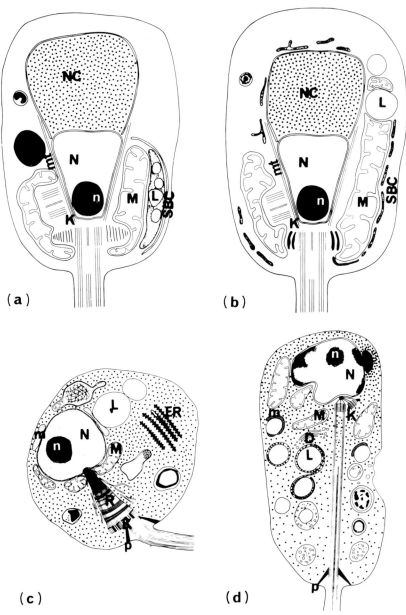

Fig. 13. Schematic drawings of zoospore type 4(a), *sensu* Lange & Olson (1979), (b) the zoospore of *Physoderma maydis*, (c) the zoospore of *Olpidium brassicae* and (d) the zoospore of *O. radicale*. The drawings in (c) and (d) illustrate the general rule that zoospores of plant pathogens are very difficult to place in groups (compare to Fig. 12), (b) represents the exceptional case where it fits perfectly well into a group (type 4). Abbreviations: N, nucleus; n, nucleolus; M, mitochondrion; L, lipid body; m, microbody; K, kinetosomes; D, dictyosome; ER, endoplasmic reticulum; mt, microtubules; r, rumposomal structure; R, striated rhizoplast; p, props; Nc, nuclear cap area.

Type 3 (Fig. 12c) is the zoospore type found in Monoblepharidales. Type 4 (Fig. 13a) is the blastocladian type of zoospore and for comparison a schematic drawing (Fig. 13b) of the *Physoderma maydis* zoospore (Blastocladiales) is shown. Type 5 (Fig. 12d) is the anteriorly uniflagellate zoospore type.

Within the biflagellate zoospores the only grouping which can be made on the presently available ultrastructural data is a grouping corresponding to the primary and secondary zoospore type, the latter being both morphologically and ultrastructurally more complex than the former (Grove and Bracker, 1978).

Too few species have been studied within the Plasmodiophorales to draw any conclusions on the type of spore formed.

III. The Use of Ultrastructural Studies

A. WHAT CAN WE LEARN FROM ELECTRON MICROSCOPE STUDIES OF ZOOSPORIC FUNGI?

The zoospore, this naturally occurring protoplast which for some species may be manipulated in the laboratory, provides excellent material for model studies of the relation between structure and function. For example, Murphy and Lovett (1966) and Lovett (1975) have attempted to elucidate the possible function of the nuclear cap in the blastocladian spore, while Kazama (1972) and Lange and Olson (1977) have attempted to correlate the phototactic response and the ultrastructural appearance of the eye-spot-like organelle in zoospores of *Phlyctochytrium* species.

A few of the most promising areas for experimental studies should be mentioned here.

(a) The study of the control of, and the individual steps in, microtubule protein synthesis.
(b) The various ways in which a naked cell survives in fresh water.
(c) The nature of the special features such as adhesion vesicles and adhesive cyst coatings developed in parasitic species.
(d) How specific recognition is made between the zoospore and the susceptible plant part.
(e) How the ultrastructural development of the zoospore or functional conidia in some plant parasitic species is influenced by the environment.
(f) The sporogenesis in zoosporic fungi as model systems for the study of cell differentiation and developmental control.

Taxonomy and phylogeny are two other main areas in which ultrastructural observations may prove to be of value. The possible phylogenetic relationship among the zoosporic fungi and the zoosporic algal groups and protozoa has classically been a field of interest. The suggestion that there was a biflagellate origin for the uniflagellate Phycomycetes was made by Olson and Fuller (1968) and was based on ultrastructural evidence. The key character here was the observation that in an uniflagellate zoospore a second "vestigial" kinetosome, surrounded by a system of non-functional props was adjacent to the functional kinetosome which was surrounded by a system of functional props.

Assuming that one accepts a biflagellate origin for the uniflagellate Phycomycetes, it is striking how different the posteriorly and the anteriorly uniflagellate Phycomycetes really are. The anteriorly uniflagellate zoospore is, however, ultrastructurally much more closely related to the biflagellate zoospore than to the posteriorly uniflagellate fungi. As noted above, the type of mitochondria, the ribosome arrangement, the axonemal terminal plate, and the mastigonemes are quite similar in the anteriorly uniflagellate and the biflagellate fungal zoospore.

In virtually every criterion already mentioned the fungi grouped in the order Plasmodiophorales are very different from other fungal groups. The available information on ultrastructural organisation in the Plasmodiophorales, which admittedly is rather sparse, does not seem to bridge the gap which exists between these organisms and the "true" fungal groups (but see also Chapters 2 and 5).

The ultrastructure of the fungal zoospore, especially the uniflagellate zoospore, has been, as noted above, the basis for the description of certain basic zoospore types (Lange and Olson, 1979), and is discussed here in relation to the taxonomic implications which would result from using this ultra-structural approach for the establishment of a new taxonomic system. In this approach Barr (1980a) has taken one further step and has divided and revised the entire order Chytridiales using the ultrastructural characteristics of the zoospore as key characters.

The placement of the obligate plant parasites (e.g. species of *Olpidium* and *Synchytrium*) in a systematic separation of the zoosporic fungi according to zoospore type has presented a real problem (Lange and Olson, 1979; Barr, 1980a). This may reflect the highly specialised nature of obligate parasites and not be merely coincidence. Here one should also note the striking diversity in zoospore ultrastructure observed within the genera *Olpidium* and *Synchytrium*; compare, for example, *O. brassicae* (Lange and Olson, 1976a,b) with *O. radicale* (Lange and Olson, 1978a) and *O. pendulum* (Lange and Olson, 1979) and compare *S. endobioticum* (Lange and Olson, 1978b) with *S. macrosporum* (Montecillo *et al.*, 1980). From these results it may be more correct to recognise these two genera as biological entities which have undergone convergent specialisation to an obligate plant parasitic way of life rather than being true taxonomic groupings. It is an open question whether similar patterns will show up when additional species of biflagellate plant parasites are studied. Special attention should here be placed on the downy mildews such as *Sclerospora*, *Sclerophthora*, *Pseudoperonospora* and *Plasmopora* and on the Plasmodiophorales such as *Spongospora*, *Polymyxa*, *Sorosphaera* and *Ligniera*.

B. FUTURE, APPLIED PERSPECTIVES OF ULTRASTRUCTURAL STUDIES OF ZOOSPORIC FUNGI

As stressed above, many of the most interesting and unpredictable ultrastructural results have been obtained from studies of the parasitic zoosporic organisms. From this, as well as from a general applied plant pathological point of view, the study of these parasites is of urgent but often neglected interest. This neglect is mainly due to technical difficulties which must be overcome or circumvented. For this, studies of closely related facultative parasites instead of the obligately parasitic species, and studies of species which may be manipulated to infect easily handled host plants may both prove to be useful approaches. Another approach which may help is the establishment of better contact between the electron microscopist and the experts who know how to handle the pathogen-host combination in question. This type of cooperation will often need to be inter-institutional and, indeed, because of the tropical nature of many of the economically important plant parasites, international collaboration must be established if these organisms are to be understood and ultimately controlled.

When considering the possible methods for the control of plant diseases caused by zoosporic fungi, one is interested in intervening at the most susceptible portion of the life cycle, and the most susceptible phase of the life cycle of zoosporic fungi is the zoospore itself. Thus far, very little has come from this type of approach for the control of the zoosporic fungi, although in a recent study of possible ways for controlling *Olpidium brassicae* in lettuce (as this fungus acts as vector for the lettuce big vein disease) various surfactants have been used successfully (Tomlinson and Faithfull, 1979a,b).

Additional information on the ultrastructural and physiological characteristics of zoosporic pathogens, along with the characteristic host response to them, could provide a base for developing improved control procedures for the diseases they cause. Careful studies of the mode of action of chemicals already known to be efficient in controlling or preventing phycomycete attacks in plant crops would be a good starting point although as Buczacki points out in Chapter 5, the polyphyletic nature of the Phycomycete grouping can sometimes lead to disappointing lines of enquiry in this respect. The thorough study of the mode of action of Ridomil on *Peronospora pisi* Sydow (Hickey and Coffey, 1980) will hopefully set the scene for many more studies of this type.

IV. The Biological Significance of the Fungal Zoospore

A. REQUIREMENTS FOR FREE WATER

The spread of a zoosporic fungus by means of zoospores is strictly dependent on the presence of water; not only for the free swimming stage but also for the final phases in the formation and the discharge of zoospores, whilst it can be of critical importance for the encystment and penetration of the host. This means that all zoosporic plant pathogens are restricted in their life cycle, in their infection cycle and epidemiologically by their requirement for free water. But this dependence on free water can be mediated in many different ways depending on a number of factors, in particular the habitat of the fungus and the part of the plant to be infected.

For plant parasites infecting the aerial portion of the plant, one can recognise two major types of water dependence: some species require heavy rainfall to give standing water in the leaf whorls whereas others can complete the infection cycle by utilising dew formed on the plant.

For the parasites found in the soil the variation on water-dependence can be just as broad. Many species can survive and spread in the soil water present between the soil particles, even in well drained soil types (e.g. species of *Pythium* and *Phytophthora*) while others require waterlogged conditions. A typical example of the latter group is *Synchytrium anemones* (DC.) Woronin (primarily infecting *Anemone nemorosa*) and *Sclerophthora macrospora* Sacc. when initiating an attack in a finger millet (*Eleusine coracana*). However, when interpreting the requirements for a parasite to initiate infection in its host it must be remembered that host susceptibility may vary so that more severe attacks are observed under conditions adverse for general plant growth.

In the classical concept, the conidial species of the lower fungi are recognised to be more advanced phylogenetically than the zoosporic species. The conidia-producing species have freed themselves from a water-requirement and have become further adapted to a terrestrial life, in parasitising the aerial part of the plants. And if one compares zoosporic plant parasites infecting the aerial plant parts with conidia-producing parasitic species, the latter group generally has a significant epidemiological advantage. On the other hand one should not overlook the possibility that there may also be an advantage in the fact that zoospore formation and discharge is only completed when sufficient free water is present; and this may be a sort of guarantee that satisfactory conditions exist during the spread of the zoospores, their encystment and host penetration. The presence of free water may also facilitate the active spread of the zoospores to a susceptible part of the plant where its chance is higher of establishing an extensive infection rather than only a local lesion.

The formation of zoospores and zoospore discharge also place certain requirements on the water potential of the micro-environment. This and related aspects of the influence of water potential on the Phycomycetes have recently been reviewed by Duniway (1979).

For many zoosporic plant parasites it is claimed that infection can only be initiated on certain susceptible parts of the plant, typically on meristematic tissue or at least on very young tissue where the epidermal wall layers are still delicate. However, from laboratory experiments it appears that these statements may need some modification particularly with reference to the role of humidity in the infection process.

If maize plants are kept under constant high humidity, a spore inoculum of *Physoderma maydis* may give rise to infections even on fully mature leaves (Olson *et al.*, 1980). This stands in contrast to the observations by Tisdale (1919) who stated that only meristematic tissue could be invaded by this pathogen. In the Spieckermann procedure for establishing infection of *Synchytrium endobioticum* in potatoes (Spieckermann and Kotthoff, 1924), it is said that only tubers which have buds that are less than 3–4 mm in length are suitable for inoculation. However, by keeping the zoospore inoculum in a persistent drop of water for 24–48 hours, infection was readily established on tomato cotyledons (Lange, 1978). On the basis of these superficially contradictory observations, it would appear that by securing high humidity many parts of the plants will be susceptible to infection by these parasites. Nonetheless, observations on the restricted nature of the susceptible plant parts are still valid when one considers field conditions.

It is tempting to look for additional factors which can be correlated with the separation of the zoosporic plant parasites from the conidia-producing species. A general characterisation of these two groups is that the conidia-producing plant parasites have overcome a strict requirement for free water; they are less restricted by the need to reach a specific susceptible plant part as the conidia simply germinate where they land. In general they have the potential to produce a long germ-tube and finally to make an appressorium which may form an infection peg; they also often start infection in the host plant through "natural openings" such as the leaf stomata. The conidium has obviously lost the possibility for motility in seeking a specific area of the plant but they have also in many cases lost the ability to obtain entry to the host cell mechanically or enzymically. When one considers the obligate zoosporic plant parasites the zoospores typically enter the host cell by mechanically or enzymically forming a hole in the host cell wall (see Fig. 9).

Do these differences between the zoosporic and the conidia producing fungal parasites reflect a more specific host–parasite relationship among zoosporic fungal parasites? At present it is not possible to give a definitive "yes" to this question but many observations point towards this general pattern. It should, however, be noted in this connection that a number of the zoosporic peronosporalean plant parasites have the ability to initiate an infection in various ways: either with or without appressoria, and by penetration either through the cuticle or through the stomata (Adegbola and Hagedorn, 1969; Kim *et al.*, 1974).

B. SPREAD OF ZOOSPORIC PLANT PATHOGENS

Zoospores may be present in the soil either as a result of sporangia falling on to the ground or as a result of the sporangia being formed directly in the soil. In both cases the spread of the fungus on the host plant and between potential host plants is achieved by movement of the flagellated zoospore. The distance covered as a result of active zoospore movement is dependent on several factors, including host-specific characteristics, soil structure and the water content of the soil. The latter aspect is reviewed in Griffin (1972, 1978).

The distance travelled by the individual zoospores is usually relatively short. For zoospores of *Aphanomyces euteiches*, *Synchytrium endobioticum*, and *Olpidium brassicae* in soil it is reported to be less than 1–2 cm (Lockwood and Ballard, 1959; Esmarch, 1927; Chupp, 1917). Pfender *et al.* (1977) reported that zoospores of *Phytophthora megasperma* Drechsler could swim up to 6 cm in an upward direction in saturated soil. However, where there exists either a natural flow of water or water flow through the soil due to irrigation, the zoospores may be spread over entire fields or even to neighbouring fields. The latter case is well illustrated by an indirect effect observed in an irrigated lettuce field in California (Grogan, personal communication). Here big vein disease, a virus-like problem transmitted by *Olpidium brassicae*, was seen to be spread with incredible efficiency throughout the entire field due to the forced spread of the zoospore inoculum by flooding of the field.

The aerial spread of zoosporic plant parasites can be mediated in three ways; through direct contact between infected leaves, through rain splash, and by air currents. In the cases of rain splash and direct contact between infected leaves, both detached zoosporangia (or resting sporangia in the case of *Physoderma maydis*) and droplets containing either zoospores (Thorold, 1955) or sporangia can be transmitted from plant to plant over short distances. This pattern is most probably the reason for the epidemiological characteristics of these diseases in causing them gradually to spread from an initial infection site until the entire field is contaminated. The spread of an inoculum by water splash can take on large-scale proportions when overhead irrigation is used (Rotem *et al.*, 1962). It should also be noted that zoosporangia may be carried in the air; those of *Sclerospora graminicola* (Sacc.) Schröter have been isolated from air samples meters above an infected pearl millet field (K. M. Safeeulla, personal communication). And Van der

Zaag (1956) reported airborne zoosporangia of *Phytophthora infestans* trapped at a distance of up to 11 km from their source. It appears that sporangia may be spread in the air to even greater distances, but it is not known whether such sporangia retain their viability.

C. RAPIDITY OF ZOOSPOROGENESIS AND THE LENGTH OF THE MOTILE PERIOD

One of the main characteristics of zoospore formation is the speed with which the process can be accomplished. When favourable humidity and temperature occur, zoospores may be discharged within a few hours. This is a part of the adaptation of these otherwise aquatic organisms to semi-terrestrial conditions. It is, in general, explained physiologically by the fact that the formation of the initial sporangial stages can take place under low humidity. The sporangial primordia may firstly wait for favourable conditions to occur and then be able to complete sporogenesis and spore discharge very rapidly.

After discharge, the fungal zoospore may remain motile for several hours. For the uniflagellates we estimate the period to be about 24 hours under favourable conditions; similarly the period of mobility of the biflagellate zoospore stage has been estimated generally to be between 20 and 30 hours (Hickman and Ho, 1966). The many reports of extremely short motile periods for zoospores may be explained by them being exposed to slightly adverse conditions during the observation period.

The duration of the active swimming period of zoospores of plant parasites is determined primarily by three factors: (1) motility may cease due to the presence of a suitable host which induces the zoospore to encyst and start infection; (2) the energy supply/potential carried by the zoospore is exhausted; or (3) the zoospore encounters extreme conditions which in some way lead to its lysis.

One could call the second of these three factors the biochemical limit for the length of the motile period of a fungal zoospore. By extrapolating from experiments with uniflagellate zoospores (Lovett, 1975), it appears that protein synthesis does not occur during the motile stage. The early stages of encystment and germination appear to be directed by long-lived messenger RNA, and one could expect that there are certain limits to the period in which even long-life messenger RNA will be functional. Additional biochemical limitations may be restricted to the amount of storage material that may be carried in the zoospore, or the amount of ATP, i.e. energy, which can be realised by utilising this material (Bernstein, 1968).

D. EPIDEMIOLOGICAL PATTERNS

In order to understand and thus explain the epidemiological characteristics of the various zoosporic plant parasites it is necessary to consider their dependence on the presence of water in order to spread and also the length of the motile period of the zoospore. But other factors are also of epidemiological importance, such as how the parasite has adapted to produce the highest zoospore inoculum with the least waste of energy, what the factors are that govern the penetration process, and whether the zoospore cyst functions as a type of resting structure or at least as a structure which can "wait for better times".

1. *Maximising the Zoospore Inoculum*

It is a characteristic of holocarpic Phycomycetes that the entire vegetative thallus develops into a single zoosporangium. Among the plant parasites, examples of this are found in the genus *Olpidium*, *O. brassicae* and *O. radicale*, for example, both infect roots while *O. viciae* Kusano and *O. trifolii* Schröter attack the aerial parts of their host plants. The causal agent of the potato wart disease, *Synchytrium endobioticum* has holocarpic sporangium formation (along with other species of this genus of obligate parasites), but here the fungus also develops a sorus of sporangia. In neither of these genera with individually produced sporangia or with sporangia produced in sori, are auxilliary structures formed and so no energy is used to form bodies that do not directly play a role in the production of zoospores.

In species of *Physoderma*, among the polycentric Phycomycetes, only the delicate rhizoids, which ensure the spread of the sporangia inside the host, do not take part directly in spore formation. Among the hyphal Phycomycetes several growth characteristics can be considered to serve in maximising the number of zoospores produced; for example, the ability to form sporangia by internal proliferation could be interpreted as an ingenious way of forming a new generation of zoosporangia without wasting energy (and time) on new vegetative structures. Examples of this may be found among *Pythium* spp.; while Thirumalachar (in Safeeulla, 1976) also reported internal proliferation in *Sclerospora graminicola*. Perhaps the most advanced pattern observed for efficient zoospore production is found in those species which form a sporangiophore which appears to be very similar to a typical multibranched conidiophore. A delicately branched sporangiophore with a single, rather small, zoosporangium formed apically on each and every tip is found in species of *Pseudoperonospora*, *Plasmopara* and in *Sclerospora graminicola*.

The role of sequential zoospore emergence (i.e. the formation of primary and secondary zoospores) is not obvious. However, it is striking that plant parasitic species typically form only the secondary zoospore stage. From an evolutionary point of view it could be argued that plant parasitic species represent an advanced group among aquatic Phycomycetes, as there has been a partial or total deletion of the primary zoospore stage.

2. *Encystment*

The active role of the zoospore in the spread of zoosporic fungal organisms is obvious. However, in order to understand the "success" of the zoosporic plant pathogens one must also consider the role of zoospore encystment followed by host penetration.

Zoospore encystment and host penetration represent a sophisticated pattern of processes, and it is difficult to analyse the various steps individually as they represent a chain of events which must be successfully completed to insure the survival of the parasite. Among the conidia-producing species of plant parasitic fungi the conidium can germinate directly to produce a hypha; thus the "uncertainty" of the zoospore-encystment sequence of events is eliminated. The elimination of these "uncertainties" doubtless represents adaptation to a terrestrial existence and can be interpreted as an advance towards a more successful terrestrial parasite.

Zoospore encystment, as well as being the endpoint of the motile period for the zoosporic pathogen, at the same time serves as the initiation of host penetration. However, it may be postulated that the encysted zoospore in some species also represents a resting structure, although generally of a rather short-lived nature. It could be that zoospore encystment independent of a suitable host may serve as a type of resting structure by which the fungus can survive a sudden adverse change in conditions, for example, free water no longer being present in sufficient quantity.

Many apparently contradictory observations on the longevity of zoospore inoculum in soil have been reported. Hickman (1958) observed a rapid loss in infectivity of *Phytophthora* zoospores in the absence of a suitable host. Luna and Hine (1964) reported that zoospores of *Pythium aphanidermatum* may survive in soil for up to seven days (without growing saprophytically); *Phytophthora cactorum* zoospores were shown by Meyer and Schönbech (1975) to survive for about three months in soil; while Turner (1965) observed survival of zoospore inoculum of *Phytophthora palmivora* in soil for up to 18 months. These results may be explained by different species—specific characteristics or the possibilities of the fungi surviving as encysted zoospores (see above) or at the germ tube stage (Zan, 1962).

3. *Host Penetration*

All plant cells have walls and, to be successful, a plant parasitic fungus must be able to overcome this mechanical barrier in order to enter the host cytoplasm. Although it is known that several plant resistance mechanisms are associated with variations in the composition of the cell wall (as well as the stimulation of host cell wall synthesis in response to parasite penetration), it is astonishing how little is known about how plant parasites actually overcome the cell wall barrier and thereby complete host-wall penetration. For the zoosporic plant parasites (including the uniflagellates, biflagellates and some species of the Plasmodiophorales) little information is available on how the zoospore penetrates its host. Is it an enzymic process, a mechanical process or a combination of the two? Are the processes similar among the different groups and genera of parasites? And can the same mechanisms for establishment of host infection also be found in conidia-producing species?

Although these and related aspects of host penetration are of greatest importance in understanding the success of obligate plant parasites perhaps the best way to investigate such problems will be via studies of facultative parasites, where the necessary laboratory experiments for elucidating these problems can be made.

E. TACTIC RESPONSE IN ZOOSPORIC FUNGI

The tactic responses in zoosporic fungi can be divided into two major groups, phototaxis and chemotaxis; although responses which could be characterised as autotaxis (Porter and Shaw, 1978), rheotaxis (Katsura, 1971) and geotaxis (Cameron and Carlile, 1977; Palzer, 1975) have also been observed for zoosporic fungi.

Observations on a positive phototactic response in zoosporic fungi were made early in the century (for the uniflagellate zoospore of *Synchytrium fulgens* Schröter see Kusano (1930); for the biflagellate zoospore of *Phytophthora cambivora* (Petri) Buisman see Petri (1925)). The more recent information on photoaxis in zoosporic fungi has come predominantly from studies of non-parasitic forms (e.g. *Phlyctochytrium* sp. (Kazama, 1972); *Polyphagus euglenae* Nowak (Powell, 1978); *Coelomomyces punctatus* Couch (Martin, 1971); *Allomyces reticulatus* Emerson and Robertson (Robertson, 1972)).

The above mentioned examples are all of a positive response to a light stimulus. While making infection experiments with certain root-inhabiting zoosporic fungi such as *Olpidium* spp. it is well known that a necessary precaution is to protect the experiment from exposure to light. This is, however, not to be interpreted as the existence of a direct phototactic influence on infection but rather to guard against the possibility that light may inhibit zoospore encystment and host penetration.

Chemotaxis can be divided into hormonal chemotaxis, chemotaxis as response to exudates from the potential substrate (such as leaf stomata, roots or seeds), and chemotaxis as a specific response to experiments *in vitro* with ions, chemical compounds and other agents. Among the uniflagellate Phycomycetes, hormonal chemotaxis has been shown for *Allomyces*, where the male gamete is attracted to the female gamete as a result of the secretion of the sexual hormone sirenin, and for *Monoblepharis* where the male gamete is attracted to the egg by a hormonal concentration gradient. Similarly, in the biflagellate fungi, a complex set of hormonal-developmental processes have been demonstrated in the development of the sexual organs of *Achlya* (Raper, 1957). Most of the chemotactic responses observed in zoosporic plant parasites could be characterised as apparent active movements (due to attraction) to potential infection sites on suitable hosts (Hickman, 1970). As the location of the infection sites varies from leaf stomata to roots or shoot tips, the chemotactic responses are also commonly subdivided.

The attraction of zoospores to leaf stomata, observed as an accumulation of zoospore cysts specifically around the stomata, has been observed for many plant parasites, especially among the peronosporalean species (e.g. *Plasmopara viticola* (Gregory, 1912; Arens, 1929a); *Pseudoperonospora humuli* (Arens, 1929b); *P. cubensis* (Iwata, 1957); *Phytophthora syringae* (Noviello and Snyder, 1962)). In contrast, *Phytophthora infestans* was observed to encyst uniformly over the leaf surface (Hori, 1935).

With regard to the chemotactic attraction of zoospores to roots, an interesting, and perhaps generally applicable pattern, was described by Royle and Hickman (1964) from their work with *Pythium aphanidermatum* in pea where the zoospores were specifically attracted to the elongation zone behind the root tip and to root wounds.

Another interesting pattern of chemotactic response of plant pathological relevance has been observed with species of *Phytophthora*: the zoospores are attracted particularly to the roots of susceptible host plants, much less to resistant host plants and, in general, not at all to non-hosts (see, for example, work on *P. cinnamoni* and *P. citrophthora* (Zentmyer, 1961) and on *P. palmivora* (Turner, 1963)).

A rather non-specific attraction to cut shoot surfaces of a wide spectrum of both host and non-host plants was observed for zoospores of *Olpidium viciae* and *O. trifolii* (Kusano, 1932).

Another type of host substrate, pine seeds, has been shown to stimulate zoospore germination of *Pythium afertile* Kanhouse & Humphrey (Agnihotri and Vaartaja, 1968).

Experiments with chemical compounds, to test their relative values as zoospore attractants, have been used as a useful approach for elucidating the nature of the chemotactic response (Cameron and Carlile, 1978). In a recent study of *Phytophthora palmivora*, Cameron and Carlile (1980) demonstrated a negative chemotactic response to low molecular-weight cations. It was suggested from these observations that the net negative charge found at the zoospore cell surface is instrumental in the negative response of the zoospore.

Only sparse information is available on the phenomenon of autotaxis (see Fig. 14a; Porter and Shaw, 1978), but further studies may show that this

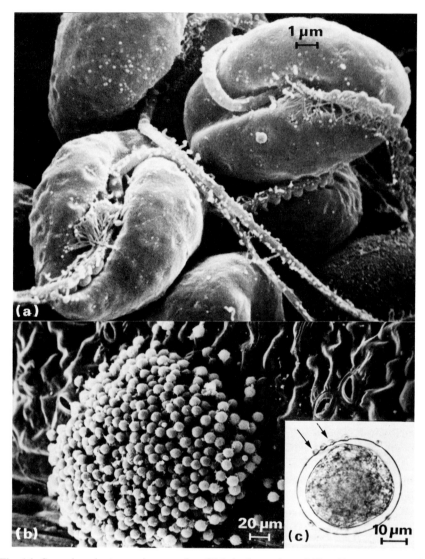

Fig. 14. Scanning electron micrographs of (a) the heterokont biflagellate zoospore of *Phytophthora infestans* and (b) aggregation of encysted zoospores of *Phytophthora drechsleri* on host surface (*Capsicum* leaf). The aggregation is interpreted to be the result of an autotactic response. By courtesy of D. S. Shaw. (c) Encysted zoospores of *Synchytrium endobioticum* (arrows) on the surface of a neighbouring zoosporangium. As in (b), this may be interpreted as being the result of an autotactic response. Light micrograph of living material.

phenomenon is quite widespread; the specific attraction of zoospores of *Synchytrium endobioticum* to the extruded sporangial wall (see Fig. 14c) for example may be categorised as a type of autotaxis.

From the examples listed above it appears that there is substantial information available on zoospores showing tactic responses. However, almost nothing is known about the nature, structure, or even existence of receptor sites for the chemotactic response; or indeed for the autotactic response. Various proposals have been presented for the location and structure of the receptor organelles for the phototactic response observed in some uniflagellate zoospores (Lange and Olson, 1979; and see also the above sections on ultrastructure but no substantial evidence has been presented. Future studies, especially with plant parasitic fungi, to elucidate some of these problems should include studies of the zoospore membrane and possible labelling techniques which could be used to tag the receptor sites.

F. ZOOSPORES AS VECTORS FOR PLANT VIRUSES

The recognition of the fact that fungi can act as vectors for plant viruses, is rather recent. The first report was by Teakle (1962); since then considerable knowledge has been gathered but many aspects of the fungal–viral infection cycle have remained unsolved.

So far all fungal species which have been demonstrated to be acting as vectors for plant viruses belong to the group of zoosporic fungi, and it is the zoospore stage which is the critical step in the transmission. It is by the active movement of the zoospore that the virus is spread, and it is by the penetration of the fungal zoospore into the host tissue that the virus establishes its infection.

Among the unsolved problems of fungal virus transmission are the basic questions of whether the virus replicates in its fungal vector, whether the virus particles or naked viral RNA are in or merely on the surface of the zoospore, and how the virus genome comes to be present in the host cytoplasm. Several attempts have been made to elucidate these problems but so far with little success.

During a televised discussion in Sweden among a group of Nobel prize winners, it was stated that the basic ingredient for a scientific break-through is to be able to foresee the problems that are mature for solution. There are several aspects of the interaction between the zoospore and virus particles which are certainly not ready for solution at the present time; additional knowledge must first be gained about the membrane chemistry of the zoospore and the viral receptor sites on the membrane, about encystment, host penetration and cytochemistry for example. These are the types of problem which are not easy to study with obligate fungal parasites, but the whole subject is treated in depth in Chapter 7.

V. Grouping of Zoosporic Plant Pathogens According to Zoospore Behaviour

The plant pathologist tends to classify fungal parasites from the viewpoint of the host or host range (especially of cultivated plants) that the pathogens attack, and the type of symptom that they produce. Conversely, the fungal taxonomist categorises the zoosporic Phycomycetes from the viewpoint of their type of zoospore flagellation, their developmental pattern and such morphological characteristics of individual life cycle stages as can be used in a taxonomic description. What follows, however, is an attempt to conceive a different system of classification in which the division of the zoosporic plant parasites is derived from the viewpoint of the zoospore itself.

How can a zoospore initiate an infection? Here many criteria could be considered for one of the most striking delineations among the zoosporic plant parasites is the separation of those organisms with an obligate zoosporic stage from those species where the "zoosporangia" are able to germinate directly and produce a germ tube. The latter group would include most of the peronosporalean plant parasites (especially many species of *Pythium* and *Phytophthora*, as reviewed by Duniway (1979)), while the former group would include all uniflagellate plant parasites. Where one would place the plant parasites belonging to the Plasmodiophorales in such a separation is still dubious—but exciting to contemplate (see Chapters 2 and 5).

How does the zoospore find a suitable host? The feature common to all zoosporic species is the potential use of active zoospore movement to spread the fungus to a new host. However, a separation of the plant parasites could be made on the basis of whether the zoospore is attracted specifically to a potential host plant or even to an especially susceptible part of such a host. The specific attraction of the zoospore to its host must be viewed as an important character for the competitive success of a plant parasite. Before we can investigate the type of chemotactic response that exists we must first determine those species in which a tactic response is involved. Then, once a chemotactic response has been demonstrated, we can begin to approach the central question of just how this response is perceived by the zoospore and how it is converted into a directed movement.

Are there different modes of zoospore encystment? Koch (1968) described and categorised different encystment models for the uniflagellate zoosporic fungi. It is tempting to apply these or similar criteria to the classification of the zoosporic plant parasites. The merit of such an approach is seen in the encystment process in plant parasitic organisms being closely associated with the recognition of a suitable host surface; this then being followed by encystment and host penetration. However, the information necessary for such a grouping is not available, for the observations required are at the limits of what can be achieved with the light microscope. Moreover, ultrastructural information is almost non-existent due to the technical problems of working with obligate plant parasites where the encystment process is very rapid, cannot be synchronised and thus is extremely difficult to locate in electron microscope preparations.

Are there different modes by which the fungus penetrates the host cell? Does the encysted zoospore have an enzymic apparatus facilitating a direct penetration of the host wall and if so what enzymes are involved? Alternatively, does the zoospore encyst and penetrate intercellularly into its host either mechanically or also by an enzymic degradation of the host wall? Or is the encysted zoospore only able to penetrate through natural openings of the host, such as the stomata or through wounds?

Typical examples of the first group, those species which penetrate the host wall directly are found in the genera *Olpidium* and *Synchytrium* (of the Chytridiales) and also in at least two genera of the Plasmodiophorales. The primary zoospores of *Polymyxa betae* Keskin and *Plasmodiophora brassicae* each have an intriguing organelle, the stachel, which appears to facilitate host wall penetration (Keskin and Fuchs, 1969; Aist and Williams, 1971). A separation of species which enter through natural openings on the host plant is difficult to make as several zoosporic peronosporalean species are reported to penetrate in any of three ways; to enter through the stomata, to encyst in the depression between the epidermal cells, or to penetrate through the cuticle (see the descriptions of *Pythium aphanidermatum* and *P. debaryanum* by Kim *et al.* (1974) and Adegbola and Hagedorn (1969) respectively).

What type of response does the host make to parasite attack? One of the most obvious criteria for grouping the zoosporic plant parasitic fungi, when considered from the viewpoint of the zoospore, would be the type of response that the host makes when confronted with the penetrating zoospore protoplast. Does this response take the form of callus production, hyper-sensitive reaction, abnormal nuclear or cellular enlargement, rapid cell division, degradation of host cytoplasm, production of phytoalexins or is there apparently no reaction at all?

What are the different types of infection that can be started by a single zoospore? Fungal zoospores are all flagellated, naked and unicellular, but in spite of this uniformity the zoospores of fungal plant parasites may be the initiators of very different types of infection. With regard to thallus development, the type of infection can be separated into three distinct groups. (1) The simplest type of infection is the development of a holocarpic thallus, where one zoospore gives rise to a single zoosporangium (or a sorus of sporangia) and nothing else (e.g. *Olpidium* spp., *Synchytrium* spp., *Lagenidium* spp. (Sparrow, 1960)). The development of a single plasmodium from a single zoospore such as occurs in *Plasmodiophora brassicae* belongs functionally to this group. (2) A polycentric type of growth such as that of *Physoderma* spp., must be considered to be one step more complex than holocarpic thallus development; here, many sporangia, all interconnected by fine, delicate rhizoids develop from a single zoospore. (3) The most advanced type of thallus development is found among the hyphal or filamentous fungi. Here a single zoospore may lead to the development of numerous sporangia (and sex organs) spread widely in the host and interconnected by hyphal threads (e.g. species of *Pythium*, *Phytophthora*, *Sclerospora*, *Sclerophthora*, *Plasmopara* and *Pseudoperonospora* in the Peronosporales, and of *Aphanomyces* in the Saprolegniales). The relevance of this type of classification, on the basis of infection type, is emphasised by the fact that another important character is correlated with this grouping: only the last two types of thallus development, polycentric growth and hyphal growth, may lead to a true systemic infection of the host. The spread of a parasite belonging to the first group, having holocarpic thallus development, is totally dependent on establishing new infection sites. This restriction imposed by holocarpic thallus development is partly overcome epidemiologically by the development of very efficient primary infection systems, often having numerous simultaneous infection sites (almost the entire surface of young host plants of *Anemone nemorosa* may be covered by sporangia of *Synchytrium anemones*) and a rapid second spread, each generation taking only a short time to be completed, such as is seen in *Olpidium brassicae* and *Synchytrium endobioticum*. Another interesting way in which the fungus can spread in the host is when the protoplast of a penetrated zoospore becomes amoeboid and then moves from one cell to another often thereby entering the subepidermal cell layers of the host. This has at various times been suggested to occur with *Plasmodiophora brassicae*, *Olpidium brassicae* and *O. radicale*. On the other hand *Synchytrium endobioticum* becomes deeply entrenched within its host tissue simply through being buried by the rapid proliferation of host cells which the fungal infection stimulates.

It would appear that the holocarpic parasites have developed certain compensatory tactics to ensure the primary spread of the fungus in its host, but it does not change the fact that of the holocarpic fungi on the one hand, and the polycentric and hyphal-forming parasites on the other, it is only the latter that can establish a true systemic infection. It is striking that a separation of zoosporic plant pathogens according to thallus development brings together such plant parasitic genera as *Olpidium* and *Synchytrium* (both uniflagellate), and *Lagenidium* (biflagellate) with *Plasmodiophora* (isokont biflagellate) in one group while species of *Physoderma* (uniflagellate) and the peronosporalean biflagellate types fall into a second group.

VI. Exclusion of Zoosporic Fungi from Plant Pathological Tests

If the attack of a zoosporic fungus on its host's roots does not lead to symptoms of fungal infection in the aerial portion of the plant, and does not produce distinctive symptoms in the root system, there is a good chance, or rather a high risk, that the fungus will not be recognised as a pathogenic organism, even though infection by it may lead to a reduction in the capacity of the root system and to give rise to a similar loss in crop yield.

Hawker (1962) made a very interesting study of the role of endotrophic mycorrhiza developed by zoosporic plant parasites such as *Pythium* spp. However, after the recognition of *Endogone* as the major cause of the development of non-septate endotrophic mycorrhiza, interest in the involvement of zoosporic fungi in mycorrhizal development decreased significantly. Quite probably many endotrophic mycorrhizal associations of economic importance have yet to be discovered.

Only a few cases are known of zoosporic fungi being vectors for plant viruses (see Chapter 7). A systematic search in nature by thorough root examinations for the correlated appearance of fungal and viral infections might lead to the recognition of additional examples of this type of associated infection.

The zoosporic fungi are in general not strong competitors on an agar surface and in many cases they will not even start development as insufficient free water is present. The conidia-producing species, however, are very strong competitors, and the relative competitive characteristics of zoosporic and conidia-producing species have far-reaching implications. Let us consider how many soil analyses have been performed without scoring a single zoosporic organism, these organisms having been excluded simply by the choice of the culturing method used; or imagine the number of seed health tests in which the potential inoculum of zoosporic pathogens has not been revealed, again because of the exclusion of this group of organisms by the nature of the test methods.

Sparrow (1960) recommended that for the detection of zoosporic fungi the substrate to be examined must be totally submerged in water and that various kinds of baits should be added to enrich for the species in question. By submerging the substrate one creates an environment which favours the zoosporic fungi. Conidia-producing fungi are suppressed by this treatment, and algae are kept down by keeping the samples in darkness.

Recently another approach for testing for the presence of species of *Pythium* and *Phytophthora* has been developed. It involves the use of selective agar media containing polyene antibiotics (Tsao, 1970). However, these media have the disadvantage of only allowing vegetative fungal growth, while polyene antibiotics and their solvents are costly and hazardous to handle.

Perhaps a possible solution to the problem of recognising the presence of zoosporic pathogens is to be found in the use of the old baiting approach combined with the use of selective media and antibiotics. But first it is necessary to broaden awareness of the restrictions consequent upon the techniques that are generally used, and then to develop suitable new techniques which are also usable in routine testing for the presence of zoosporic fungi.

2

The Zoosporic Grouping of Plant Pathogens

Entity or Non-Entity?

DONALD J. S. BARR

Biosystematics Research Institute, Agriculture Canada,
Central Experimental Farm, Ottawa, Ontario, Canada

I. Introduction	. . . .	43
II. Development of Taxonomic Theory.	. . . .	45
A. Historical Highlights	. . . .	45
B. The Flagellar Apparatus	. . . .	47
C. The Thallus and Life Cycle	. . . .	52
D. Chemotaxonomy	. . . .	59
III. Phylogeny and Classification	. . . .	70
A. The Taxonomic Groupings	. . . .	70
IV. Synopsis	. . . .	81
Acknowledgements	. . . .	83

I. Introduction

Zoosporic fungi are often considered difficult to study, especially from a taxonomic standpoint. Many taxa do not make satisfactory herbarium specimens because so much essential information is lost. Thus, there is excessive dependence on literature descriptions that are often deficient in essential details. In spite of this, there is considerable consensus among mycologists on the classification of many groups. Moreover, the advent of culturing techniques for some taxa has substantially influenced phylogenetic and taxonomic concepts. Many of these have not yet been incorporated into classifications, and even fewer have reached mycology text books. The purpose of this chapter, then, is to analyse recent studies on zoosporic fungi and related eukaryotic organisms, and show how research is influencing the classification of zoosporic fungi and thoughts about their evolution.

ZOOSPORIC PLANT PATHOGENS
ISBN 0 12 139180 9

The name **Phycomycetes**, which includes the present Mastigomycotina and Zygomycotina (Table 1) no longer has any taxonomic status. The term **lower fungi** is usually used synonymously with Phycomycetes even though authors occasionally extend it to include slime moulds also. Nevertheless, it is useful and contrasts with "higher fungi" which includes the Ascomycotina, Deuteromycotina and Basiodiomycotina. The term **aquatic fungi** should be avoided in taxonomy for many zoosporic fungi are not aquatic—many occur in soil, and some have been isolated from desert sand. Moreover, there are many non-zoosporic aquatic fungi. In a broad sense the term **water moulds** includes all aquatic fungi, but in a restricted sense only the orders Saprolegniales and Leptomitales of the Oomycetes, some of which are not aquatic! **Chytrid** is the common name for species belonging to the order Chytridiales. Other members of the class Chytridiomycetes are *not* chytrids and use of the name in any sense other than for the order is incorrect. In this chapter the term **phytopathogens** is used solely to indicate pathogens of vascular plants.

Theories on fungal phylogeny and classification cannot be restricted to phytopathogenic taxa. Much of the theoretical and practical work has been based on saprophytes which are generally easier to study, while many zoosporic phytopathogens are obligate parasites which do not lend themselves to biochemical investigation. Some of the subject matter of this chapter therefore concerns zoosporic fungi other than phytopathogens and there are further reasons why this is desirable. All organisms have their rôle in the ecosystem and some saprophytes may be crucial as biodegraders of organic material. Others are parasites of nematodes, and yet others may be found to be important in the practice of plant pathology. The best examples of this are

Table 1. The position of zoosporic fungi in the classification of the Kingdom Fungi

Kingdom Fungi
Division Eumycota (True Fungi)
  Subdivisions: Basidiomycotina
               Ascomycotina
               Deuteromycotina
               Zygomycotina
               Mastigomycotina (Zoosporic Fungi)
                  Classes: Oomycetes
                          Hyphochytriomycetes*
                          Chytridiomycetes
                          Plasmodiophoromycetes
Division Myxomycota (Slime Moulds)

* Originally written as "Hyphochytr*id*iomycetes" by Sparrow (1958); however, this is etymologically incorrect because the stem of the word is based on *Hyphochytrium* and Hyphochytriaceae (see Appendix I).

the Hyphochytriomycetes, a small class of zoosporic fungi which are often thought of as curiosities. Some, such as *Hypochytrium catenoides* Karling and *Rhizidiomyces* spp. are abundant in a wide range of soil types and may be cosmopolitan in their distribution. They have been reported as parasites on Oomycetes and may be responsible for the natural decline of phytopathogenic species in soil (Ayers and Lumsden, 1977; Sneh *et al.*, 1977; Wynn and Epton, 1979). The possible phylogenetic relationship of the Hyphochytriomycetes to the Oomycetes is of considerable importance when fungicides are claimed to have activity only against Oomycetes (see Chapter 6) for the destruction of their natural parasites may merely intensify disease problems later. It is for these reasons, and for systematic completeness, that the Hyphochytriomycetes are freely discussed here. Conversely, some fungal taxa classified with the zoosporic fungi are not known to produce zoospores, and at least some have probably lost this capability permanently. In genera such as *Pythium*, a few species seldom, if ever, produce zoospores whereas others produce them normally; in other genera such as *Peronosclerospora* and *Peronospora*, none of the species produces zoospores.

The phylogenetic relationship of Oomycetes to heterokont algae* is a long established concept, generally supported by the evidence presented here. Moreover, other evolutionary associations between zoosporic and other fungi, algae or protozoa are often suggested from experimental data. Therefore, frequent reference will unavoidably be made to lower eukaryotic organisms in addition to zoosporic fungi. Literature citations will largely be confined to review papers or very recent studies.

II. Development of Taxonomic Theory

A. HISTORICAL HIGHLIGHTS

Are aquatic fungi degenerate algae? This question has been debated since aquatic fungi were first discovered about 150 years ago. Although such discussions may be regarded as harmless and useless philosophical exercises, they are fundamental to taxonomic theory as it is difficult to conceive of a good system of classification without regard to evolution. Historically, ideas on fungal evolution can be divided into two camps; those that supported a monophyletic, and those that argued for a polyphyletic origin. In the early

---

* In the original sense, heterokont algae were those with motile spores having two flagella of unequal length but the term has come to mean those with motile spores that have flagella of different external morphology. Thus, those with one whiplash (naked) flagellum and one tinsel flagellum, regardless of length, are now generally considered heterokont. The main classes that could have a relationship to Oomycetes are the Chrysophyceae, Phaeophyceae and Xanthophyceae.

1800s fungi were generally thought to be a closely related group of organisms. However, on the discovery of aquatic forms, their relationship to algae became much debated. Braun (1847) considered fungi merely as parasitic groups of algae and lichens while Cohn (1854) claimed that the chief difference between fungi and algae was their habitat and mode of life, but he thought there should not be any systematic separation between them. Pringsheim (1858) then published a comparative account of the sexual organs and oospores of the Saprolegniaceae and Vaucheriaceae, and concluded that there was a close relationship between these families. Perhaps as a natural succession to these ideas, Sachs (1874) proposed that different fungal groups had different origins among the algae; the Chytridiales were considered to be derived from the Protococcideae, the Oomycetes from the Oosporeae (Oedogoniales), the Zygomycetes from the Conjungatae, the Ascomycetes and Basidiomycetes from the Floridean algae. The idea that the fungi were derived from Protozoa was also proposed at an early time (Dangeard, 1901; and others).

The natural relationship of the Plasmodiophorales to other organisms has been the subject of lively debate among mycologists (see Chapter 5). Until recently (Karling, 1968; Sparrow, 1973a; and others) they were placed in the Myxomycetes. Some mycologists claimed that the Plasmodiophorales evolved in a separate line from the Protozoa, and others with equally definite views claimed they evolved from the Chytridiales. Karling, who thoroughly reviewed the historical literature on this subject (Karling, 1968), concluded that the zoospores and zoosporangia of the Plasmodiophorales are more fungus-like than animal-like, and proposed that they be classified with the Phycomycetes. This proposal has not ended the debate on their origin but at least it provided a reasonable place for them in classification.

The polyphyletic origin of the Phycomycetes was not accepted by all early mycologists; De Bary (1884) was opposed to Sach's views and revived the monophyletic theory. He thought there was a link between *Pythium* or *Monoblepharis* to what he considered morphologically similar algal taxa such as *Vaucheria* or *Oedogonium*. He favoured the idea that the Chytridiales were derived from the Oomycetes by degenerative influence of parasitism and he considered the Phycomycetes (excluding the Plasmodiophorales which he thought were evolved from the Protozoa), a monophyletic group. Mez (1929) came to similar conclusions regarding the degenerative state of unicellular, endobiotic Phycomycetes except that he suggested two distinct origins: that *Monoblepharis* was related to *Codium* and degenerated into the Olpidiaceae, whereas the Saprolegniales arose from the Siphoniales and degenerated into the Olpidiopsidaceae. It is of particular interest to the development of the taxonomic theory that Mez rejected outright any idea that the biflagellate fungi were related to the uniflagellates.

Atkinson (1909), and later his student Fitzpatrick (1930), supported the monophyletic theory but opposed De Bary's view that the Chytridiales were degenerate Oomycetes and considered it more likely that the opposite was true. He based this on the fact that diplanetism is unique to the Phycomycetes but was clearly mistaken in assuming that diplanetism could be traced in a primitive state into the Chytridiales. He critically compared the various morphological forms of sexual reproduction and traced what he considered a primitive form in the Chytridiales through to the Oomycetes and to a more advanced form of development in the Zygomycetes. He thought the absence of starch in fungi was evidence against the theory that they evolved from algae and supported views that they evolved from flagellates. Furthermore, he supported those who believed the Ascomycetes evolved from the Phycomycetes, not the red algae.

Ernst Bessey (1942) also supported the idea of a monophyletic origin for Phycomycetes and suggested that they evolved from unicellular heterokont algae. The basis of his theory was that the uniflagellate Chytridiales and Hyphochytridiales each evolved after the loss of one flagellum; the Chytridiales following the loss of the tinsel flagellum and Hyphochytridiales following the loss of the whiplash flagellum. He regarded the Plasmodiophorales as an entirely different group, related to the Myxomycetes, and evolved from the Protozoa.

The modern theory on the classification of zoosporic fungi is derived from concepts based on the flagellar types of the zoospore that were long used in algal classification. Lotsy (1907) and Scherffel (1925) each proposed that the uniflagellate and biflagellate fungi constituted distinct and separate series but it was Sparrow (1935, 1943, 1960, 1973a) who firmly entrenched this idea into the present-day system. Thus, the Oomycetes have a tinsel and a whiplash flagellum which are generally unequal in length. The Plasmodiophoromycetes have two whiplash flagella of unequal length. The Chytridiomycetes have a single, whiplash flagellum and the Hypho-chytriomycetes a single, tinsel, flagellum.

B. THE FLAGELLAR APPARATUS

1. *Overview*

Because flagellar type is so fundamental to the classification of zoosporic fungi, it is discussed first although the issue is complex because it involves the ultrastructure of the zoospore (covered in detail in Chapter 1) and, in particular, the flagellar apparatus, i.e. the flagellum proper, the transition zone, the kinetosome and the rootlet system. These are considered conservative characteristics, little affected by environmental pressures, and have been retained during the course of evolution. They are therefore very useful for developing phylogenetic and taxonomic concepts. The $9 + 2$ pattern of microtubules in flagella and cilia and the architecture of centrioles and kinetosomes (Wolfe, 1972), are examples of extreme conservatism because these patterns occur in almost all eukaryotic organisms. However, modifications have occurred to the rootlet system, the transition zone, and to the external surface of the flagellum proper during the course of evolution. These modifications will be discussed in relation to the taxonomic theory.

2. *Flagellar types*

Among zoosporic fungi there are two types of flagella. The whiplash (smooth or naked) type trails behind the zoospore and provides forward thrust by a wave-like motion, whereas the tinsel (flimmer) flagellum, so called because of hair-like mastigonemes on its surface, is directed forward and pulls the zoospore. The mechanics of flagellar propulsion were explained by Sleigh (1982). The mastigonemes are formed during zoosporogenesis in cisternae of the endoplasmic reticulum that buds from the nuclear envelope (Heath *et al.*, 1970; Bouck, 1972). It is not known how they are prevented from attaching to the whiplash flagellum in heterokont biflagellated organisms.

The mastigonemes are attached in two opposite rows to the sheath of the anterior flagellum in at least some taxa of the ·Saprolegniales and Peronosporales and in many heterokont algae. However, in algae there is some variation in this arrangement, and Bouck (1972), stated that the classification of organisms on the basis of mastigoneme distribution has not withstood the test of time. Whether this statement will be applicable to Oomycetes and Hyphochytriomycetes remains to be resolved.

The flagellar type is important in fungal classification rather than the position of its attachment relative to the movement of the zoospore. In primary, pear-shaped zoospores of the Oomycetes, the flagella are generally attached anteriorly, whereas in secondary, kidney-shaped zoospores, they are attached laterally. In the Chytridiomycetes, the single whiplash flagellum is usually posterior but it may be lateral or even anterior. When it is anterior in the Chytridiomycetes, it trails around the side of the zoospore and is directed backwards. Ultrastructural examination of these zoospores shows that the whole body is turned around so that what is structurally the posterior is in the forward part of the zoospore relative to the direction of its movement.

3. *Kinetosomes and Centrioles*

The term kinetosome is synonymous with basal body or blepharoplast, and is generally used in mycological literature. Each flagellum is attached to a kinetosome which is, in fact, a flagellated centriole. The primary function of the centriole is probably for the assembly of microtubules, particularly those of the flagellum (Pickett-Heaps, 1969). In all four classes of zoosporic fungi there are two centrioles, and, at zoosporogenesis in the Oomycetes and Plasmodiophoromycetes, both function in flagellum formation. In the Chytridiomycetes and Hyphochytriomycetes, one centriole remains non-functional in that it does not produce a flagellum. It has been suggested that the nonfunctional centriole is a vestigial kinetosome and indicative of a biflagellate origin for the uniflagellate zoospore (Koch, 1956; Olson and Fuller, 1968). In zoospores and gametes of heterokont algae, the tinsel flagellum is longer than the whiplash flagellum, which, in many, is short or vestigial. In some it is absent but in these there is a nonfunctional centriole (Manton, 1959; Hibberd and Leedale, 1972) as in the Hyphochytriomycetes. Thus, there is good reason to suppose that the nonfunctional centriole in the Hyphochytriomycetes is also vestigial. However, the same argument cannot be made for the Chytridiomycetes because there is no other group of organisms that is obviously related to them for this kind of comparison. The Chytridiomycetes may be among the most primitive eukaryotic organisms and the nonfunctional centriole may therefore be an embryonic rather than a vestigial body.

The relative position of the centrioles prior to zoosporogenesis, and the rearranged position of the kinetosomes in biflagellated zoospores, or kinetosome and nonfunctional centriole in the uniflagellate zoospores, may be important to taxonomic theory because it relates to locomotion. In the Oomycetes the centrioles are aligned at 180° before zoosporogenesis (Heath, 1980). In the primary zoospore the angle is close to 90° and in the secondary zoospore about 130–150°. The 130–150° arrangement and position on the side of the zoospore facilitates movement whereas the narrower angle of the kinetosomes in the primary zoospore is distinctly awkward for movement and suggests that the ancestral type had two similar flagella. In the Hyphochytriomycetes the centrioles are at approximately right angles before zoosporogenesis (one report, see Heath, 1980), but during zoosporogenesis the nonfunctional centriole shifts to a 130–150° position (Barr, 1981). This suggests that the second flagellum was whiplash at some earlier time in evolutionary history because it would have been directed backwards. In the Chytridiomycetes, the centrioles are at right angles prior to zoosporogenesis but there is considerable variation in arrangement in the zoospores between the different taxa. In the Blastocladiales, the angle remains at 90° whereas in the Chytridiales the nonfunctional centriole shifts to a parallel position for which there is no clear explanation. In the Plasmodiophoromycetes, the centrioles are aligned at 180° before zoosporogenesis but shift to about 120–150° in *Polymyxa graminis* Ledingham (Barr and Allan, 1982), to 45° in *Plasmodiophora brassicae* Woronin (Aist and Williams, 1971), to 30–35° in *Woronina pythii* Goldie-Smith (see Chapter 8), and 55° in *Sorosphaera veronicae* Schroeter (Talley *et al.*, 1978). In *Polymyxa* zoospores, the shorter

of the two flagella is usually turned backwards so that both flagella beat in a
similar direction; however, motion of these zoospores is seemingly awkward
and the lesser angle in other taxa may be more efficient.

4. *The Transition Zone*

In contrast to the largely uniform structure of kinetosome as well as flagella
or cilia in eukaryotic organisms, the transition zone between them displays
unique characteristics that distinguish each of the four classes. While the
phylogenetic association of green algae and higher plants is long established,
it is not widely appreciated that the flagellar transition zone of Chlorophyta
zoospores and spermatozoa of primitive vascular plants have a unique star
shaped structure (Manton, 1965). A different structure, described as a helix,
is found in the core of the transition zone in heterokont algae, except the
Phaeophyceae (Hibberd, 1979) and this may be similar to the concertina-like
structure in the Oomycetes (Heath and Greenwood, 1971). Recent studies on
Oomycetes and Hyphochytriomycetes (Barr, 1981) suggested this structure
is a series of appressed rings alternating in diameter and that the structure
in heterokont algae may be a series of rings with equal diameter, instead of a
helix. If it proves to be correct that all these organisms have a series of rings
in the transition zones of their flagella, it will be a powerful indicator of a
phylogenetic relationship.

The transition zone in the Chytridiomycetes is quite different and includes
a fine spiral structure that is attached to the A-tubule of the flagellar axoneme
(Barr and Hadland-Hartmann, 1978). There are also distinctive kinetosome
props that attach the C-tubule to the plasmalemma and as far as it known the
particular architecture of these props is unique to the Chytridiomycetes
among eukaryotic organisms (Barr and Hadland-Hartmann, 1978).

In the Plasmodiophoromycetes details of the transition zone have been
reported in *Polymyxa graminis* (Barr and Allan, 1982). It lacks the
arrangements found in the other zoosporic fungi and in its simplicity is
similar to the transition zone found in flagellated and ciliated protozoa.

5. *The Rootlet Complex*

It is generally accepted that the rootlet complex, which consists of systems of
microtubules and fibrils, provides an anchoring mechanism for the flagellum
and also cytoskeletal support for the zoospore and its organelles. Manton
(1965), stressed the phyletic significance of rootlets and said it represented a
new field for investigation. Numerous investigations, for example, in the
Chlorophyta (Markowitz, 1978; Moestrup, 1978; Melkonian, 1979), the
Chrysophyceae (Bouck and Brown, 1973; Hibberd, 1976a; Schnepf *et al.*,
1977), colourless flagellates (Hibberd, 1976b), in swarm cells of slime moulds

(Haskins, 1978) and in Protozoa (Lynn, 1981) have confirmed that the rootlet complex is varied, but within groups, certain similarities and presumed phylogenetic traits exist.

The different arrangements of rootlet systems among the four classes of zoosporic fungi have been reviewed elsewhere (Barr, 1981). The Oomycete zoospore has the most complex arrangement of microtubules, comprising six systems. At least some of these systems are also found among heterokont organisms such as *Ochromonas* (Bouck and Brown, 1973) and *Poterioochromonas* (Schnepf *et al.*, 1977). The considerable uniformity in the rootlet systems of Oomycetes, except the Thraustochytriales, is a strong indication that the class has a recent origin (Barr, 1982). The minor variation found in rootlet morphology of the Thraustochytriales indicates that this order is only distantly related to other Oomycetes, as also suggested on other evidence by Perkins (1974).

Hyphochytriomycete zoospores also have rootlet systems that are somewhat similar to those in heterokont algae (Barr, 1981) but the overall arrangement is different from rootlets in Oomycetes. It is unlikely, on this evidence, that the Hyphochytriomycetes evolved recently from the Oomycetes following "loss" of one flagellum. However, if the rootlet microtubules associated with the second kinetosome were also lost, and the remaining rootlet components rearranged, evolution from a common ancestor would conceivably be possible (Barr, 1981).

The variation of the rootlet systems among the Chytridiomycetes provides further strong evidence that the group has a very long evolutionary history. Certain patterns can be traced in different taxa and these show the Monoblepharidales are probably evolved from the Chytridiales, and the Blastocladiales from the Spizellomycetales. With the exception of the Blastocladiales, there is considerable variation in microtubule arrangement within each of the other orders. In the Spizellomycetales, these differences have been used formally to describe genera (Barr, 1980a) and there is enough evidence to suspect this may eventually be possible for the Chytridiales and Monoblepharidales. The use of ultrastructural characteristics in classification does not mean, however, that specimens must be examined in the electron microscope for routine identification for once the naturalness of a grouping has been established, one finds other characteristics to use for an identification key.

Among the Plasmodiophoromycetes, rootlet systems have been examined in detail only in *Polymyxa graminis* (Barr and Allan, 1982); however, studies on the zoospore of *Sorosphaera veronicae* (Talley *et al.*, 1978) indicate it may have similar rootlets. The *Polymyxa* rootlet system is very simple compared with the Oomycetes and does not bear any resemblance to rootlets in any other class of zoosporic fungi. Although somewhat similar to the protozoan rootlets such as those in the ciliate *Cyrtolophosis* (Didier *et al.*, 1980), it is simpler than any protozoan rootlet system thus far studied, and the zoospore lacks pellicular microtubules.

C. THE THALLUS AND LIFE CYCLE

1. *Overview*

Thallus morphology and life cycles in zoosporic fungi have been the subject of intensive investigation for over 100 years. In spite of this, many details are still unclear because of technical difficulties in culturing and in synchronising life cycle stages. Furthermore, although the resolution of the light microscope often proves inadequate, ultrastructural examination is more time consuming and expensive. Classification has been based primarily on thallus morphology but the discussion here will concentrate on recent studies that are influencing phylogenetic and taxonomic thinking.

2. *Evolution of the Thallus*

Thallus morphology is an important basis for distinguishing different families and orders of zoosporic fungi. There are two principal forces determining the type of thallus development. One is the evolutionary pressures that tend to result in greater complexity; the other is the degenerative influence of parasitism and of ecological specialisation in general that tends to result in morphological simplicity. However, the wide assortment of thallus forms is not difficult to understand if the evolutionary process is followed. It has been assumed that the place and movement of the nucleus in the thallus is all important in this process (Barr, 1978) and this has been part of the basis for a major reclassification of the Chytridiomycetes (Barr, 1980a) although it is applicable to other groups of organisms. Basically, there are four main steps in the evolutionary process as it relates to the nucleus, and resulting thallus morphology.

The most primitive type of thallus development is where the nucleus remains in the zoospore cyst which enlarges into a sporangium (**endogenous sporangium development**). The development of anucleate rhizoids is explained as a means of enlarging the absorptive surface and in many forms the sporangium and its rhizoids lie amongst the substrate. In others such as *Rhizophydium graminis* Ledingham, the sporangium is positioned on the host surface and a host cell is penetrated by rhizoids. This has the advantage of providing a stable food supply for the short time it takes the fungus to complete its development; however, the basic type of development is simple because the nucleus remains in the zoospore cyst which develops into a single sporangium.

The next step in the evolutionary process is migration of the nucleus from the zoospore cyst into the germ tube (**exogenous sporangium development**). In the simpler forms, a single sporangium develops in the germ tube. In some endoparasitic species such as *Olpidium brassicae* (Woronin) Dang, there are no rhizoids but there is an intimate association of host and parasite. In saprophytic species such as *Entophlyctis variabilis* Powell, rhizoids generally develop. The fact that the nucleus moves into the germ tube, thus enabling these organisms to penetrate a host or substrate, is the important indicator of the evolutionary process, not the mere presence or absence of rhizoids. The advantages that these organisms have over those that develop sporangia external to their host cells are that their reproductive body (sporangium or resting spore) is both closer to its food supply and protected by the host cell. It is unlikely by chance alone that the nucleus is able to move out of the zoospore cyst in species with this type of sporangium development. However, with the exception of the Plasmodiophorales, the cytological changes that permit this to happen have not been adequately studied. It is most unlikely that endogenous sporangium development would at any time evolve from the exogenous type of development among heterotrophic organisms because there would be no advantage gained, and much to lose.

The major step in evolution came when nuclei were able to behave independently following mitosis, thus giving rise to polycentric development. In the **colonial polycentric** type such as occurs in the Synchytriaceae, daughter nuclei can behave independently, resulting in a number of sporangia developed in a single sorus; in the **filamentous polycentric** pattern such as in *Physoderma*, daughter nuclei are able to migrate independently in the rhizomycelium. The fungus is thus able to grow to new food supplies and increase its chance of survival by producing many fruiting bodies, and also to survive for a long period with the capability of fragmentation and without the necessity of zoosporogenesis.

All four types of development are found among the Chytridiomycetes. However, the evolution of the thallus in other classes of zoosporic fungi is not understood because the morphologies of their immediate ancestor are not known. The Oomycetes, for example, may have evolved directly from a filamentous autotrophic alga, from a unicellular colourless flagellate, or possibly from some other morphological form. In Oomycetes moreover, evolutionary pressures have resulted in two main adaptations. One is in response to the terrestrial environment. Here sporangia are able to resist unfavourable conditions such as drought, and are able to germinate by germ tubes instead of zoospores. In other species the degenerative influence of parasitism results in a morphologically simple thallus. The space limitations within a host cell encourage monocentric development. The evolution of the various patterns of thallus development have occured in phylogenetically different groups of organisms and there are many examples of morphological parallelism and convergent evolution. The classical example of the latter is found in *Olpidiopsis*, *Anisolpidium* and *Olpidium* which are holocarpic, monocentric obligate parasites belonging to the classes Oomycetes, Hyphochytriomycetes and Chytridiomycetes, respectively. When morphological degeneration occurs as a result of parasitism, it is unlikely that there would subsequently be a reversal of the process because of natural genetic selection. Instead, the organism would either die out, or evolve in a coevolution sequence with its host as has been well documented for the Uredinales (Savile, 1979). However, coevolution has not yet been critically considered as a factor among lower fungi.

3. *Mitosis*

In reviews on mitosis in fungi (Fuller, 1976; Heath, 1978, 1980; Kubai, 1980; and others) most authors have speculated on the evolution of the mitotic process, arguing that it is fundamental to all eukaryotic organisms, and not likely to have arisen through multiple endosymbiotic events (Oakley, 1978). It has been argued that the characteristics of mitosis are more reliable evolutionary markers than any other criterion and there are a number of interesting mitotic characteristics that vary in different organisms (conveniently tabulated with references by Heath, 1980). Unfortunately many reports on individual species do not include all pertinent details so that few comparisons can be made. Moreover, only *Rhizidiomyces* has been examined in the Hyphochytriomycetes (Fuller, 1976), and only two species of the Plasmodiophoromycetes, *Sorosphaera veronicae* (Braselton *et al.*, 1975; Dylewski *et al.*, 1978), and *Plasmodiophora brassicae* (Garber and Aist, 1979a). Mitosis in the Chytridiomycetes is better understood, but it is only the Oomycetes, including many phytopathogenic species, that have been examined in sufficient numbers to give reasonably conclusive information on the mitotic characteristics for the class.

The mitotic characteristics that can be compared in the various classes of zoosporic fungi at this time are: (1) the state of the nuclear envelope; (2) the angle that the pair of centrioles assume; (3) the state of the nucleolus; and (4) the arrangement of the chromosomes at metaphase. In many lower eukaryotic organisms the nuclear envelope remains more or less intact, or "closed", during mitosis. In the Blastocladiales and Oomycetes it is closed with the spindle apparatus formed within the nucleus. In the Chytriodiomycetes excluding the Blastocladiales, and in the Hyphochytriomycetes and Plasmodiophoromycetes there are polar fenestrae that allow entry of spindle microtubules. In the Oomycetes and Plasmodiophoromycetes the centrioles are aligned at 180° whereas in the Chytridiomycetes and Hyphochytriomycetes they are placed at 90°. The nucleolus is persistent during mitosis in the Oomycetes and Plasmodiophoromycetes. In the Hyphochytriomycetes the nucleolus is dispersed, whereas in the Chytridiomycetes it becomes indistinguishable from the ground nucleoplasm, or it is expelled from the nucleus during telophase (Powell, 1980), (although in one species, *Catenaria anguillulae* Sorokin (Ichida and Fuller, 1968) it is persistent). The chromosomes are not aligned along the metaphase plate in the Oomycetes, whereas in the other three classes, alignment is conspicuous.

The striking cruciform-like nuclear division in the Plasmo-diophoromycetes can be seen clearly and photographed in the light microscope (Miller, 1958) and may prove to be a useful class characteristic. It is, in fact, a combination of characteristics 3 and 4 (above) and has been examined ultrastructurally (Keskin, 1971; Dylewski *et al.*, 1978). The nucleolus is stretched out perpendicularly to the chromosomes aligned along the metaphase plate.

There may be much less variation in mitosis among the Oomycetes than there is among the various orders of the Chytridiomycetes. Powell (1980) reviewed some subtle differences between the Chytridiales (*Phlyctochytrium irregulare* Koch and *Rhizophydium sphaerotheca* Zopf) and the Spizellomycetales (*Entophlyctis*), whereas the absence of polar fenestrae in the Blastocladiales sets this order apart from the other Chytridiomycetes. The one species of the Hyphochytriomycetes examined (*Rhizidiomyces apophysatus* Zopf) (Fuller and Reichle, 1965) is much more like the Chytridiomycetes than the Oomycetes in mitosis. On present evidence, however, it is difficult to make any predictions about the relationship of any of the zoosporic fungi with other fungi, or other lower eukaryotes.

4. *Asexual Reproduction*

The advantage of sporangial reproduction in zoosporic fungi is that it is ideally suited to heterotrophic organisms with absorptive nutrition. While the food supply lasts, the organism undergoes rapid growth. When the food supply becomes exhausted, there is rapid maturation of sporangia, zoo-sporogenesis and dispersal. The large number of zoospores produced at one time favour survival. The presence of sporangia in all four classes is an example nonetheless of parallel evolution and not an indicator of relation-ships between them.

Only aspects of asexual reproduction with a significant impact on biosystematics will be discussed here. Firstly, sporangium morphology is highly variable in many taxa. Size varies not only between different substrates or hosts, but even under standard conditions. Moreover, because of limited information on sexual reproduction in the Hyphochytriomycetes and Chytridiomycetes, there has been an overdependence on asexual reproduction as a taxonomic criterion.

The presence of an operculum in many chytrids has long been considered important (Braun, 1856), and was the basis for subdividing these fungi in two series (Sparrow, 1943, 1960, 1973b) although numerous authors now oppose this view (see, for example, Karling, 1977; Barr, 1980a).

A vesicular type of zoospore release is found in certain taxa of the Oomycetes, Hyphochytriomycetes and Chytridiomycetes and, in the latter, fully formed zoospores are released with gelatinous matter. At first they do not move, but within a few minutes there is flagellar movement and, soon, the vesicle bursts. The method of discharge may be useful for distinguishing species and possibly genera among the chytrids (Barr, 1975). In the Oomycetes and Hyphochytriomycetes, undifferentiated protoplasm is exuded from the sporangium and the zoospores are differentiated in the vesicle. The presence of the vesicular type of release is the formal characteristic for distinguishing *Pythium* from *Phytophthora* (see Chapter 4), and is also used for distinguishing taxa of the Hyphochytriomycetes, although there it is of unproven reliability.

5. *Sexual Reproduction and Life Cycles*

Among zoosporic fungi sexual reproduction is only relatively well understood in the Oomycetes for there it can often be demonstrated in culture. In the other classes, with few exceptions, it has not been easy to demonstrate. Sexual reproductive morphology is important, however, because it is generally regarded as a conservative characteristic and for this reason much stress has been placed on it in the development of taxonomic theory in all groups of eukaryotic organisms. The life cycle is also important because it indicates the progress of evolution within a group with a long evolutionary history; those with a gametophyte (haploid) generation are generally considered to be more primitive than those where the sporophyte generation dominates.

The oogamous type of reproduction in the Oomycetes is their major distinguishing characteristic, and morphological variations in the process are important taxonomic criteria. Although oogamous reproduction occurs in many eukaryotic organisms, including certain of the Chytridiomycetes, in the Oomycetes it is unique because there is no motile stage; the oosphere is fertilised within the oogonium either by fusion of gametic nuclei generated in the oogonium or as a result of transfer of a gametic nucleus from an antheridium. The non-motile means of fertilisation would have facilitated adaptation to the terrestrial environment and may in part explain how a group of organisms with an aquatic origin have become successful plant parasites. However, the same argument would not explain any advantage for the strictly aquatic members although there would not necessarily be any disadvantage other than to discourage heterothallism. Much of the early discussion on Oomycete evolution was based solely on a gross morphological comparison with oogamous reproduction in the heterokont algae, the Monoblepharidales or the Blastocladiales where it is also found.

Dick (1969) emphasised the importance of the distinctive internal oospore structure and the developmental patterns during oosporogenesis and proposed three groups based on sexual reproductive morphology. He concluded that there was a very wide gulf between the Saprolegniaceae (group 2) and other Oomycetes (groups 1 and 3) and suggested that during the course of evolution, homothallism gave way to heterothallism and the suppression of morphological features associated with sexuality.

Meiosis in the Oomycetes has been a subject of much controversy (see Chapter 3), but in recent years the consensus of opinions (see Dick, 1972) indicates that it is gametangial in the Saprolegniaceae and Peronosporaceae. The vegetative and asexual stages are therefore diploid, or possibly polyploid in some species (Win-tin and Dick, 1975). The contention that meiosis is gametangial is supported by genetical studies on *Phytophthora cactorum* (Lebert & Cohn) Schroeter (Elliott and MacIntyre, 1973), *P. drechsleri* Tucker (Khaki and Shaw, 1974) and *Pythium aphanidermatum* (Edson) Fitzpatrick (Dennett and Stangehellini, 1977). Dennett and Stanghellini (1977) further suggested that the 1:5.83 segregation ratio in the sexual progeny of a drug-resistant mutant of *P. aphanidermatum* could be explained by fusion of heterozygous gametangia. In this case, the oospores would be tetraploid and at meiosis would produce diploid hyphae.

There is no cytological information on the root parasitic Oomycete *Lagena radicicola* Vanterpool & Ledingham which is the only known phytopathogen in the Lagenidiales. The time of meiosis in other Lagenidiales is also far from clear at present. In *Lagenisma coscinodisci* Drebes, a parasite in a marine diatom, Schnepf *et al.* (1978a,b) found some evidence suggesting that meiosis occurs in the zoosporangium whereas in *Lagenidium callinectes* Couch, a parasite on crab ova, Amerson and Bland (1973) observed synaptonema-like complexes in zoospores. The position of meiosis in the holocarpic Lagenidiales is rather fundamental to evolutionary concepts of the Oomycetes for if this order has a haploid life cycle, it is more primitive than the Saprolegniales and Peronosporales, which are generally believed to have a diploid thallus. The Lagenidiales may then be the ancestral group in the Oomycetes but such a process would require a long evolutionary history or a simultaneous change of position in the life cycle of both karyogamy and meiosis. If the Lagenidiales prove to be diploid, with meiosis occurring in the gametangia as in the Saprolegniales and Peronosporales, they may be merely morphologically reduced, as already suggested.

There is no conclusive evidence that sexual reproduction occurs in the Hyphochytriomycetes. However, in the marine species *Anisolpidium ectocarpii* Karling, Johnson (1957) observed apparently fortuitous pairing of uninucleate spore protoplasts inside cells of *Ectocarpus*, followed by zygote development. If this proves to be sexual reproduction, then it is very different from oogamous reproduction in the Oomycetes and indicative of a somewhat different ancestral origin.

Sexual reproduction in the Chytridiomycetes is extremely varied and has been extensively reviewed by Sparrow (1960); however, opinions differ on the type of sexual reproduction in the phytopathogenic species. Biflagellate or multinucleate zoospores are often seen, particularly in culture, but in most instances this is most likely due to incomplete differentiation of zoospores at zoosporogenesis. Moreover, although once it was generally assumed that thick-walled spores had a sexual origin, it is now known that asexual sporangia may become thick walled and survive desiccation. In the root parasite *Rhizophydium graminis*, Ledingham (1936) reported that epibiotic resting spores were sexually produced following fusion of rhizoids inside root hairs, but Barr (1973) claimed thar resting spores were merely resistant sporangia formed in the absence of rhizoid fusion. However, Moore and Miller (1973) have shown conclusively that resting spores can result in culture following rhizoid fusion in the saprophyte *Chytridiomyces hyalinus* Karling. In *Olpidium* it has generally been assumed, since the classical work of Kusano (1912) on *O. viciae* Kusano, that sexual reproduction results from fusion of isogamous gametes. In *O. radicale* Schwartz & Cook (*O. cucubitacearum* Barr & Dias) zoospores may come together at the time of encystment (Barr, 1968) but this act was not proved to be sexual since the fate of these cysts was not followed. In *O. brassicae* the resting spore is more likely to be asexual in origin and sexual reproduction has not been demonstrated. In *Synchytrium* various forms of sexual reproduction have been reported (Karling, 1977) but only fusion of isogamous gametes has been adequately verified. Apparently, in both *Olpidium* and *Synchytrium*, zoospores may behave as gametes, or encyst without fusion and reproduce new sporangia asexually. It is generally accepted that resting spore formation in *Synchytrium* follows gamete fusion and the haploid state is restored during resting spore development or germination but much less is known about the life cycle of *Olpidium*.

In the Blastocladiales, the life cycles have been critically studied (Emerson, 1941; Olson and Borkhardt, 1978; Olson and Reichle, 1978) and meiosis occurs during resting spore development or germination. *Physoderma* is the only phytopathogenic genus and in many species two distinct life cycle phases are reported (Sparrow, 1947; Lingappa, 1959a,b; Karling, 1977); an ephemeral, epibiotic state that produces zoospores capable of acting as gametes, and a polycentric, endobiotic state that produces resting spores. Cytological evidence on *P. maydis* Miyabe (Lange and Olson, 1980a) suggests that the epibiotic state is haploid and the polycentric one diploid with meiosis occurring prior to resting spore formation. In contrast to the Blastocladiales, the more primitive unicellular Chytridiales and Spizellomycetales are generally presumed to be haploid.

Until recently, evidence for sexual reproduction in the Plasmodiophoromycetes was largely indirect (see Chapter 5). However, critical observations by Kole (1954) on *Spongospora subterranea* (Wallr.) Lagerheim and by Keskin (1964) on *Polymyxa betae* Keskin have confirmed that fusion of "zoospores" actually takes place. In zoospore suspensions of *Polymyxa graminis* the proportion of larger than normal, quadriflagellate zoospores would not account for the large number of cystosoral plasmodia that develop two days later (D. J. S. Barr, unpublished observations). One explanation is that cell fusion occurs at more than one stage of development, but perhaps in some isolates would occur predominantly at one stage. Recent ultrastructural evidence in *Plasmodiophora brassicae* has shown that karyogamy takes place prior to sporogenesis in the cystosoral plasmodium (Buczacki and Moxham, 1980). The widely held assumption that meiosis occurs concurrently with cleavage into cysts (resting spores) has been supported by electron microscopy on *Sorosphaera veronicae* (Harris *et al.*, 1980) and *Plasmodiophora brassicae* (Garber and Aist, 1979b). The diploid phase in Plasmodiophoromycetes then is very short and characteristic of a simple type of life cycle. The zoosporangial plasmodia are haploid and zoospores may, in some species, reproduce the zoosporangial cycle, or act as gametes that initiate the sexual and cystosoral cycle (Kole and Gielink, 1963).

D. CHEMOTAXONOMY

1. *Overview*

Chemotaxonomic approaches have their impact at widely different levels of classification. For example the presence of specific metabolic pathways helps develop concepts on the interrelationships among major groups of eukaryotic organisms and the theory on evolution. Serology, conversely, has its impact essentially on the development of species concepts.

It is all very well to devise biochemical approaches to classification, but quite a different matter to implement these into a working system. A complex and expensive test is generally impractical for routine identification of a species. Indeed, difficulties arise when studies other than those connected with light microscopy are undertaken. However, a biochemical examination may be useful in establishing concepts of the relationships among taxa and this has been done when there is uncertainty about the position of a species, or a higher taxonomic group that lacks some of the essential characteristics for its inclusion in the classical system. A good example is the absence of sexual reproduction in some Oomycete-like fungi. Such tests generally need to be done only once. Alternatively, a simple test such as response to temperature may be reliable and is usefully applied as an aid for routine identification of certain species of zoosporic fungi. The application of biochemical or physiological tests moreover, has its limitations in the systematic study of zoosporic phytopathogens because most are obligate parasites, and, for the foreseeable future they will only be applicable to *Achlya, Aphanomyces, Pythium* and *Phytophthora.*

2. *Physiology and Nutrition*

Extensive reviews of the relevance of physiology to phylogeny have been made by Cantino (1950, 1955), and Cantino and Turian (1959). They proposed that in heterotrophic organisms there is a gradual loss of the ability to synthesise cellular substances, and an increase in dependence on the substrate or host. Moreover, once the synthetic capacity is lost, it is not regained. The characteristics used in their scheme included: (1) loss of capacity to reduce sulphate and to use it as the sole source of sulphur for growth; (2) loss of capacity to synthesise one or more vitamins; (3) loss of capacity to utilise nitrate and ammonium salts as the sole sources of nitrogen. Gleason (1976) reviewed more recent literature as far as the lower fungi are concerned.

The requirement for organic sulphur appears to be common to all members of the Blastocladiales and Saprolegniales. The loss of the ability to synthesise one or more vitamins has occurred in several species scattered among the various orders of the Oomycetes and Chytridiomycetes and there does not appear to be a trend among taxa at the ordinal rank. However, among species and genera it may be of taxonomic interest. The most relevant studies on vitamins among these fungi have been with *Pythium* and *Phytophthora* (Roncadori, 1965; Ridings *et al.*, 1969). All species of *Phytophthora* required an exogenous source of thiamine as does *Pythium vexans* De Bary but 14 other species of *Pythium* required only the pyrimidine moiety of thiamine and 22 other species of *Pythium* did not require thiamine in any form. However, among *Pythium* there was no morphological correlation between those species requiring pyrimidine and those without this requirement.

Data on nitrogen requirement (Gleason, 1976) are difficult to assess in relation to Cantino and Turian's concepts on physiological phylogeny for so few species have been examined although some generalised observations can be made. The Oomycetes, including *Pythium* and *Phytophthora*, can utilise ammonium nitrogen and some species can also utilise nitrate nitrogen (Roncadori, 1965; Kraft and Erwin, 1967). It is possible that the Saprolegniales cannot utilise nitrate nitrogen, in which case they would be phylogenetically more advanced than the Peronsporales according to this scheme.

Very limited nutritional studies have been made with the Hyphochytrio-mycetes. *Hyphochytrium catenoides* Karling is unable to utilise ammonium and nitrate nitrogen, but can use inorganic sulphur (Barr, 1970a), but no definite phylogenetic conclusions can be drawn from this.

Many species of the Chytridiales and Spizellomycetales can utilise both ammonium and nitrate nitrogen; some require organic nitrogen. Many grow better on organic than they do on inorganic nitrogen, but this could be related to changes in pH values during growth, or the presence of essential trace substances. The Monoblepharidales and Blastocladiales, which are phylogenetically more advanced (Barr, 1978, 1981) and probably represent two different evolutionary lines, may have lost the ability to use nitrate nitrogen, while some of the Blastocladiales require organic nitrogen. Nonetheless, Cantino and Turian's theory is probably better regarded as a phylogenetic trend with no immediate taxonomic use.

The ability to utilise complex organic materials such as cellulose and chitin could be regarded as an evolutionary advance. However, fungi with chitin in their cell walls produce endochitinase, just as Oomycetes produce endo-cellulases. These enzymes are necessary for softening the hyphal tip and wall, both for tip elongation and branching (Gooday and Trinci, 1980). Their presence as extracellular enzymes is therefore not surprising. Nonetheless, for the very reason that Oomycetes have cellulose in their walls, it was assumed until recently that they were incapable of utilising cellulose as a carbon source. Extracellular cellulase activity has, however, been detected *in vitro* in species of *Pythium*, *Phytophthora*, *Achlya*, *Aphanomyces* and *Thraustotheca* (Unestam, 1966; Nemec, 1974; Park, 1975; Berner and Chapman, 1977; Miele and Linkins, 1978), and *in vivo* by *Pythium* and *Sclerospora* (Vidhyasekaran, 1971; Sharma and Wahab, 1976). Interestingly, *Phytophthora cinnamoni* Rands has very low (Weste, 1978) or no cellulase activity (McIntyre and Hankin, 1978) whereas some other species produce extracellular cellulases. Comparative studies on other Oomyceyes indicate there may be specific differences in extracellulase activity (Nemec, 1974; Berner and Chapman, 1977). Among the Hyphochytriomycetes, *Hypochytrium catenoides* degrades cellulose fibres *in vitro* (Barr, 1970a) and it is well known that certain Chytridiomycetes grow on and degrade cellulosic substances (Sparrow, 1960).

Chitin decomposition has been demonstrated in *Chytridiomyces* (Reisert and Fuller, 1962), which presumably contains chitin as a wall constituent. The utilisation of N-acetyl-D-glucosamine, the component of chitin, can be employed as a test to distinguish certain unicellular Chytridiales and Spizellomycetales (D. J. S. Barr, unpublished observations).

Miele and Linkins (1978) demonstrated *in vitro* that extracellular cellulase from *Achlya bisexualis* Coker & Couch binds as an active hydrolase to both the substrate and to its own wall which is then disrupted. However, an organism with enzymes potentially capable of breaking down its own walls, must have some protective mechanism if it is to survive in nature. It is therefore the evolution of protective mechanisms that are of interest rather than the mere presence of cellulases in Oomycetes, or chitinase in other fungi.

Any test for detecting the presence of extracellular enzymes must be qualitative and simple to be of routine use in identification. Ho and Foster (1972) demonstrated such a test using an artificial starch agar medium and iodine to show the radial extent of starch utilisation in the medium. They grew 11 *Phytophthora* species on this medium and were able to classify species into three categories.

Response to temperature in *Pythium* was tested by Middleton (1943), and was suggested as a useful characteristic for separating morphologically similar species of *Phytophthora* in culture (Ribeiro, 1978). *Olpidium brassicae* and *O. radicale* also have different responses to temperature (Barr, unpublished observations), but it is not practical to use this as a taxonomic criterion because they are obligate parasites. Among isolates of any species there is some variation, not only in the range of temperatures responded to, but also in the growth rate. The most reliable and simple criterion is derived by testing only the maximum temperature at which a fungus will grow, but even this may only be useful for distinguishing taxa with maxima 5°C or more apart.

3. Cell Walls

Early work on fungal cell-wall composition (see Aronson, 1965) was primarily cytochemical and was imprecise. Although Wettstein (1921) believed that lower fungi consisted of two groups, one with cellulosic and the other with chitinous walls, it was not until walls were examined by X-ray and other analysis that an accurate inventory on carbohydrate composition of various fungi was assembled (Bartnicki-Garcia, 1968). It was partly on this evidence that Bartnicki-Garcia (1970) supported earlier claims by Gäumann (1964) and others that the Chytridiomycetes are probably the ancestral group to the Zygomycotina, Ascomycotina and Basidiomycotina. The presence of cellulose (which was never precisely defined) in Oomycetes had long been considered an indication that they evolved from the algae.

The problems with any taxonomic theory based on cell wall composition are the exceptions that apparently do not fit into the taxonomic position traditionally assumed for them. Thus, evidence that the Hyphochytriomycete *Rhizidiomyces*, and the Oomycetes *Apodachlya* and *Leptomitus* contain both cellulose and chitin (Fuller and Barshad, 1960; Lin *et al.*, 1976; Aronson and Lin, 1976, respectively) has thrown confusion into the matter. More recently, in a comparative study on cell-wall sugars in 41 Oomycetes, Vaziri-Tehrani and Dick (1980a) reported glucosamine (which may indicate but does not prove the presence of chitin) in widely scattered taxa, particularly in *Achlya*, *Apodachlya* and *Leptomitus*. The presence of cellulose and chitin in walls of the ascomycetes *Ceratocystis ulmi* (Buisman) Moreau (Rosinski and Campana, 1964), as well as chitin in certain algae (Mackie and Preston, 1974;

Pearlmutter and Lembi, 1978) suggests that cell-wall composition may be a pliable characteristic related to environmental needs.

The presence of cellulose in *Ceratocystis* was explained by LéJohn (1974) who suggested that because the fungus is a plant parasite, it borrowed the host plant enzymes at some early time in evolution. Subsequently, *Ceratocystis* or its ancestor adapted its own genome to code for the enzyme necessary for cellulose synthesis. If this is correct, then cellulose may be found in walls of closely related taxa, or in some other phytopathogens.

LéJohn's theory explains the presence of cellulose in species that are not expected to have it, but it does not explain the presence of chitin in certain Oomycetes and Hyphochytriomycetes. Moreover, chitin, when present, is a major component of the fungal wall, whereas cellulose never is. Analytical work done on *Pythium* (Cooper and Aronson, 1967; Manocha and Colvin, 1968; Sietsma *et al.*, 1975) and *Phytophthora* (Tokunaga and Barnicki-Garcia, 1971) cell walls has shown that so-called cellulose is a poorly crystallised hexose polymer and actually comprises a rather small proportion of the carbohydrate content. The Oomycete cell wall consists largely of a $\beta(1,3)$-linked glucan; however, there is some difference of opinion as to the nature of the microfibrils (Sietsma *et al.*, 1975; Hegnauer and Hohl, 1978).

Present evidence shows that chitin synthesis, in spite of its requirement for at least seven enzymes (LéJohn, 1974), has been evolved a number of times (e.g. in insects, algae and fungi). Presumably, environmental pressures determine the carbohydrate content of walls; however, the evolutionary process for chitin synthesis would be slow because of the number of enzymes involved. The logical explanation for the presence of "cellulose" in the Oomycetes and the Hyphochytriomycetes is that it is a vestigial characteristic, selected against in favour of chitin. In the Chytridiomycetes, which have a long evolutionary history, cellulose synthesis, if it ever existed, has been completely lost whereas in the Oomycetes and Hyphochytriomycetes, which have a relatively short evolutionary history, it is in the process of being lost.

Vaziri-Tehrani and Dick (1980b) found that amino acid ratios in 39 Oomycetes varied widely, and concluded that differences could be related either to mating types or their ecological niche. However, they found (Vaziri-Tehrani and Dick, 1980c) that differences in amino acid ratios of valine, leucine, isoleucine, lysine and proline provided characteristic patterns for different Oomycete families.

4. *Fatty Acids*

Erwin (1973) proposed an intriguing scheme for evolution based on the polyunsaturated fatty acid content of the cell. He suggested that the volvocid green algae were the ancestral group for eukaryotic organisms because they contain the greatest number and diversity of long-chain fatty acids including both α and γ linolenic acid; the ability to synthesise these fatty acids was presumed to be suppressed and gradually lost during the course of evolution. If this is so, then both the Oomycetes and Chytridiomycetes may be evolved in separate lines from the Chrysophyta, which in turn are evolved from the Chlorophyta. Both forms of linolenic acid and other long-chain fatty acids are present in all four groups. The Ascomycotina and Basidiomycotina, which have lost their ability to synthesise γ linolenic acid, are thus evolved in one line from the Chytridiomycetes. The Zygomycotina, which have lost their ability to synthesise α linolenic acid, are also evolved from the Chytridiomycetes, but in an independent line. Although the fatty acid composition of cells deserves further consideration as a phylogenetic marker, insufficient zoosporic fungi have yet been tested (Shaw, 1965; Bean *et al.*, 1972; Southall, 1977) to support adequately Erwin's scheme. Moreover, technical problems in fatty acid analysis make it necessary to view existing results cautiously.

5. *Sterols*

Ergosterol or chemically related sterols dominate in many Zygomycotina, Ascomycotina and Basidiomycotina although there are numerous exceptions (Weete, 1974). In the Oomycetes and Hyphochytriomycetes, fucosterol and 24-methylene cholesterol are the main sterols (McCorkindale *et al.*, 1969; Bean *et al.*, 1972). Fucosterol is also the dominant sterol in the Phaeophyceae (Patterson, 1971), and has been detected in other eukaryotes including certain Zygomycotina (see Ragan and Chapman, 1978). *Pythium* and *Phytophthora* do not synthesise sterols although they require an exogenous source of them for sexual and asexual reproduction (Hendrix, 1970). Among the Chytridiomycetes only *Allomyces macrogynous* (Emerson) Emerson & Wilson and *Rhizophlyctis rosea* (DeBary & Woronin) Fischer, have been examined for sterols; neither contains ergosterol but *A. macrogynous* contains traces of fucosterol (Bean *et al.*, 1972). The taxonomic value of sterols must be viewed with much caution because there are many interlocking pathways in sterol synthesis; for example, 22-dehydrocholesterol, which is a less common sterol, has been found in *Rhizophlyctis rosea* and in a red alga which are presumed to be quite unrelated organisms. Moreover, as far as the zoosporic fungi are concerned, relatively few species have been analysed for sterols and none of those tested were phytopathogens; most phytopathogenic species either do not produce sterols or are obligate parasites. *Plasmodiophora brassicae* resting spores contain the same sterols as the host plant (Knights, 1970), which suggests an exogenous requirement for sterols in this obligate parasite.

6. Nucleic Acids

Cytoplasmic ribosomes of eukaryotes sediment by centrifugation at 80S and comprise 40 and 60S subunits. The subunits are composed of poly-ribonucleotide molecules; the 40S is made up of one rRNA species of 17–18S, and the 60S of three rRNA species with values of 5S, 5.8S and 25–28S. The molecular weights of these can be determined by polyacrylamide-gel electrophoresis and measured as 10^6 daltons (Mdal). The molecular weights of the various rRNA species in different organisms has been used to suggest relatedness. The 17–18S and 25–28S species are the ones generally studied. The assumption is that the more complex organisms have higher molecular-weight rRNA but in fungi the reverse is the situation because lower fungi have the higher molecular weights at both 17–18S and 25–28S (Lovett and Haselby, 1971; Porter and Smiley, 1979). Similarity in molecular weights of the rRNA species between organisms could be mere chance, but a distinct difference is a powerful indication of phylogenetic unrelatedness. Thus Lovett and Haselby (1971) argued that the Oomycetes had an origin independent of the Chytridiomycetes, Zygomycotina, Ascomycotina and Basidiomycotina. *Rhizidiomyces* was the only Hyphochytriomycete tested and it did not show a clear connection to either the Oomycetes or Chytridiomycetes. The significance of recent studies on rRNA in *Plasmodiophora brassicae* is discussed in Chapter 5.

Considerably more analysis has been made of fungal DNA than of RNA, but the taxonomic interpretations of the results are conflicting. Relatively little is known about the DNA sequence, and chemotaxonomic analysis has been based on the ratio of guanine plus cytosine (GC) to the total DNA. Storck and Alexopoulos (1970) tested a large number of fungi from all subdivisions and found that the GC per cent of the total DNA varied widely. For example, in 12 species of the Saprolegniaceae the ratio ranged from 40 to 62 % and in four isolates of *Phytophthora cinnamomi* from 49 to 58 %. Mandel (1968) tested three Chytridiomycetes and reported a range of 44–66 %. More recent studies on the Saprolegniaceae, and the discovery of a satellite DNA component that could have led to erroneous results, have demonstrated that the technique is useful for developing generic concepts (Green and Dick, 1972; Clark-Walker and Gleason, 1973; Neish and Green, 1976, 1977). Moreover, Ojha *et al.* (1973) reported three fractions of 62, 53 and 48 % in *Allomyces* and *Blastocladiella*, and two (55 and 41 %) in *Saprolegnia*. A re-evaluation of the DNA base composition in *Pythium* and *Phytophthora* may be useful in light of these findings.

Hybridisation experiments between DNA and RNA have not been applied to zoosporic fungi. However, DNA–DNA hybridisation results between *Allomyces* and *Blastocladiella* indicate a close similarity between members of the Blastocladiales, and a more distant relationship between these and

Saprolegnia (Ojha *et al.*, 1973). Among other organisms there has been some success in applying hybridisation as a phylogenetic criterion (Ragan and Chapman, 1978) but there appear to be many technical problems to overcome before this technique can be used routinely.

7. *Electrophoresis*

Electrophoresis is one of the more promising techniques for distinguishing species of fungi and has been used effectively with *Pythium* and *Phytophthora*. The mobility of proteins allows them to separate in gels of low porosity into distinctive banding patterns. Individual enzymes or the total protein content may be tested in this way although some may be more suited than others to this type of analysis. Present evidence shows that certain enzymes are suited for distinguishing species in certain genera but not in others. Thus, in *Phytophthora*, esterase patterns showed up as distinct differences whereas differences in peroxidase patterns were less clear (Kaosiri and Zentmyer, 1980). In *Pythium* there are five distinct peroxidase isozymes that are indicative of a generic characteristic but various combinations of these are indicative of species (Clare *et al.*, 1968). Oxidase patterns may be a generic characteristic in *Pythium* but a species charac- teristic in *Phytophthora* (Clare *et al.*, 1968).

In *Phytophthora*, various workers have been able to distinguish *P. cinnamomi*, *P. citrophthora* R. E. Smith & E. H. Smith, *P. palmivora* (Butler) Butler, *P. fragariae* Hickman, *P. cactorum* and *P. megasperma* Dreschsler var. *sojae* Hildebrand either by their total protein or esterase patterns (Clare and Zentmyer, 1966; Gill and Powell, 1968a,b; Hall *et al.*, 1969; Gill and Zent- myer, 1978). The most striking impact of electrophoresis on taxonomic theory in *Phytophthora* is on the *P. palmivora* complex that causes the destructive black pod and canker disease of cocoa. Painstaking measurement showed that there were three morphological forms, two of which have a limited geographic distribution (Zentmyer *et al.*, 1977; Kaosiri *et al.*, 1978; Brasier and Griffin, 1979) and two of which can also be distinguished cytologically (Sansome *et al.*, 1979). Protein and esterase patterns have confirmed differences between these morphological forms and indicate that they are sufficiently different to be classified as distinct species (Kaosiri and Zentmyer, 1980). Similarly, Vaartaja (1965) used protein patterns as supportive evidence for three new species of *Pythium*.

There is every indication therefore that enzyme or total protein patterns are stable characteristics for species determinations. Moreover, isolates of any one species from different hosts and geographic location have similar patterns. For this reason the technique does not appear promising for distinguishing races, although some success has been reported in defining races in higher fungi. Among lower fungi, Gill and Powell (1968b) examined total protein patterns in eight races of *Phytophthora fragariae* but they were unable to distinguish the races involved.

8. *Serology*

The application of immunodiffusion and immunofluorescent techniques has also been useful in establishing species limitations in various organisms, and has been applied to the study of lower fungi. *Phytophthora cinnamomi*, *P. cactorum* and *P. erythroseptica* Pethybr. can be distinguished (Burrell *et al.*, 1966) using this technique as can *P. cinnamomi*, *P. cambivora* Petri (Buisman) and *P. cryptogea* Pethybr & Lafferty (Halsall, 1976). The limitation of serological techniques is that every species must be tested directly against every other; thus, in the above examples, there is no serological evidence of the relationships between *P. cactorum* or *P. erythroseptica* with *P. cambivora* or *P. cryptogea*. Nevertheless, the serological similarities between *P. cryptogea* and *P. drechsleri* Tucker (Halsall, 1976) supported the claim made by Bumbieris (1974), based on morphological grounds, that these were in fact one species. In addition, serological studies on different morphological forms of *P. palmivora* have shown a relationship of these with other less well studied species (Merz *et al.*, 1969).

Immunofluorescence studies have also contributed to species concepts in *Pythium*. White (1976) showed that *P. graminicola* Subram. could be distinguished from 17 other species but not from *P. aristosporum* Vanterpool, which is morphologically similar to *P. graminicola*, while the application of immunodiffusion tests to 15 isolates of *Pythium* showed that they could be separated into four groups (Krywienczyk and Dorworth, 1980). It is of particular interest that these groups correspond more to groupings arranged according to asexual than to sexual characteristics.

9. *Enzymes and Metabolic Pathways*

Vogel's (1960, 1965) studies on lysine synthesis were hailed for their significance in evolutionary theory. They indicated a relationship of the Chytridiomycetes to the Zygomycetes and higher fungi, and supported other evidence that the Oomycetes are related to the heterokont algae. Vogel reported the diaminopimelic acid (DAP) pathway to be present in bacteria, algae, vascular plants, the Oomycetes and Hyphochytriomycetes and the aminoadipic acid (AAA) pathway in *Euglena*, the Chytridiomycetes, Zygomycotina, Ascomycotina and Basidiomycotina. These pathways are quite different and it was argued that one could not have evolved spontaneously from the other. LéJohn (1974), however, pointed out that lysine degradation proceeds along a route similar to the reversal of the AAA synthesis pathway. Thus, he argued, the latter may merely be a retrograde evolutionary process. This conveniently explains the apparently anomalous placement, based on lysine synthesis, of *Euglena* with the Chytridiomycetes. It would also explain the apparent relationship of the Chytridiomycetes to the higher fungi *if* other evidence was weighed against this relationship. If lysine synthesis by the AAA pathway is a retrograde process, it certainly has not happened very often during the evolutionary history of eukaryotic organisms, because there are no known examples of organisms with both pathways, nor any anomalous situations, with the possible exception of *Euglena*. LéJohn's theory on the possible origin of the AAA pathway therefore does not diminish the significance of lysine synthesis in the development of evolutionary concepts; it merely explains how the difference may have evolved.

The final biosynthetic pathways leading to tryptophan as the end product consist of the shikimate pathway to chorismate, and the tryptophan pathway from chorismate to tryptophan. Analysis by ammonium sulphate fractionation and sucrose density gradient techniques of the enzymes involved in these pathways show that those extracted from some organisms separate, but in others they aggregate. In bacteria (Crawford, 1975) and fungi (Ahmad *et al.*, 1968; Rines *et al.*, 1969), it has been demonstrated that the aggregations relate to clustering or cotranscription of genes. The tryptophan pathway is particularly interesting because the five enzymes involved sediment in a variety of patterns called "types" (Hütter and DeMoss, 1967). It is argued that the types of sedimentary patterns are not likely to have been caused by simple reversible mutations, and, because there is no evidence of severe selective forces on tryptophan enzymes, all descendants after a change will possess the same new pattern. Hütter and DeMoss (1967) proposed that the types of sedimentation reflect phylogenetic relationships; however, there is some controversy over the interpretation of their results (LéJohn, 1974). The work of Hütter and DeMoss (1967) and others was reviewed by Ragan and Chapman (1978) who listed all fungi as comprising five types. Of relatively few zoosporic fungi tested, two Oomycetes (*Pythium* sp. and *Saprolegnia* sp.) have one type, whereas two Chytridiomycetes (*Allomyces macrogynus* and *Rhizophlyctis rosea*) have another type in common with certain Ascomycotina and Basidiomycotina.

LéJohn (1971, 1974) studied dehydrogenases in lower fungi with the object of relating their regulatory properties to fungal phylogeny. He argued that because the citric acid cycle occupies a strategic place in metabolism, and must therefore have a long evolutionary history, the regulatory factors are presumed to be indicative of various phylogenetic parameters. The higher fungi produce both NAD^+- and NADP-linked glutamic dehydrogenase whereas the Myxomycota, Mastigomycotina and Zygomycotina possess only NAD^+-linked glutamic dehydrogenase. Among the more primitive Chytridiomycetes and most Zygomycotina, LéJohn found the NAD^+-linked glutamic dehydrogenase lacked any mode of regulation whereas in the Blastocladiales there is a complex mode of regulation. The Oomycetes have NAD^+-dependent glutamic dehydrogenase allosterically regulated by NADP. Wang and LéJohn (1974a) studied 50 Oomycetes and found that, by and large, species in the same genus possessed isozymes that were similar in electrophoretic mobility and regulatory patterns. Moreover, the Peronosporales had a greater complexity than the Saprolegniales in the regulatory properties in their isozymes. The authors cautiously suggested that increased complexity could reflect a high environmental adaptation rather than an indication of evolutionary advancement. The activators and inhibitors in the Hyphochytriomycetes' glutamic dehydrogenase are only slightly different from those in the Oomycetes (LéJohn, 1974). This suggests that the Oomycetes and Hyphochytriomycetes have a common ancestry far closer than either does with the Chytridiomycetes. However, LéJohn (1971) examined only three species of primitive Chytridiomycetes and the possibility of other species possessing both NAD^+ and NADP-linked forms cannot be discounted at present.

The Oomycetes and Hyphochytriomycetes contain only the $D(-)$ form of NAD^+-linked lactic dehydrogenase and it is allosterically controlled by GTP (LéJohn, 1971). There is considerable variation in the degree of inhibition among different species of Oomycetes (Gleason, 1972; Wang and LéJohn, 1974b); however, the Chytridiomycetes and Zygomycotina are distinctive because they have a second class of lactic dehydrogenase that is uncontrolled except by ATP (LéJohn, 1971). Another difference is the isocitric dehydrogenase which in the Chytridiomycetes and Zygomycotina is NAD^+-linked, and usually allosteric, whereas in the Oomycetes and Hyphochytriomycetes NAD^+-linked isocitric dehydrogenases are absent (LéJohn, 1971).

The possible value to the taxonomic theory of other enzymes and metabolic processes in eukaryotic organisms was reviewed and discussed by Ragan and Chapman (1978) although there are no data for the lower fungi.

10. Cytochrome

The potential of cytochrome analysis on the taxonomy of lower fungi was suggested by LéJohn (1974) and, judging by work on other organisms (Dayhoff et al., 1975; Fitch, 1976), it may prove to be a chemotaxonomic characteristic of prime importance. Cytochromes a, b and c have been found in the Oomycetes and Chytridiomycetes (Gleason and Unestam, 1968; Ribeiro et al., 1976), and differences in the absorption spectrum of any one of these cytochromes in different fungi is indicative of structural differences of the protein. It is thought that cytochromes have not undergone any radical change during the evolutionary history of eukaryotic organisms because of their specialised function as electron transferring proteins. Critical studies of cytochrome *c* (Dayhoff *et al.*, 1975) have shown that among the higher fungi examined, the imperfect species appeared closely related to the Ascomycotina and more distantly related to the Basidiomycetes. Mycologists may find it profitable to follow the lead of those in other disciplines who have made strides in this exciting area of chemotaxonomic research.

III. Phylogeny and Classification

A. THE TAXONOMIC GROUPINGS

1. *Oomycetes*

The generally held assumption that the Oomycetes are related to heterokont algae has been supported by the ultrastructural and biochemical evidence documented in the preceding sections (see Fig. 1). Ultrastructure of the zoospore (Chapter 1) and its microtubule rootlet system (this Chapter) provide the best evidence for this relationship. The primary Oomycete zoospore (Holloway and Heath, 1977a,b) has, in common with zoospores of the Chrysophyceae such as *Ochromonas*, (Bouck and Brown, 1973), a similar symmetry; a pyriform nucleus with the narrow end close to the flagellar apparatus and a Golgi apparatus nearby; similarities in their rootlet systems; and, most probably, a similar structure in the transition zone of the

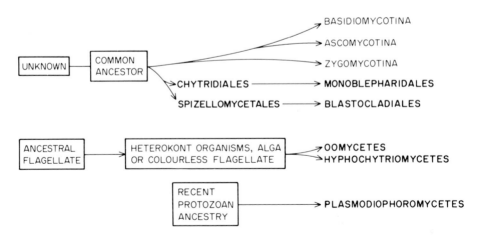

Fig. 1. The probable phylogenetic relationship of the various taxa of zoosporic fungi (**bold face**) with other eukaryotic organisms.

flagellum. The problem in comparing zoospore ultrastructure between the Oomycetes and other lower eukaryotic organisms is that most Oomycetes do not have an active primary zoospore. In fact, because the primary zoospore is a poor swimmer, or encysts without swimming (or in many taxa is not even known to exist), it can be considered a degenerate stage. The secondary zoospore is structurally more complex and organised, and the mere existence of secondary zoospores in Oomycetes suggests a forward step in evolution from an ancestral past, when only zoospores of the primary type were present. In early evolutionary history the development of secondary zoospores would have enhanced dissemination and survival. This would have been particularly desirable for the early heterotrophic forms which may have had poor competitive ability with other saprophytes. The evolution of chemotactic systems in response to host exudates, as found in many Oomycetes (Chapter 1), would largely replace the advantage diplanetism provides an organism for its dispersal.

Probably the most significant events in the evolution of the Oomycetes were the development of the secondary zoospore and the change in the angle of the kinetosomes to about 130–150°, accompanied by a shift in their position to the side of the spore, so that one flagellum could be directed forward and the other backward for efficient swimming. In zoospores of most heterokont algae, the angle of the kinetosomes is about 30–90° and thus both flagella are directed more or less forward as they emerge from the zoospore. An explanation for this awkward arrangement is that the heterokont algae are evolved from the Cryptophyceae which have two, anteriorly directed, tinsel-type flagella (Dodge, 1979). It is not surprising that among the heterokont algae the whiplash flagellum had a diminished importance in locomotion and became shorter than the tinsel flagellum. The Oomycetes have taken a different route; that of changing the angle of the kinetosomes. However, the whole architecture of the primary zoospore does not favour efficient movement and the evolution of the secondary zoospore with laterally attached flagella was the outcome.

The evolutionary time scale encompassing the ancestral relationship with heterokont algae is more difficult to analyse. *Ochromonas* is considered to be among the more primitive of the chrysophycean algae (Bourrelly, 1957). It is therefore of interest that the *Saprolegnia* primary zoospore is structurally simpler than the *Ochromonas* zoospore although the primary Oomycete zoospore may now be structurally as well as functionally degenerate. However, the form of mitosis in Oomycetes is considered to be more primitive than that in any of the heterokont algae (Heath, 1980). Moreover, cell-wall composition, the presence of secondary zoospores, a non-motile type of sexual reproduction and the type of nutrition cannot support any possibility of a very recent connection, in terms of evolution, with the heterokont algae. It seems probable therefore that the Oomycetes evolved from a heterokont alga morphologically and cytologically more primitive than any existing group.

A reasonably factual account of the course of evolutionary history in the Oomycetes might be within our grasp if studies of zoospores and their rootlet systems were applied to a wider spectrum of genera in different orders, for ultrastructural studies have proved valuable in assessing the phylogeny of the Chytridiomycetes and many other lower eukaryotes. Already, studies on the ultrastructure of *Thraustochytrium* and *Schizochytrium* zoospores (Perkins, 1976; Kazama, 1980; Barr, 1981) have provided good evidence that the Thraustochytriaceae is only distantly related to other Oomycetes. Nothing is known about the ultrastructure of Leptomitales zoospores, nor, with the one exception of *Lagenidium* (Gotelli, 1974), the Lagenidiales zoospore. Nor has the ultrastructure of zoospores of the Albuginaceae or Peronosporaceae been reported on and knowledge of the fine structure of zoospores is confined to *Saprolegnia*, *Aphanomyces*, *Pythium*, *Phytophthora* and *Lagenidium* among the "true" Oomycetes. The greater similarity between these zoospores than among those of the Chytridiomycetes, indicates they are a relatively closely related group of organisms with a short evolutionary history. Those endobiotic, holocarpic Oomycetes in the Lagenidiales, with typical kidney-shaped secondary zoospores, will probably prove to be morphologically degenerate Oomycetes which at one time had a filamentous saprophytic ancestry, whereas those with other shaped biflagellate zoospores may not even be true Oomycetes. At present, biochemical evidence supports the relationship of the Oomycetes to heterokont algae but does not give many clues on the course of evolution among the Saprolegniales and Peronosporales. Dick (1968), using a different approach, suggested that the ecological origin of Oomycetes was in the littoral zone and their morpho-logical characteristics relate to movement both into open water (stout hyphae, occasionally sexual sporulation, extended zoospore phase, and large, smooth-walled oogonia with many oospores) and into soil (slender hyphae, resting spore formation, brief zoosporic phase and small spiny oogonia with one oospore). I suggest that the stout hyphae seen in many Saprolegniaceae are a carry-over from their autotrophic past and therefore this is a primitive characteristic. The stout hyphae are probably more efficient for transporting large volumes of water and nutrients, and, as Dick (1968) pointed out, evidence for this is the drying and cracking of the agar in the advance of the mycelium when species with stout hyphae are cultured. However, stoutness has not persisted in the Oomycetes, or, for that matter, in other fungi, and smaller diameter hyphae, with a proportionately larger surface area to volume for greater absorption, appear to have been favoured during evolution.

During the course of Oomycete evolution, the following trends seem to have occurred: (1) movement from an aquatic to a terrestrial environment;

(2) reduction in the diameter of the hyphae; (3) suppression of cellulose as a cell wall carbohydrate and the development of chitin synthesis; (4) a change from a saprophytic role, first to that of an opportunistic parasite, then of a facultative parasite, then of a fully fledged parasite with a decline in saprophytic ability, and finally to that of an obligate parasite; (5) specialisation in host range; (6) formation of haustoria, holocarpic development completely inside a host cell, or systemic infection; (7) loss of the ability to synthesise sterols, and possibly vitamins, as well as a requirement for organic sulphur and nitrogen; (8) decrease in the number of oospores in the oogonium accompanied by changes from centrifugal to centripetal oosporogenesis; changes in the position of the lipid globules from peripheral to central, and a progressive development of the periplasm as outlined by Dick (1969); (9) development of the sporangiophore; (10) a reduction in the number of antheridia; (11) suppression of homothallism in favour of heterothallism; (12) development of asexually formed resistant sporangia, conidia or gemmae; (13) loss of the primary zoospore and, in advanced forms, loss of zoosporogenesis.

Accumulated evidence indicates that the Saprolegniales are the most primitive order of Oomycetes and *Achlya* is the most primitive genus of phytopathogens. *Achlya* is largely aquatic or occurs in very wet soils, has wide-diameter hyphae and is an opportunistic parasite that attacks plants under physiological stress. In the host, hyphae ramify through and between cells. Oogonia are characterised by the presence of several oospores. The primary zoospore, which is probably never flagellated, encysts immediately after release, and forms a cluster of cysts outside the sporangium. *Aphanomyces* is also primitive; indeed its sporangia are simpler than those of *Achlya*, consisting of undifferentiated hyphae, but the oogonium is single-spored and many species are clearly soil inhabiting, although their development is favoured by high soil moisture. *Aphanomyces* species are more restricted in their host ranges and *A. raphani* Kendrick may only occur on crucifers.

Oogonium–oospore morphology as outlined by Dick (1969), supported by complexity in thallus morphology, indicates that the Leptomitales and Peronosporales evolved directly from the Saprolegniales and then diverged into two lines. In the Leptomitales, the primary zoospore is retained in some species, but at least some have chitin in their cell walls, indicating that they are not ancestral to the Peronosporales. The Lagenidiales may be an artificial group derived from any of these orders, as well as from other lower eukaryotes as the result of the degenerative influence of parasitism. *Lagena*, because it has a vesicular type of zoospore discharge, may be evolved from *Pythium*.

Pythium has characteristics that are generally more primitive than *Phytophthora* and many species are aquatic and saprophytic (also see

Chapter 4). Among the parasitic species there is a tendency towards host specialisation, particularly towards the Gramineae. As in all Peronosporales, the oospores are single. The number of antheridia, on the other hand, varies between different species, but in some, such as *P. ultimum* Trow, it is usually single. Most species are homothallic, but there are some important exceptions such as *P. sylvaticum* Campbell & Hendrix. Sporangia are borne singly on hyphae and there is considerable morphological variation. In some they are formed from undifferentiated hyphae, in others they occur as lobulate or globular intercalary swellings of the hyphae, and yet in others they are terminal or lateral with simple sporangiophores. As in other Peronosporales, no primary zoospores are produced; however, the vesicle in which secondary zoospores are formed may be equivalent to an undifferentiated mass of primary zoospore protoplasm. In many species no zoospores are formed at all and "sporangia" freely germinate by germ tubes.

Phytophthora is clearly more advanced than *Pythium* although there is enough similarity between them, and enough morphological variation in

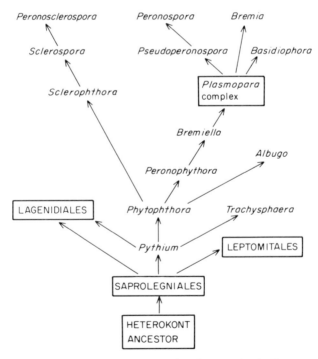

Fig. 2. The probable phylogenetic relationship of genera in the Peronosporales and other Oomycetes. Part of this scheme is based on a simplified proposal of Shaw (1978).

Pythium, to suspect that *Phytophthora* evolved directly from a *Pythium*-like ancestor. There are few aquatic species of *Phytophthora* whereas at the other ecological extreme some, such as *P. infestans* (Mont.) DeBary, occur on leaves and fruit. They are well adapted to their role as parasites and most have poor or no saprophytic ability. Many species have wide host ranges whereas others are restricted to certain families or genera, and there are physiological races in some. Many species penetrate host cells with haustoria, many are heterothallic (Savage *et al.*, 1968; Boccas, 1981) or have a chemically controlled heterothallic-like behaviour (Ko, 1978; Barr, 1980b) and secondary zoospores are formed inside the sporangium.

The Peronosporaceae, in contrast to the Pythiaceae, are terrestrial organisms that can attack parts of plants above ground and are known as downy mildews. They have evolved to the level of, essentially, obligate parasites, although some can be cultured on special media. They are distinguished from the Pythiaceae by their well developed and conspicuous sporangiophores. Inside host tissue they have well developed haustoria while some grow systemically. Shaw (1978) considered that two distinct lines evolved from *Phytophthora* (Fig. 2). One is confined in host range to the Gramineae and progressed from *Sclerophthora* to *Sclerospora* and *Peronosclerospora*. The other is largely confined to non-graminaceous Angiosperms and centres around the *Plasmopara* complex. He stressed the importance of the apical region associated with zoospore discharge during the evolutionary process. Loss of the operculated papillum, which occurred in both lines, resulted in permanent loss of zoosporogenesis. Such "sporangia" are structurally and functionally conidia. All species of *Peronosclerospora* and *Peronospora* have conidia, and are therefore the most advanced members of the Peronosporales.

2. *Hyphochytriomycetes*

The presence of a nonfunctional centriole (vestigial kinetosome) in zoospores of the Hyphochytriomycetes suggests a biflagellate origin for this group and the angle of the nonfunctional centriole relative to the kinetosome is a strong indication that the second flagellum was of the whiplash type and directed posteriorly. The presence of rings of alternating diameters in the transition zone of the flagellum indicates a relationship to the Oomycetes, but differences in the rootlet systems, in zoospore ultrastructure in general, and in the mitotic process, show that one class probably did not evolve from the other, but had a distant common ancestor. Chemotaxonomic data support the concept of a general relationship between these classes and the heterokont algae. Thus, it is reasonably conclusive that the Hyphochytriomycetes, Oomycetes and heterokont algae are related, but present knowledge is not precise enough to indicate an exact relationship among any of them; if indeed species which are in existence today have retained enough ancestral characteristics to make such indications of a relationship possible.

There are no phytopathogens in this small group of fungi but several are destructive parasites of Oomycete oospores (Ayers and Lumsden, 1977; Sneh *et al.*, 1977; Wynn and Epton, 1979) and may be responsible for the natural decline of phytopathogenic Oomycetes in soil. Sparrow (1973b) recognised six genera of Hyphochytriomycetes and in all there are only about 15 species, some of which are poorly described and of doubtful validity. It is also very probable that most of the characteristics used for distinguishing taxa are unsound because of morphological variation. For example, reports of *Anisolpidium* in pine pollen may merely be *Hyphochytrium catenoides* which is generally holocarpic when grown in this substrate (Barr, 1970a). Also, the method of zoospore release needs evaluation as a taxonomic criterion. If further study proves that these organisms are ecologically important as parasites on Oomycetes, a morphological and taxonomic study must be a prerequisite to any other investigation.

3. *Chytridiomycetes*

There is little factual evidence to account for the origin of the Chytridiomycetes. The simplicity of many unicellular saprophytic forms, and in particular the ultrastructure of their zoospores, indicates that they are very primitive. There are no eukaryotic algae that have comparable zoospores. The *Chlamydomonas* zoospore, which is considered similar to the ancestral type of the Chlorophyta by many phycologists, has a more complex rootlet system (Ringo, 1967; Cavalier-Smith, 1974) than that found in primitive Chytridiomycetes. The rootlet systems in flagellated Protozoa, Plasmodiophoromycetes (Barr and Allan, 1982) and Myxomycetes (Haskins, 1978), are also more complex than those in primitive Chytridiomycetes. The amoeboid movement of Chytridiomycete zoospores and the fact that zoospores of some saprophytic species can swim for up to three days without encysting, suggests that the zoospore is the primordial body. In primaeval times, the ancestral Chytridiomycete zoospore probably reproduced by binary fission, and the sporangium developed sometime later. The prolonged stationary period during sporangium development provides an advantage to an organism dependent upon absorptive nutrition. As the Chytridiomycete thallus evolved, and in particular when it become filamentous and polycentric, the importance of the vegetative thallus increased at the expense of the zoospore. It is interesting that in my culture collection, certain polycentric Chytridiomycetes have been growing for many years without producing zoospores. The wide diversity in developmental morphology, zoospore ultrastructure and habitat among members of this class shows that it has a very long evolutionary history. The unicellular members of the group may in fact be among the most primitive nucleated organisms living and are without any ancestral group in the eukaryotes that has modern representatives.

The Chytridiomycete mitochondrion contains flattened cristae as do those of the Zygomycotina, Ascomycotina, Basidiomycotina, certain flagellated protozoa, the Chlorophyta, higher plants and animals. Taylor (1978) argued that the type of cristae (flattened or tubular) is correlated with chloroplast and flagellar features and in combination suggest certain evolutionary lines. Whether this argument will hold true after more organisms are examined remains to be seen, but it is reasonable to use mitochondrial characteristics negatively; that is to claim a lack of relationship between organisms with different mitochondria. Thus, the Chytridiomycetes, Zygomycotina and higher fungi are phylogenetically distinct from the Oomycetes, Hypho-chytriomycetes and Plasmodiophoromycetes as well as the Myxomycetes on the basis of this feature.

Similarly of the transition zone in the flagellar apparatus (Barr and Hadland-Hartmann, 1978), supported by studies on cell wall carbohydrates (Bartnicki-Garcia, 1970), provides convincing evidence that the four orders of the Chytridiomycetes are a monophyletic group. Admittedly, too few have been examined in sufficient detail to make it certain that *all* posteriorly uniflagellate fungi are true Chytridiomycetes but enough representative species for most of the larger genera have been critically examined to justify this claim as a generalisation.

Zoospore ultrastructure, and in particular the microtubule rootlet systems, shows that extant species evolved along two separate lines; one from the Spizellomycetales to the Blastocladiales, and the other from the Chytridiales to the Monoblepharidales (including the Harpochytriales). There is a substantial number of cytologically and morphologically simple species in existence to justify the construction of a plausible evolutionary scheme (Barr, 1978, 1981). Morphological and developmental evidence show that there is considerable parallelism between these lines. Ecologically, the Chytridiomycete–Monoblepharidales line includes fungi which are largely aquatic, whereas the Spizellomycetales–Blastocladiales orders are soil inhabitants, although there are a number of exceptions as would be expected in a group with a long evolutionary history.

Physoderma and *Synchytrium* are strictly confined as obligate parasites to vascular plants and therefore must be at least as recent in origin as the most primitive plant they parasitise. *Physoderma* has a typical blastocladiaceous zoospore, but it could also have evolved from a fungus related to *Kochiomyces* in the Spizellomycetales that has a zoospore with certain similar characteristics (Barr and Allan, 1981). Moreover it is interesting, although possibly coincidental, that the epibiotic state of *Physoderma* often has dichotomous rhizoids that are morphologically similar to the *Kochiomyces* rhizoids.

The genus *Synchytrium* presents an enigma because zoospores of the two species that have been examined ultrastructurally are somewhat different (Chapter 1). *S. macrosporum* Karling has characteristics of the Chytridiales–Monoblepharidales line, whereas *S. endobioticum* (Schilberszky) Percival is somewhat like the Spizellomycetales zoospore. A long coevolutionary history with plants may account for this diversity. It might also be helpful if critical studies were made on zoospores of *Micromyces* or *Endodesmidium*, which are the only chytrids presumed to be closely related to *Synchytrium*. At present, it is not possible to classify this genus in an entirely appropriate position in the class, and its taxonomic grouping with the Chytridiales is a tenuous position.

Present evidence on zoospore ultrastructure of *Olpidium brassicae* and *O. radicale* (Chapter 1) and their developmental morphology indicates a phylogenetic relationship close to *Karlingia*, a genus of soil saprophytes that can digest cellulose, and *Entophlyctis variabilis*, a soil saprophyte that has similar morphology to *Olpidium* when it grows inside plant substrata. These taxa all belong to the order Spizellomycetales. It would be a mistake to assume that all species of *Olpidium* are closely related, or that they evolved by the process of coevolution. On the contrary, light microscopic observations of those growing on algae indicate that they may be related to the Chytridiales. Thus, it is best at present to consider the likelihood that *Olpidium* is an artificial group of look-alike Chytridiomycetes.

Rhizophydium graminis is the only species within the large and complex genus *Rhizophydium* that is parasitic on vascular land plants. Most others grow saprophytically or parasitically on algae and other substrates in aquatic habitats. Although the classification of this genus is, at present, in need of study, many have been grown and studied in culture and examined in the electron microscope, and there is no reason to suggest that *R. graminis* does not belong naturally here.

The Chytridiomycetes are often considered a difficult group to understand because they are morphologically and ecologically very diverse. To compound the problem, many species, and even some genera, have been inadequately described. However, all phytopathogens belong to four genera, and so the complexities of the classification are of little interest to the plant pathologist. Those interested can refer to Karling's excellent compilation of the classical genera (Karling, 1977) and to my treatment (Barr, 1980).

4. *Plasmodiophoromycetes*

This class of obligate parasites is found in vascular plants, Oomycetes and *Vaucheria*. The catholic host range suggests two phylogenetically distinct groups that merely look similar; however, reports of cruciform-like mitosis in *Octomyxa achlyae* Couch and *Woronina polycystis* Cornu (Couch *et al.*, 1939; Goldie-Smith, 1954) which are parasitic in Oomycetes, are a good indication that the Plasmodiophoromycetes are in fact one natural group. The fact that green algae are not parasitised by these fungi adds to the mystery of why such unrelated organisms as vascular plants and heterokont eukaryotes are hosts. The most likely explanation is that Oomycetes are the ancestral hosts of the Plasmodiophoromycetes, and at some earlier time in evolutionary history, a plasmodiophoromycetous parasite of a phytopatho-genic Oomycete adapted to a vascular plant as a secondary host.

The very simple zoospores and microtubule rootlet system show that the Plasmodiophoromycetes are very primitive. The zoospore has some charac-teristics of a protozoan zoospore and is quite unlike those of the Chytridiomycetes or Oomycetes. The means of root penetration by *Polymyxa betae* and *Plasmodiophora brassicae* (Keskin and Fuchs, 1969; Aist and Williams, 1971) is another indication of their protozoaceous origin. In these species, a bullet-like stachel is formed in a tubular cavity within the zoospore cyst. The stachel punctures the host cell wall and the fungus is rapidly injected into a root cell. This apparatus appears to be a modified trichocyst; it is most like some of the simpler trichocysts found in many ciliated protozoa, for example in *Paramecium* and *Frontonia* (Yusa, 1963, 1965). Trichocysts are also found in certain algae (Dodge, 1973) but algal trichocysts are somewhat more complex in their structure.

The evolution of *Polymyxa* as a genus of two economically important phytopathogens, of sugar beet and cereals, can be attributed to agricultural practices. *Polymyxa* is morphologically similar to *Ligniera*, a poorly understood genus most of which infect aquatic and semi-aquatic flowering plants (Cook, 1926). In *Ligniera* the sporangial plasmodium is multicellular, and each cell becomes a sporangium that can release its zoospores independently of the adjacent sporangium on the thallus. This independence of the sporangia is a desirable feature for an aquatic organism because it ensures a constant supply of inoculum. The *Polymyxa* sporangial plas-modium segments into separate cells but at maturity the internal walls largely dissolve and the whole acts as a single sporangium. As such, it is well adapted as a soil organism because it can take advantage of sudden rains when many zoospores are produced at one time. *Ligniera pilorum* Fron and Gaillat also occurs in agricultural soils and may be the species bridging the two genera. It has a wide host range including the Chenopodiaceae and Gramineae (Barr, 1979). The well established principle that intensive cropping encourages the development of new strains of phytopathogens is applicable to *Polymyxa* for the intensive cropping of cereals has encouraged the selection of strains pathogenic to these plants. While there is no known natural distribution of *P. graminis* among weeds, it grows abundantly well on cereals and rather poorly on a few related grasses. The natural host for *P. betae* is probably *Chenopodium album*, a weed that has flourished as a result of agricultural practices and *P. betae* apparently infects sugar beet after this crop is grown for a number of years in one area.

Karling (1968) recognised one order, one family and nine genera of Plasmodiophoromycetes, seven of which are phytopathogens (Appendix 3). The most important species without any question is *Plasmodiophora brassicae*, the cause of club root of crucifers (Chapter 5). Other important pathogens include *Spongospora subterranea* which causes powdery scab of potatoes, and *Polymyxa* species which transmit plant viruses (Chapter 7).

5. *Myxomycota*

The zoosporic, phytopathogenic organism *Labyrinthula* appeared suddenly in epidemic proportions in the 1930s on the marine eel grass *Zostera* along the eastern coast of North America, and in Atlantic coastal waters of Europe. Loss of *Zostera* in some years reached 90% with ecological repercussions to wildlife, and to fisheries resources (Cottam, 1945), however, the epidemics were probably the result of a combination of factors (Pokorny, 1967). The labyrinthulas are a small number of organisms with uncertain phylogenetic affinity. Mycologists tend to classify them with the slime moulds (Myxomycota), whereas protozoologists (Levine, 1980) place them with the genus *Thraustochytrium*, in the protozoan phylum Labyrinthomorpha. The most authoritive account of the labyrinthulas is by Olive (1975), who classified them with *Thraustochytrium*, and allied genera, in the subphylum Labyrinthulina. However, he recognised the classification by Whittaker (1969) and placed them, with the slime moulds, in the Kingdom Protisa, as a distinct entity from fungi and therefore they are not considered here in detail.

6. *Algae*

Many algae grow epiphytically in plants without causing any appreciable damage and, when damage is reported, it is believed to be due to loss of water and dissolved minerals, a shading effect and loss of photosynthetic area, or possibly by secretion of toxins. *Cephaleuros virescens* Kunze is the only alga that ranks as having any importance as a phytopathogen. It is the cause of red rust of tea and black fruit disease of pepper vines in Asia. It also occurs elsewhere in warm temperate and tropical countries, and in glasshouses, on citrus and ornamentals. Some algae such as *Rhodochytrium* and *Chlorochytrium* are colourless endophytes and Lingappa (1956) demonstrated that an organism which had previously been described as *Synchytrium borreriae* Lacy was in fact a colourless, endophytic alga and it is very possible that other reports of organisms identified as chytrids may in fact be algae. Most parasitic algae produce motile spores that act either as gametes, or behave as zoospores and reproduce the vegetative thallus asexually. All phytopathogenic algae belong to the Chlorophyta; *Phyllosiphon* has been considered a member of the Xanthophyceae, but it contains chlorophyll *a* and *b* and is therefore a green alga (Leclerc and Couté, 1976); however, it is one that is not known to produce motile spores. Additional information on parasitic algae may be found in a review by Joubert and Rijkenberg (1971).

7. *Protozoa*

Phytomonas is a genus of flagellated protozoa that occurs in plant sap. Members of this genus cause wilt to coffee (Vermeulen, 1963), hartrot disease of cocoa palm (Parthasarathy *et al.*, 1976; Waters, 1978) and a possibly destructive disease to the African oil palm in Latin America (Thomas *et al.*, 1979). *P. elmassiani* (Migone) is transmitted to plants of the family Asclepiadaceae by lygaeid bugs (*Oncopeltus* spp.) (Vickerman, 1962; McGhee and Hanson, 1964), and can multiply in their salivary glands but it is uncertain at present what other species of insects transmit phytomonads. *Phytomonas* is a member of the Trypanosomatidae which are better known for diseases of man and domestic animals, such as sleeping sickness, although most species occur in the guts of such invertebrates as annelids and arthropods. Baker (1974) concluded that trypanosomatids exploited the opportunity offered by their insect hosts to develop as plant and blood parasites, and these parasites developed secondary hosts. There is no recent classification of *Phytomonas* although the status of this genus and related taxa has been reviewed by Vickerman (1976). McGhee and McGhee (1979) claimed that there is a strict family-host specificity among species of *Phytomonas*, and that this serves as a means of classifying them.

IV. Synopsis

The Mastigomycotina are a group of eukaryotic organisms with an absorptive mode of nutrition and most species have a zoosporic stage at some stage in their life cycle. However, they are collectively an artificial group because the Oomycetes and Hyphochytriomycetes have an ancestral affiliation with the heterokont algae, the Chytridiomycetes have no known ancestral origin but are probably the progenitors of the Zygomycetes and higher fungi, whereas the Plasmodiophoromycetes are probably related to protozoa. They are all nevertheless "fungi" because the Kingdom Fungi is an artificial group of eukaryotic organisms that have in common the absorptive mode of nutrition, just as algae and protozoa are artificial groups based on their respective modes of nutrition.

In order for the fungi to be a natural group, as is implied by the divisional name Eumycota (True Fungi), the Chytridiomycetes would be retained as the subdivision "Chytridiomycotina" and the other three zoosporic classes despatched to positions within the algae and protozoa. The Oomycetes have in fact been accepted by Leedale (1974) and other phycologists as a part of the heterokont algae, and the Plasmodiophoromycetes are classified by the committee on systematics and evolution of the Society of Protozoologists (Levine, 1980) as the class Plasmodiophorea in the subphylum Sarcodina. There is, however, the human element that will resist these moves. The majority of those who study zoosporic fungi may continue to feel more comfortable as "mycologists" because of their traditional affiliations in universities, institutions and professional societies, and because much of their literature is in mycological treatments.

An alternative approach has been taken by Margulis and Schwartz (1982) who classified zoosporic fungi with algae and protozoa in the Kingdom Protista, separate from the Zygomycotina and higher fungi that form the Kingdom Fungi. This is also an artificial system because the protists are a very heterogeneous group of organisms. To circumvent the artificiality of the system, each natural group of organisms was given the status of a phylum; thus the groups of zoosporic fungi became Oomycota, Hyphochytr[id]io-mycota, Chytridiomycota and Plasmodiophoromycota. There is obviously some logic to this system but it creates far too many high-ranking taxa for systematic convenience and many of these contain relatively few species.

The fact remains that however desirable it is to strive towards a natural system of classification, it will to some extent remain artificial in order to accommodate miscellaneous groups of organisms for the convenience of the user. Taxonomy must be adaptive to practical needs, as well as reflecting, as far as possible, the natural relationships of organisms; however, it is important that it should identify the unnatural associations. The alternative of dual classifications, one attempting to show natural relationships, and the other striving for the ultimate in practicality, is not appealing. At present there is no compelling reason for transferring zoosporic fungi to the Protista or any other Kingdom. Moreover, if the studies of protozoa, algae and fungi are recognised primarily as disciplines, there is no practical advantage in removing zoosporic fungi from mycology. The zoosporic fungi, collectively classified as the subdivision Mastigomycotina, are a taxonomic entity for convenience, but phylogenetically they are a non-entity.

Acknowledgements

The author extends his sincere appreciation to Drs. M. W. Dick, M. J. Powell and G. A. Neish for many useful suggestions they made in their review of this paper. However, the author accepts sole responsibility for viewpoints that are of a controversial nature. The author is appreciative of a preliminary reading of the manuscript by Mrs. P. Allan-Wojtas. He also thanks Dr. T. W. Johnson for use of his unpublished manuscript on "The Saprolegniaceae". He is appreciative of many interesting discussions on phylogenetic theory which his colleague Dr. D. B. O. Savile.

3

The Peronosporales

A Fungal Geneticist's Nightmare

D. S. SHAW

*School of Plant Biology, University College of
North Wales, Bangor, UK*

I.	Introduction	86
II.	A Cytologist's Nightmare	86
	A. Arguments about Meiosis—Hyphal Nuclei are Haploid	86
	B. The Renaissance of Diploidy: More Controversy	88
	C. Confirmation	90
III.	The Genetics of the Peronosporales Begins	91
	A. Variation in *Phytophthora infestans*	91
	B. A Preface to the Interpretation of Genetical Data	94
	C. The First Genetical Analyses	95
	D. Matings with Induced Mutants: Conclusive Evidence of Diploidy	96
IV.	The Struggles Continue—To Understand the Basic Mechanisms Generating Variation	98
	A. Mutation and Heterozygosity	98
	B. The Influence of Chromosomal Rearrangements and Tetraploidy	99
	C. Selfing in Matings of Heterothallic Species: An Added Complication or an Unexpected Advantage?	101
	D. Mitotic Crossing-Over	103
	E. Somatic Recombination	105
V.	The Inheritance of Mating Type (Compatability Type)	107
	A. *Phytophthora* spp.	107
	B. *Peronospora parasitica* and *Albugo candida*	108
	C. *Bremia lactucae*	109
	D. *Pythium sylvaticum*	110
VI.	The Genetics of Aggressiveness and Virulence: Can We Breed Better Crop Plants?	111
	A. The *Phytophthora infestans*–Potato Interaction	111
	B. The *Bremia lactucae*–Lettuce Interaction	116
VII.	In Conclusion: A Little More Speculation	118
	Acknowledgements	121

ZOOSPORIC PLANT PATHOGENS
ISBN 0 12 139180 9

I. Introduction

If you wish to learn about a complex biochemical pathway, about the molecular mechanisms of recombination, the behaviour and expression of chromosomal or mitochondrial genes or what happens when such genes are introduced into a bacterium, you reach for the nearest convenient culture of a eukaryotic microbe, most probably a fungus, and put it through its paces along well trodden paths. The fungus you would choose would be a yeast, a *Neurospora* or perhaps an *Aspergillus*; it would grow easily on defined media, have a rapidly completed sexual cycle and perhaps meiotic products held together as tetrads or octads. The genome of your fungus would be well mapped and an extensive range of mutants would be available for all sorts of sophisticated experiments.

In contrast, members of the order Peronosporales, and other lower fungi, offer little or nothing to those searching for a model system. These are delicate and fickle organisms, difficult or impossible to isolate and grow in pure culture; zoospores are fragile and oospores notoriously dormant. Who would choose to do genetics with this temperamental tribe? The Peronosporalologist, interested in the fungi themselves and the diseases they cause, accepts the challenge and enjoys the slow and sometimes lonely advance through the many preliminary problems. He must be able to dream of marked chromosomes, of understanding the genetics of pathogenicity, virulence and mating behaviour and of one day being able to perform some of the tricks which frequently give pleasure to the microbial geneticist.

The zoospore does not figure prominently in this genetical account of the Peronosporales as it is not an obligate stage in the life cycle (*Bremia* and *Peronospora* form no zoospores). Although not involved in karyogamy or associated with meiosis, the uninucleate zoospore is nonetheless a boon in genetical analyses. Clones can be quickly and easily established from physical mixtures of genotypes, from heterokaryons and from heteroplasmons while genotypes present in a heterogeneous mycelium can be sampled using a population of zoospore isolates.

II. A Cytologist's Nightmare

A. ARGUMENTS ABOUT MEIOSIS—HYPHAL NUCLEI ARE HAPLOID

Cytology with these organisms is no less difficult. It is no wonder that the behaviour of their tiny nuclei confused several generations of cytologists and that a healthy controversy developed more than once over the nature of the nuclear divisions. The last years of the last century and the first years of this were ones of intense activity amongst cytologists. Detailed observations were being made on different stages of the life-cycle of a great variety of zoosporic plant pathogenic organisms including *Pythium*, *Phytophthora*, *Peronospora*, *Plasmopara*, *Sclerospora* and *Albugo*, as well as the related saprophytes *Saprolegnia* and *Achlya* (reviewed by Dick and Win-Tin, 1973). Although these first descriptions were often meticulously made and accompanied by extensive and beautiful drawings of the development of gametangia and zygotes, interpretation was not consistent and was often confused. Many cytologists were more interested in details of fertilisation than in whether meiosis was gametangial or zygotic; the significance of meiosis was at that time only just beginning to be realised. Some workers suggested that each large nucleus in the gametangium went through two (meiotic) divisions to produce four (gametic) nuclei (e.g. Wager, 1900; Trow, 1899; Stevens, 1899a,b; Rosenburg, 1903; Ruhland, 1902) and that divisions of the zygote nuclei were similar to those of vegetative nuclei in hyphae. Others, including Berlese (1898), Hartog (1899), Claussen (1908) and Murphy (1918) were convinced that, if division occurred in the gametangia, it was mitotic and that meiosis took place in the zygote. Occasionally the debate became so heated that non-conforming colleagues were accused of interpeting their results according to their own preconceptions (Murphy, 1918).

It is now rather easy to suggest how these different interpretations came to be made.

(1) The very small size of the nucleus makes the observation of chromosome number and behaviour difficult in all but the most favourable material.

(2) It is now believed that much better preparations can be made using the simple squash method than by the sectioning technique used by these early workers.

(3) A complete developmental sequence, essential for the detection of stages of all nuclear divisions in gametangia and zygote is not easy to obtain from axenic cultures and even more difficult to obtain from the field material embedded in host tissue which was so often used in the past. Thus, many were convinced that only one division, inferred to be mitotic, occurred in gametangia. It is now generally agreed that a single mitotic division does take place in the very young oogonium of most species but that the two divisions of a conventional meiosis follow. It appears that the case for zygotic meiosis rested almost wholly on the apparent absence of recognisable meiosis in the gametangia and only rarely in claims of reduction of chromosome number or nuclear size after division of the zygote nucleus (see, for example, Berlese, 1898). The claims for gametangial meiosis were much more substantial and included evidence of synapsis and meiotic products in tetrads (Rosenburg, 1903), of distinctive metaphase chromosomes (Ruhland, 1902), of half the number of chromosomes at anaphase of the first division and of distinctive

differences in chromatin appearance between the first and second division (Stevens, 1899a,b).

How then, did it become generally accepted that the Oomycetes had a haploid vegetative phase like most other fungi? After the first period of lively debate concerning the site of meiosis, little further evidence was produced. Several reviewers (Dick and Win-Tin, 1973; Caten and Day, 1977) have emphasised that the idea of gametangial meiosis was not given proper consideration and was becoming increasingly unpopular as mycologists became convinced that the higher fungi were haploid (Shear and Dodge, 1927; Cutter, 1951; Olive, 1953). Surely since the Oomycetes were perfectly good fungi, they too must have a zygotic meiosis. Opinion must also have been swayed by the sporadic appearance of new evidence, mostly against gametangial meiosis but also for zygotic meiosis (e.g. Arens, 1929b; Tsang, 1929; Saksena, 1936a,b,c; Blackwell, 1943; Damle, 1943; Thirumalachar *et al.*, 1949). McDonough (1937), Bosc (1946) and Ziegler (1953) did produce convincing evidence that the early divisions in zygotes were distinctive and different from those in actively growing hyphae. Apparently the work of Trow (1899), Stevens (1899) and their colleagues was all but forgotten. Had it been appreciated that this class of organism has less affinity with the higher fungi and more with mycetozoa, ciliates, dinoflagellates and brown algae (Stewart and Mattox, 1980; Margulis and Schwartz, 1981), false assumptions might not have been so readily made.

B. THE RENAISSANCE OF DIPLOIDY: MORE CONTROVERSY

It was in this climate that renewed claims for vegetative diploidy were made in the early 1960s. The renaissance began following the efforts of Sansome, a dedicated and highly experienced cytologist. Using improved squashing techniques and undoubtedly drawing on her vast experience of chromosomal cytology with such diverse organisms as *Pisum*, *Datura* and *Neurospora*, she re-examined the division in gametangia of *Pythium*, *Achlya*, *Sclerospora* and homothallic species of *Phytophthora* (Sansome, 1961, 1963a, 1963b, 1965; Sansome and Harris, 1962; review by Sansome, 1966).

Most characteristics of a classical meiosis were found, such as the large nuclei of a prolonged first prophase giving rise to metaphases with very small but distinctive bivalents (in some material, multivalents) which were not observed in dividing nuclei of hyphae or germinating oospores. Sansome noted that products of the first division did not increase in size but instead entered immediately into a second division whose products were smaller than any products of division in hyphae or in germinating oospores. Abortion was noted in oogonia before oospores were mature. It was suggested that segregation of recessive lethals was taking place as a result of meiosis.

Amazingly, this work confirming that of 60 to 70 years before, met with scepticism and considerable resistance. Had this evidence arrived in the 1920s, acceptance would probably have been immediate but by this time a whole generation had been educated against such an idea. Sansome's work served to re-open the old debate and stimulate a long overdue effort to analyse these organisms genetically.

The hope that the pattern of Mendelian inheritance of simple characters would immediately confirm or refute Sansome's claim (and those of Trow, Stevens and others) was not to be realised easily or quickly. The first genetical results came from species of *Phytophthora* and were almost totally inconclusive (see the review by Gallegly, 1968). However, two characteristics seemed to stand out: that traits segregated in the F_1 progeny, and that some propagations from single germinated oospores of *Phytophthora infestans* (Mont.) de Bary showed somatic segregation (Galindo and Zentmyer, 1967; Romero and Erwin, 1969; Satour and Butler, 1968; Laviola, 1968; Timmer *et al.*, 1970). It was argued that F_1 segregation was characteristic of haploid organisms and that the somatic instability was a product of zygotic meiosis where more than one meiotic product could survive.

In its turn, this unsatisfactory genetical work stimulated yet another flush of cytological activity in many laboratories. Nuclear divisions were examined in a wide range of Oomycetes including members of the Saprolegniales: (see, Barksdale, 1968; Bryant and Howard, 1969; Flanagan, 1970; Howard and Bryant, 1971; Win-Tin and Dick, 1975) and Peronosporales: *Phytophthora drechsleri* Tucker (Galindo and Zentmyer, 1967), *P. nicotianae* van Breda de Haan var. *parasitica* (Dastur) Waterhouse (Huguenin and Boccas, 1970), *P. capsici* Leonian (Maia *et al.*, 1976), *Bremia lactucae* Regel (Tommerup *et al.*, 1974; Sargent *et al.*, 1977), *Peronospora parasitica* (Fr.) Tul. and *Albugo candida* (Gmelin ex. Pers.) Kunze (Sansome and Sansome, 1974) and *Pythium* spp. (Win-Tin and Dick, 1975). In all these cases, evidence was that the divisions were meiotic.

However, the old controversy continued a little longer. Laviola and Portacci (1974) argued that their cytological work with *Phytophthora infestans* was consistent with their genetical results; meiotic divisions could not be found in gametangia but were thought to occur in germinating oospores. Similarly, Stephenson *et al.* (1974b) presented evidence for meiotic divisions in oospores of *Phytophthora capsici* which they claimed supported Timmer's genetical work with the same fungus (Timmer *et al.*, 1970). Could it be, as Gallegly (1970) had suggested, that homothallic *Phytophthora* species were diploid yet heterothallic species were haploid?

Sansome returned to the scene and in a series of papers on heterothallic species of *Phytophthora* (Sansome, 1976, 1977; Sansome and Brasier, 1973, 1974; Brasier and Sansome, 1975) confirmed that these fungi, too, were diploid. Laviola's isolates of *P. infestans* and Stephensons's of *P. capsici*, together with isolates of *P. drechsleri* and *P. cinnamomi* Rands were shown to have gametangial meiosis with the same features found in her earlier studies. Chiasmata were observed at prophase and, due to slow terminalisation, were still apparent at metaphase. At the end of the second division the meiotic products could frequently be seen in tetrads (cf. Rosenburg, 1903). In addition, rings and chains of four chromosomes were frequently identified at first metaphase indicating that heterothallic species were structural hybrids as a result of reciprocal translocation. Frequent gametic abortion in an isolate of *P. infestans* appeared related to non-disjunction of an association of six chromosomes. Multiple associations of meiotic metaphase chromosomes were also observed in the homothallic *P. megasperma* Drechsler var. *megasperma*. These are thought to be multivalents resulting from multiple pairing in a tetraploid variety (var. *megasperma*) which has twice as many chromosomes as the diploid variety (var. *sojae* Hildebrand) (Sansome and Brasier, 1974). Similar multivalents have been reported in *Pythium debaryanum* Hesse and *P. echinulatum* Matthews which have $n = c$ 20 chromosomes but not in *P. multisporum* Poitras, *P. ultimum* Trow and *P. torulosum* Coker & Patterson which have $n = 10$ chromosomes (Win-Tin and Dick, 1975).

C. CONFIRMATION

The very last phase of the haploidy–diploidy debate called in ultrastructural and microchemical evidence and was finally closed by the production of results from several extensive and rigorously conducted genetical analyses.

Ultrastructural studies of *Saprolegnia terrestris* Cookson (Howard and Moore, 1970) and of *Achlya ambisexualis* Raper (Ellzey and Huizar, 1977) confirmed many of the observations made on dividing gametangial nuclei with the light microscope and, further, claimed to have identified axial elements associated with synaptonemal complexes or the complexes themselves. Michelmore (personal communication) has now produced detailed evidence of synaptonemal complexes in oogonia of *Bremia lactucae*. Meiosis must be taking place in these gametangia.

Using DNA specific stains or fluorochromes, the relative DNA contents of nuclei at various stages of the life cycle can be compared. Light absorbed or emitted by stained nuclei is proportional to their DNA content and can be measured using a microscope fitted with micro-densitometer or photomultiplier respectively. DNA contents of Feulgen stained nuclei in gametangia of *Saprolegnia terrestris* (Bryant and Howard, 1969) and of *Apodachlya brachynema* Pringsheim (Howard and Bryant, 1971) showed the expected reduction $(4C \rightarrow 1C)$ during the two divisions. Similar proof of meiosis was obtained by Mortimer and Shaw (1975) who measured light emitted from nuclei of *P. drechsleri* stained with the fluorochrome BAO (2,5-bis-3-4[4'-aminophenyl-(1')]-1,3,4-oxadiazole). Only the gametic nuclei in gametangia had the basal 1C DNA content. Prophase nuclei in gametangia had 4C and 2C DNA contents whereas the vegetative nuclei were mainly 2C, but higher levels of DNA typical of nuclei undergoing replication were also detected. These results were obtained for isolates 6500 and 6503 of *P. drechsleri*. The same isolates were shown to have gametangial meiosis (Galindo and Zentmyer, 1967; Brasier and Sansome, 1975) and Mendelian inheritance typical of a diploid (Khaki and Shaw, 1974).

These and other genetical analyses, together with the accumulating cytological and microchemical data make the conclusion unavoidable: that the Peronosporales and other orders of the Oomycetes have a diploid life cycle (Raper type g; Raper, 1966; Burnett, 1975). Such a cycle, having karyogamy immediately following meiosis, is rare in the fungi but typical of animals and algae of the orders Fucales and Siphonales. The lack of even a short haploid phase means that recessive lethal genes will be less often eliminated unless able to act in gametic nuclei within the multinucleate gametangium.

III. The Genetics of the Peronosporales Begins

The higher fungi, shown to exhibit Mendelian patterns of inheritance in the 1930s, soon become favoured material for the geneticist who wanted to study the structure and function of genes. Organisms like yeast and *Neurospora* with a rapidly completed life cycle and large numbers of easily cloned progeny offered clear advantages over flies, plants and animals. Nutritional mutants (auxotrophs) were easily selected in the extended haploid phase of the life cycle; analysis of meiotic products, conserved in tetrads, simplified mapping and led towards an understanding of gene fine structure and of the molecular mechanism of crossing over. Well-mapped chromosomes provided ideal material for studying the basis of somatic variation; chromosomal and extra-chromosomal systems were characterised.

In comparison, the oosporic fungi offered few of these advantages. The plant pathologist was left to struggle on with material which, with rare exceptions, could not even be made to complete its life-cycle. Ignorance of the mating system and of the physiology of sexual reproduction meant that zygote production in many species when grown in the laboratory was rare and erratic. Oospores, when produced, were dormant and could not be germinated.

A. VARIATION IN *PHYTOPHTHORA INFESTANS*

Pending the development of techniques for establishing sexual progeny, much was learnt about the variation in response of field isolates to different host genotypes and about asexually generated (somatic) variation, particularly in species such as *Phytophthora* spp. which were easily grown in pure culture (see the review by Erwin *et al.*, 1963).

The devastating effect of *P. infestans* on potato crops made this fungus a popular subject for the study of variation and it became clear that isolates from tomato and potato were often better adapted for growth on their own host (Giddings and Berg, 1919; Berg, 1926). Soon after resistant material derived from *Solanum demissum* was introduced into *S. tuberosum*, potato isolates were found to vary in their ability to attack different resistant genotypes of the host. This variation in the pathogen was referred to as pathological specialisation (Schick, 1932). Introduction of several potato genotypes, each with a unique single dominant gene for resistance (R gene), enabled the identification of many physiologic races able to overcome the effects of one or more resistance genes. An international system was developed (Black *et al.*, 1953) in which races were classified by numbers corresponding to the number(s) of R gene(s) they were able to overcome. There would seem to be a gene-for-gene relationship governing the reaction of this pathogen with host genes comparable to the relationship identified in flax and flax rust (Flor, 1956). An isolate able to attack plants with resistance gene R_1 is supposed to carry a virulence (pathogenicity) gene, p_1. If the fungus is also able to attack differential genotypes with resistance genes R_2 and R_3, it is supposed to carry p_2 and p_3 genes and is classified as race 1,2,3, although these hypothetical p genes have yet to be identified unambiguously in genetical experiments.

The race composition of *P. infestans* was determined in many countries using a series of differential hosts and was found to be heterogeneous despite the fungus being asexual outside Mexico (e.g. Howatt and Grainger, 1955; Graham, 1955). In some laboratory experiments, growth of an isolate (e.g. race 0) on senescent or succulent leaves of a resistant host, R_4, produced a race (race 4) with increased virulence (reviewed by Black, 1954). Mills (1940a) suggested that the new races resulted from a physiological adaptation during passage through resistant leaves similar to the increase in aggressiveness of potato races when grown serially on tomato, a phenomenon he termed "adaptive parasitism". In other experiments no increases in virulence were detected (Müller, 1933; Black, 1952). However, new virulence characteristics were known to have arisen in populations never exposed to new resistant genotypes (Mastenbroek and de Bruin, 1955; Schick *et al.*, 1958); these could not have arisen physiologically by "training" but, it was thought, could have resulted from gene mutation. Thus gene mutation became the preferred explanation of race change and, coupled with strong selection pressure, offered an acceptable mechanism for the origin of new races during passage on resistant material (Gallegly and Niederhauser, 1959).

There was clearly a need to understand more fully how new potato cultivars with blight resistance genes were being overcome by genetical changes in fungus populations. The apparent absence of any sexual stage precluded analysis using the sexual cycle until oospores resulting from gametangial fusions were discovered and found to be common in the Toluca Valley of Mexico (Smoot *et al.*, 1958; see Fig. 1). This work led to a re-examination of the mating system of self-sterile *Phytophthora* spp. *P. infestans* and other self-sterile species were shown to be truly heterothallic, having two clearly defined compatibility or mating types, A1 and A2 (Galindo and Gallegly, 1960). Although first defined in *P. infestans*, these mating types were found to occur in all heterothallic species throughout the genus; for example, pairings of A1 or A2 isolates within or between species were sterile; intra- or inter-specific pairings of A1 with A2 isolates were fertile (Savage *et al.*, 1968). Tracings of hyphae from gametangial unions established that isolates of each mating type were bisexual but varied in their preference to act as antheridial and oogonial parents; i.e. sexuality was relative (Galindo and Gallegly, 1960; Gallegly, 1968).

Zygotes could now be produced in abundance; their germination was to be the next impasse. It was well known that these oospores became dormant at maturity and showed a very low rate of germination over many months, a rate which was not greatly increased by any of a wide variety of environmental factors (see, for example, the reports by Blackwell and Waterhouse, 1931; Smoot *et al.*, 1958; Savage and Gallegly, 1960). However, techniques which gave only a few per cent germination allowed a few crosses to be examined between parents differing in physiological race (Smoot *et al.*,

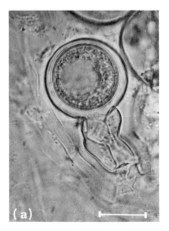 

Fig. 1. Oospores of *Phytophthora infestans* resulting from a mating of A1 and A2 isolates. The fertilisation tube emerges from the antheridium (a) and swells at its apex (b) near to the point of penetration into the oosphere (now the oospore). Scale bar = 20 μm. By courtesy of C. Laviola.

1958; Niederhauser, 1961). Sansome's revival of the ploidy debate (Sansome, 1961) provided an added stimulus for genetical experiments with *P. infestans* (Romero and Erwin, 1969; Laviola, 1968), *P. drechsleri* (Galindo and Zentmyer, 1967), *P. capsici* (Satour and Butler, 1968) and *P. cactorum* (Leb. & Cohn) Schroeter (Shaw and Elliott, 1968). Would genetic data support or refute the diploid hypothesis?

B. A PREFACE TO THE INTERPRETATION OF GENETICAL DATA

Before considering any genetical results it is useful to examine what we might expect according to the alternative hypotheses. The simplest possible situation is where the segregation of a difference is due to the segregation of two alleles of a single gene. Selfed progeny from a homothallic isolate would show no segregation in a haploid but would show a 3:1 (or 1:2:1) segregation in a diploid if the parent were heterozygous at a single locus. Progeny from the F_1 generation of a mated heterothallic pair would show a 1:1 segregation in a haploid but might be heterokaryotic if more than one meiotic product survived in the germinating oospore. Segregation in a 1:1 ratio would continue in backcrosses and in F_2 progeny when the different phenotypes were crossed. In a diploid, the F_1 progeny would also show a 1:1 ratio if one parent was heterozygous but would not segregate if both parents were homozygous. Uniform F_1s on intercrossing would be expected to yield 3:1 or 1:2:1 segregations in F_2 and would yield 1:1 ratios in backcrosses with the recessive parent.

Thus, in selfed progeny of homothallic species the expected difference between a haploid and a diploid hypothesis is quite clear, but for matings of heterothallic species, the difference is only detectable in the F_1 progeny if both parents are homozygous. The 1:2:1 or 3:1 segregation in the F_2 is only to be expected in a diploid.

However, clear cut results such as these are expected only in crosses between pure breeding parents. If field isolates are used, parents may differ by alleles at many gene loci. These differences could result in heterogeneity in the F_1 progeny in both haploids and diploids and could modify the ratio of single gene segregations if modifier or lethal genes were involved. Wild outbreeding parents, if diploids, could be highly heterozygous so that each gamete produced would be likely to be unique; the possibilities for recombination in the progeny would be vast.

Further complications are added if the bisexual nature of each parent is expressed during mating to yield a proportion of single parent progeny (selfs). In a mating of two compatible isolates, oogonia of the A1 parent might combine with antheridia of the A2 parent and vice versa. Each parent might also self. Thus, any one, two, three or four of these combinations might produce viable offspring. Obviously, for chromosomal genes, the two kinds of hybrid (reciprocal) matings will yield identical results but a proportion of one, other or both of the selfed combinations would disturb the expected ratios and might even alter the classes of progeny expected. For example, in a haploid, selfing could result in any proportion of the two phenotypes significantly different from the 1:1 proportion expected and in the extreme case where selfing replaced crossing, a uniform F_1 mimicking the F_1 in a diploid would result. Similarly, in a diploid, where parents are homozygous, selfing could give the segregating F_1 and stable F_2 expected of a haploid.

C. THE FIRST GENETICAL ANALYSES

The first sexual progenies established and analysed were from matings of *P. infestans* from Mexico (see the review by Gallegly, 1968). The few F_1s recovered from these matings showed a range of phenotypes having new combinations of the parental race characters and mating type. Segregation of some race differences was close to 1:1 while others deviated from this significantly (Romero and Erwin, 1969). Single oospore colonies yielded a single phenotype. Laviola (1968, in Gallegly, 1970) provided more extensive data mainly for the mating race 0 × race 1,2,3,4. Within 33 single oospore offspring, 13 of the 16 possible combinations of the single race characters appeared. Oospores frequently formed a single germ tube with a terminal zoosporangium. Zoospores from some of these "germ" zoosporangia, in this and in other crosses, segregated for both race and mating type characters. Mating type segregation occurred in most of these matings. Some of the ratios that were found fitted 1:1; others deviated significantly in one or the other direction. Attempts to raise F_2 progeny (Castro and Zentmyer, 1969) were thwarted due to the non-viability of germinated oospores.

What can we conclude about ploidy from these data? Firstly, let us apply Occam's razor and assume that a single race difference, i.e. virulence on R_1 plants *vs.* avirulence on R_1 plants, is determinated by a pair of alleles. Haploid progeny should show a 1:1 segregation for presence or absence of each single race character. Laviola's somatic segregation argues for survival of more than one meiotic product in some germinating oospores. If alleles determining virulence were recessive in a diploid, 1:1 segregations in progeny would result only if the avirulent parent carried a virulence allele in the heterozygous condition; otherwise F_1s would be avirulent. If virulence were dominant, progeny would be virulent or would segregate if the virulent parent carried an allele for avirulence. The finding of a race 2 character in the progeny of a mating of 1 × 1,4 (Niederhauser, 1961) is explained if parents were heterozygous for the same recessive virulence allele. Equal proportions of the two mating types in progenies could result from segregation from a heterozygote if haploid or from crosses in which one mating type is heterozygous and the other homozygous recessive.

Clearly, these matings with *P. infestans*, using race and mating type as markers do not provide the data needed to distinguish between a haploid and diploid life-cycle. Further data were produced from the mating of isolates of two other heterothallic species of *Phytophthora*. Galindo and Zentmyer (1967) used isolates of *P. drechsleri* from *Capsicum* in which gametangial meioses had been proposed. The inheritance of several natural differences between the two parents was followed in the F_1 and backcross progenies. All differences, including those of colony morphology and colour, repulsion/stimulation when paired, and sensitivity to malachite green, segregated in the F_1 and in backcrosses in which the parents differed. Some segregations fitted a 1:1 ratio, others did not. Satour and Butler's crosses of *P. capsici* showed similar segregations of natural differences such as zoospore producing ability, pathogenicity on tomato and pepper and mating type (Satour and Butler, 1968). Neither of these studies could distinguish between segregation in a haploid and that in a diploid due to heterozygosity.

D. MATINGS WITH INDUCED MUTANTS: CONCLUSIVE EVIDENCE OF DIPLOIDY

How likely was it that the data produced so far were complicated by the segregation and recombination of many pairs of allelic differences for each character being studied? Obviously, this could be avoided by following the pattern of inheritance of induced mutations giving a distinct phenotype. Using standard techniques, Timmer *et al.* (1970) selected a series of auxotrophs of *P. capsici* hoping that each would have a mutation of a single gene. However, the results were no less complicated than before. Simple segregations of prototrophy and auxotrophy were not as expected in a haploid and were confused by the segregation of additional auxotropy not shown by either parent. If diploid, the parental auxotroph must have been homozygous for a recessive auxotrophic allele or perhaps heterozygous for a gene regulating the activities of structural genes and, maybe, heterozygous for the recessive auxotrophy which appeared unexpectedly. In short, results could not easily be explained by any scheme involving a single mutant gene on either a haploid or a diploid model.

Shaw and Khaki (1971) selected drug resistance as a marker, arguing that this could result from a single dominant mutation and should be obtainable in either a haploid or a diploid fungus. The usual mutations of structural genes to auxotrophy are recessive and would not be expressed in a diploid. Mutants resistant to chloramphenicol and *p*-fluorophenylalanine were selected in the isolates of *P. drechsleri* used by Galindo and Zentmyer (1967). Although the F_1 progeny were variable for natural markers (Galindo and Zentmyer, 1967; Rutherford, unpublished work) they were uniformly drug resistant, gave a 3:1 ratio when sib-crossed and 1:1 ratios in one of the backcrosses. These results fitted exactly to the pattern expected if resistance is determined by a dominant allele in a diploid organism. Further backcrossing of F_2 progeny identified both the homozygous and heterozygous genotype within the resistant phenotypes of that progeny. It is understandable that such a result, clearly arguing for diploidy, was not immediately accepted. Perhaps all the classes and ratios of progeny were being mimicked by selfing in a haploid organism (Day, 1974).

However, supporting evidence was soon to emerge from a series of studies with homothallic species. Boccas (1972) established F_1 progeny from a single zoospore isolate of *P. syringae* (Klebahn) Klebahn. He measured the rates of linear extension of their hyphae across agar and compared them with rates of a sample of zoospore progeny from the same isolate. Zoospore propagations were relatively uniform whereas oospore progeny showed great heterogeneity for growth rate. This release of variation is most simply explained by segregation from a diploid genotype, heterozygous for genes affecting growth rate. The only variation expected in a selfed haploid would have a cytoplasmic basis; there was little evidence of cytoplasmic variation in the uniform zoospore propagation. Further evidence of segregation in progeny produced by selfing in a homothallic species was provided by Elliott and MacIntyre (1973). Zoospores from a single-spore culture of *P. cactorum* were treated with a chemical mutagen and survivors were allowed to form oospores. Although these survivors were prototrophic, it was reasoned that, in a diploid, recessive auxotrophic mutations may have been induced; mutants would then segregate on selfing. While most of the selfed progenies were wild type (prototrophic), others did show segregation of a proportion (25%) of auxotrophs. An intensive analysis of one survivor which segregated methionine requirers established that these auxotrophs and one-third of the prototrophs were pure breeding (i.e. homozygous), whereas the other two-thirds of prototrophs segregated into one-quarter auxotrophs (i.e. they were heterozygous). Third-generation cultures from selfed second-generation prototrophs confirmed the existence of homozygotes and of heterozygotes carrying the recessive allele for methionine requirement. Data conforming to this pattern are now available for *P. megasperma* var. *sojae* (Long and Keen, 1977b) and *Pythium aphanidermatum* (Edson) Fitzpatrick (Dennett and Stanghellini, 1977).

Thus it was shown that single gene markers (both dominant and recessive) can be selected after mutagenesis and, in contrast to natural markers, can be used to provide patterns of inheritance easily recognised as characteristic of a diploid in both homothallic and heterothallic species of *Phytophthora*.

The only doubt that may remain is that vegetative nuclei might be diploid but be derived from a meiotic division of a tetraploid zygote. This possibility was ruled out in *Phytophthora drechsleri* following the quantitative study of DNA in nuclei of gametangia and vegetative hyphae (Mortimer and Shaw, 1975). The fluorimetric measurements argued for a true meiosis and against a mitotic division of haploid or of diploid nuclei in gametangia.

With the accumulation of so many data from cytological, genetical and microchemical sources pointing to a gametangial meiosis, it became generally accepted, at last, that both the water moulds and the pathogenic members of the Peronosporales have a basically diploid life cycle corresponding to Raper's type g (Raper, 1966) and to haplobiontic type B (Webster, 1980), i.e. karyogamy follows immediately after meiosis without any intervening haplophase, as happens in higher animals. This is yet another feature of the Oomycetes which sets them apart from other classes of fungi.

IV. The Struggles Continue—To Understand the Basic Mechanisms Generating Variation

Any hopes of rapid progress towards an understanding of the genetics of these fungi faded when it became generally realised that the life cycle was diploid and that they did not lend themselves to the kind of analyses so easily made with haploid fungi. All the inconveniences of the diploid condition had to be grappled with.

A. MUTATION AND HETEROZYGOSITY

The isolation of convenient mutants for use in marking chromosomes is a basic requirement in genetics and has proved to be a difficult task in *Phytophthora* spp. The induction and isolation of dominant alleles conferring drug resistance seems to pose no special problems other than that of screening enough zoospores to allow the selection of rare mutations, but there is evidence to suggest that some of the more frequent drug-resistant mutations isolated in *P. cactorum* (Shaw and Elliott, 1968), *P. capsici* (Timmer et al., 1970) and *P. megasperma* var. *sojae* (Long and Keen, 1977b) could have been due to mutations in non-chromosomal genes.

The isolation of single recessive alleles for auxotrophy does pose a problem in a diploid. Although this is neatly solved by selfing of mutagen-treated parents and recovery of auxotrophic segregants in the progeny in a homothallic species (MacIntyre and Elliott, 1973) the technique would only be useful for heterothallic species if selfing could be induced. Timmer's work (Timmer et al., 1970) suggested that directly isolated auxotrophs of *P. capsici* were not homozygous for a recessive allele; their determination remained obscure but possibly involved regulatory rather than structural genes. It is interesting to note that the frequency of recovery of homozygous recessive mutants in a diploid strain of a slime mould was much higher than predicted from their frequency of recovery in a haploid strain (Williams, 1976). This suggests that homozygotes may be generated from heterozygotes by mitotic crossing over rather than by further mutation.

In the wild, we should expect our diploid fungus to carry a range of recessive mutations which would segregate in sexual progenies. Long *et al.* (1975) inbred field isolates of *P. megasperma* var. *sojae* over five generations and showed a progressive decline in variation in their progenies. Selected inbred lines from progenies showing uniform growth rate, colony morphology and aggressiveness, were used as wild-types in their genetical analyses—thus segregations due to unwanted recessive alleles were avoided. In heterothallic species such as *P. drechsleri*, persistent inbreeding to lower heterozygosity and to increase isogenicity by crossing has been attempted but has resulted in an undesirable loss in fertility (Rutherford, unpublished work).

Sansome (1965, 1966) first produced evidence for the existence of deleterious recessive alleles in her study of abortion of oogonia and oospores of *P. erythroseptica* Pethybridge. Since then, various workers have recorded abortion and loss in viability at different stages of oogonium, oospore and germling development. It seems likely that abortion of this kind will be minimal in habitual and regular inbreeders (recessive lethals would be quickly eliminated) and will increase as heterozygosity increases in outbreeders and in populations in which sexuality is rare. It would be interesting to find out if differences in oospore viability in various isolates within a species, as observed in *P. megasperma* f.sp. *medicaginis* by Förster (personal communication), are related to differences in sexual experience.

Isolates for genetical studies should be carefully chosen; those showing little or no abortion and giving a high proportion of easily germinated oospores are obviously desirable. Heterothallic pairs, isolated from the same population and thus having natural isogenicity, would seem to offer the best chances for success. So far, most workers have tended to use different species, pathogenic on different hosts; more rapid progress might be made if promising material and mutants were shared so that results could be confirmed and extended as additional useful mutants became available, as first suggested by Day (1974).

B. THE INFLUENCE OF CHROMOSOMAL REARRANGEMENTS AND TETRAPLOIDY

Recent cytological work has underlined the value of studying the karyotype of isolates to be used in genetics. Reciprocal translocations held in the heterozygous condition have been identified in all heterothallic species of *Phytophthora* and in the heterothallic *Bremia lactucae* (Sansome, 1980; Michelmore and Sansome, 1982; see Fig. 2). It has been argued that this structural hybridity is a mechanism for the maintenance of genes on the translocation complex in a heterozygous condition by preventing the survival of homozygous somatic recombinants (Sansome, 1980). Certain post-meiotic segregations and instabilities in progeny of the heterothallic *P. drechsleri* are

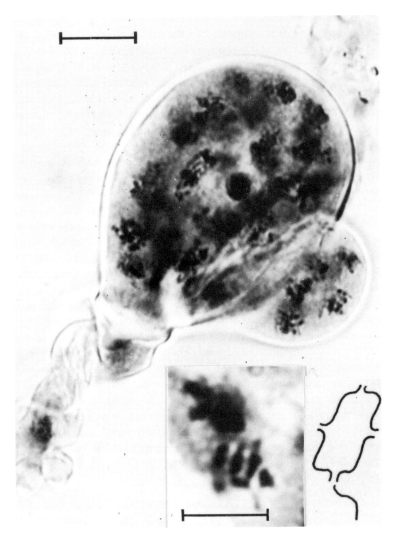

Fig. 2. The oogonium and paragynous antheridium of *Bremia lactucae* formed as a result of mating of isolates IL4 and IM44. Nuclei, stained by an aceto-orcein squash method, are mostly in late prophase 1 of meiotic divisions. Scale bar = 10 μm. Insert: metaphase I nucleus from a self-fertile isolate IM25L2 showing an association of five chromosomes. An interpretation is provided in the adjacent line drawing in which the extra chromosome of the trisomic nucleus has homology with two segments of chromosomes belonging to the reciprocal translocation complex. Scale bar = 5 μm.
By courtesy of R. W. Michelmore and Trans. Br. mycol. Soc.

thought to be the results of the aberrant disjunctions to be expected from the associations of the four translocation chromosomes at meiotic metaphase I. Meiotic products suffering deficiencies are probably eliminated or aborted but products having duplications are likely to survive to produce trisomic progeny. It has been estimated that as many as 15% of the progeny of wild-type matings of *P. drechsleri* are aneuploid and somatically unstable (Mortimer *et al.*, 1977; Sansome *et al.*, unpublished work). It can be inferred that genes linked to the translocation complex may frequently show post-meiotic segregation and aberrant patterns of inheritance. This offers one explanation for the segregation of physiological race and mating type from single oospores of *P. infestans*, segregations which were used as evidence for zygotic meiosis (Laviola, 1968, in Gallegly, 1970). Other explanations involve the rare oogonia having two oospores, observed in Laviola's material by Sansome and Brasier (1973), or dikaryotic oospores (Khaki and Shaw, 1974).

Chromosome counts have provided evidence of polyploidy in *Phytophthora* (Sansome and Brasier, 1974). Small oospored isolates of *P. megasperma* which attack herbaceous plants have a haploid complement of 9–12 chromosomes which is typical of most other species in the genus (counts are given as a range owing to the difficulty of distinguishing separate metaphase bivalents). In contrast, the large oospored form of *P. megasperma*, which is a pathogen of woody plants, has a count in the range $n = 22$–27. Thus it seems likely that a diploid form, presently referred to as *P. megasperma* var. *sojae*, gave rise to the tetraploid, *P. megasperma* var. *megasperma* as a result of chromosome doubling. It remains to be seen if the tetraploid isolates show diploid or tetraploid inheritance. Most of the synapses observed in isolates of *P. megasperma* var. *megasperma* were of bivalents although a few multivalents were noted (Sansome and Brasier, 1974); it would seem that isolates were functionally diploid (Sansome and Brasier, 1974; see also Brasier, 1983, for a review of the status of forms of *P. megasperma*).

C. SELFING IN MATINGS OF HETEROTHALLIC SPECIES: AN ADDED COMPLICATION OR AN UNEXPECTED ADVANTAGE?

Single zoospore isolates of heterothallic species of *Phytophthora* are usually self-sterile but act as A1 or A2 mating types. In pairings with an opposite mating type, an isolate is typically bisexual i.e. it contributes antheridia which fuse with oogonia from the other parent, and oogonia which fuse with antheridia from the other parent. Thus reciprocal crossing takes place to form the two kinds of hybrid oospore, A1♀ × A2♂ and A1♂ × A2♀. However, under certain environmental conditions, single isolates may become self-fertile and function as if they were homothallic. Selfing of A2 isolates of a number of species can be induced by certain antagonistic soil microbes (Brasier, 1971; Pratt *et al.*, 1972), by mechanical injury (Reeves and Jackson, 1974), by a root exudate (Zentmyer, 1979) or by a volatile fungicide (Noon and Hickman, 1974). Also, single mating types were shown to self when separated from a homothallic isolate by a cellophane membrane (Brasier, 1972) or when separated from a compatible mating type by a porous polycarbonate membrane (Ko, 1978). These observations suggest that progenies from matings might be more complex than was first imagined; heterothallic crosses might yield progeny from the selfing of each parent in addition to the expected hybrid progeny.

Is there any evidence that selfing plays a role during sexual reproduction in pairings of heterothallic isolates? Tracings of hyphae from fused gametangia (Stamps, 1953; Galindo and Gallegly, 1960) have shown that the antheridium and the oogonium arise from hyphae belonging to different parents and thus that zygotes are likely to be true hybrids. This technique is difficult and laborious and, therefore, less common, selfed gametangia may not have been detected. Another approach is to make use of the differential uptake of stain by oogonia of each parent in some matings (Brasier and Sansome, 1975; Sansome *et al.*, 1979). Work of this kind and similar work using fluorochrome labelling of one parent (Huguenin, 1973) has suggested that selfed combinations of gametangia do occur as well as hybrid combinations, at least in some matings. What is not yet known is whether viable oospores are formed from these selfings and, if they are, what proportion of the established oospore progeny are selfed. As selfed oospores of *P. drechsleri*, induced by the antagonistic soil fungus *Trichoderma* have been germinated (Brasier and Sansome, 1975), it seems likely that those produced in matings will also be found to be viable.

If a significant proportion of single oospore progeny resulted from selfing, then surely these would be detected in genetical analyses? Sansome (1970) has re-analysed the data of Galindo and Zentmyer (1967) for the segregation of natural markers of *P. drechsleri* and has concluded that the aberrant ratios fit closely to those expected in a diploid if about one third of the progeny result from selfing of one parent. But it can be argued (see, for example, Shaw, 1983) that no selfing (or very little) could have occurred in Khaki's matings (Khaki and Shaw, 1974) of parents derived from Galindo's isolates of *P. drechsleri*, since the observed frequencies in her progenies were very close to those expected. It could be that selfing competence is easily destroyed by small changes in genotype; like those that may have been induced during Khaki's mutagenesis treatments. Not only might we expect small differences in genotype to influence selfing but also small differences in environment; the same mating might yield very different selfing frequencies on different occasions. It looks as if the frequency of any selfing will need to be determined for each mating on each occasion; simple genetical methods could easily be developed for this purpose (Shaw, 1983).

Matings between A1 and A2 isolates belonging to different species of *Phytophthora* usually result in prolific oospore formation (Barrett, 1948; Galindo and Gallegly, 1960; Savage *et al.*, 1968). To find out if these oospores are interspecific hybrids or result from selfing of the parent(s), several studies have compared isozymes and other proteins of parents and progenies (Boccas and Zentmyer, 1976; Kaosiri and Zentmyer, 1980; Boccas, 1981; Erselius and Shaw, 1982). Almost without exception, progenies were found to be parental in their protein patterns. In some matings, there was evidence that selfing of both parents contributed to the progeny, in others, selfing of one parent provided most or all of the progeny. This work suggests that selfing may be common when parents are unrelated. It would be useful to find out if selfing frequency falls as relatedness of mates increases; are progeny always hybrid when parents are isogenic or nearly so?

Although selfing is troublesome for the geneticist especially when its frequency is unknown, he should be able to turn it to his advantage, just as the phytophthoras seem to do. The disadvantages of working with a heterothallic species disappear if selfing can be added to crossing as a genetical tool. Troublesome alleles in the heterozygous condition, occurring spontaneously or induced by mutagens could be quickly eliminated. Other alleles might be worth recovering as auxotrophic or temperature-sensitive markers; new mutants could be screened to identify the most useful ones showing simple patterns of inheritance on selfing. The genetics of *P. infestans* could progress in any laboratory, allowing the Mexican A2 mating type to remain safely at home.

D. MITOTIC CROSSING-OVER

Crossing-over at mitosis by breakage and rejoining of chromosomes is known to occur in diploid strains of the higher fungi and in vegetative cells of both plants and animals. It seems likely that wherever two homologous sets of chromosomes share the same nucleus, rare pairing of homologous chromosome segments is followed by crossing-over at the four-strand stage just before mitosis. This event can result in the segregation of the two kinds of homozygote from diploids having a heterozygous locus distal to the site of crossing-over.

Evidence for the same or a similar kind of mitotic crossing-over in the Oomycetes is meagre. Although there are several examples of segregation occurring during vegetative growth (Buddenhagen, 1958; Shaw and Elliott, 1968) which could have resulted from crossing-over, these could also have resulted from segregation of new mutations or from segregation of non-chromosomal genes. Examples of segregation from known heterozygotes have not been reported. However, Khaki and Shaw (1974) using *P. drechsleri* and Lasure and Griffin (1974) using *Achlya bisexualis* Coker found that their dominant drug-resistant mutants were homozygous and not heterozygous, as might have been expected. It is possible that heterozygotes produced by mutation of a single allele became homozygous during selection and characterisation before matings were made. It seems unlikely that a further mutation took place in the recessive allele during this time.

Support for mitotic crossing-over has come from cytological studies of trisomic isolates of *P. drechsleri* (Sansome, 1980; Sansome *et al.*, unpublished observations). Single oospore isolates which were self-fertile but segregated A1 and A2 types during vegetative growth were found initially to be tertiary trisomics of type I (see Fig. 3) i.e. they had an additional chromosome homologous to one of those showing structural hybridity. Segregants from a trisomic of this type were still trisomic but showed evidence of type II configurations of chromosomes at meiosis (see Fig. 3) i.e. the extra chromosome had homology with each of two pairs of chromosomes.

Evidence of mitotic crossing-over should be sought in hybrids from parents differing in a number of characters. Michelmore and Ingram (1981) crossed isolates of *Bremia lactucae* with different virulence characters (i.e. race 1,4,5,6,7,8,9,10,11 × race 1,2,3,4) to produce hybrids which were essentially avirulent on a differential series of lettuce cultivars but were presumed to be heterozygous for nine virulence factors. Michelmore (personal communication) has never observed any segregation of virulence factors during asexual propagation of hybrids over many generations.

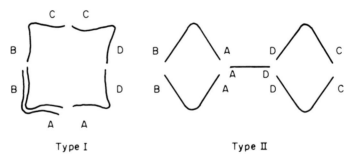

Type I   Type II

Fig. 3. Tertiary trisomics have an extra chromosome homologous with one of the chromosomes involved in structural hybridity (type I) or alternatively, the extra chromosome may have segments homologous with segments of two pairs of chromosomes (type II). Line drawings show possible configurations of meiotic metaphase chromosomes in trisomic nuclei which would be easily recognised in squash preparations of gametangia. Each line represents one metacentric chromosome with two chromatids. Homologous segments are labelled with the same letter. Mitotic crossing-over can yield type II from type I but not *vice versa*. See Fig. 2 for a clear example of a type I configuration.

However, if virulence genes were tightly linked to their centromeres or sited on the translocation complex, segregants could be rare or absent.

Mitotic crossing-over may play a most important role in the survival of diploid fungi. A rachet-like mechanism which is thought to operate in diploid organisms explains how the mutational load may increase with successive accumulations of deleterious mutations in the fittest member of the population (each additional mutation is like the irreversible click of a rachet). In theory, the rachet is most active in small populations but may not function if repair of mutational damage to DNA is particularly efficient. It can be reversed by sexual recombination or by mechanisms, like mitotic crossing-over, which result in the elimination of harmful mutations when they become fully exposed in a homozygous condition (see Muller, 1964 in Maynard Smith, 1978 for a fuller discussion of Muller's rachet). This accumulation of harmful mutations may be crucial in species which suffer such a catastrophic decline that their populations become very small during part of the life-cycle (for example, overwintering in *P. infestans*). Populations of such species would reverse their rachet during periods of sexual reproduction but if this possibility were denied (for example in *P. infestans* outside Mexico), they would be dependent on mitotic crossing-over for survival. Will the rate of mitotic crossing-over be high in predominantly asexual species with a population bottleneck or is there some other explanation for their continued success? All we need, to start to find out, is a series of known markers in the heterozygous condition in a range of species!

E. SOMATIC RECOMBINATION

Heterokaryosis can be looked upon as a mechanism adopted by the haploid fungi which offers the advantages of diploidy (the sheltering of mutations which are not immediately useful) and which, in the Ascomycetes, can respond rapidly to the environment of the moment. Not only does it allow vegetative mixing of nuclei and cytoplasms of different genotypes but provides for the rare fusions of unlike nuclei and recombination of their genes before a return to the haploid condition. This parasexual cycle appears to replace sex in imperfect fungi but is also found in those which are sexually competent.

Perhaps heterokaryosis should not be expected to have a similar importance in filamentous fungi with diploid nuclei and a well developed sexuality. However, Long and Keen (1977a) provided a convincing account of heterokaryosis in *P. megasperma* var. *sojae*. Prototrophic heterokaryons were forced when complementing auxotrophic mutants were grown together on minimal medium. A heterokaryon was confirmed when parental monokaryons segregated from single hyphal tip propagations.

Segregation from a heterokaryon has also been detected in *Phytophthora drechsleri* (Mortimer *et al.*, 1977; Sansome, 1980) and in *Bremia lactucae* (Michelmore and Ingram, 1981; Michelmore and Sansome, 1982). A proportion of the offspring from matings of *P. drechsleri* were self-fertile and were later shown to be trisomic. Similar trisomy has been found in a naturally occurring self-fertile isolate of *B. lactucae*. In both cases, trisomic cultures were unstable; segregation of single mating types occurred during vegetative growth and sporulation. This is evidence that mating type heterokaryons are formed when determinants for a single mating type segregate after mitotic crossing-over or somatic non-disjunction in the original trisomic nuclei which carry determinants for both mating types. Propagations from these heterokaryons using single zoospores (*P. drechsleri*) or single conidia (*B. lactucae*) showed segregation of the constituent monokaryons (Mortimer *et al.*, 1977; Michelmore and Ingram, 1982). It is not yet clear if heterokaryons are persistent in the hyphae of these cultures or are evanescent and resolved into monokaryons soon after their formation.

The apparent ease with which *P. infestans* generates variants having various combinations of virulence characters (Malcolmson, 1969) has stimulated a search for mechanisms allowing somatic recombination in this fungus which apparently is asexual throughout most of its range (Smoot *et al.*, 1958). Indeed, phenotype-phenotype analysis (Wolfe *et al.*, 1976) has suggested that free exchange of the genetic determinants of race occurs within a sampled population in North Wales (Shattock *et al.*, 1977).

Most workers have taken pairs of isolates having different race characteristics and grown them together either in pure culture (Wilde, 1961; Leach and Rich, 1969) or on host tissue (Wilde, 1961; Denward, 1970; Malcolmson, 1970) and have examined asexual progeny from the mixture for new race characteristics. Although the various results differ in detail, the common finding was that if the two parents were, say, race 1,2 and race 3,4 then race 1,2,3,4 could be recovered from amongst the progeny. "Recombinants" of this general kind were recovered from host leaves with resistance which selected against both parental races (e.g. Malcolmson, 1970) but were also recovered along with parental types from mixtures in pure culture when there was no selection for a recombinant and against the parents (e.g. Leach and Rich, 1969). Leach and Rich went on to show that zoospore propagations from single sporangia derived from a mixed culture were of parental as well as recombinant phenotype. Assuming that race is chromosomally determined, this suggests that nuclei of both parents associate in a heterokaryon and somehow generate "recombinant" nuclei which co-exist with parental types within the common cytoplasm until they segregate into monokaryons at zoospore formation. Thus the available evidence suggests that new races ("recombinants") are not heterokaryons of complementing parental nuclei but are stable monokaryons.

Similar findings have been made using a different approach. Dyakov and

Kuzovinikova (1974) and Shattock and Shaw (1976) mixed pairs of drug resistant mutants of *P. infestans* of known virulence and selected "recombinants" resistant to both drugs on a double-drug medium. Both groups found that "recombinants" were particularly vigorous and could be propagated without any sign of instability through single zoospores. Clearly these growths were not heterokaryons. Also, in both cases, the virulence characteristics of only one of the parents were detected in the "recombinant". In a comparison with the parental isolates, Dyakov found that his "recombinant" had nuclei containing twice as much DNA, had an increased resistance to ultra-violet irradiation and segregated single zoospores after irradiation which showed various combinations of parental race and drug resistance phenotypes (Dyakov and Kulish, 1979; Kulish and Dyakov, 1979). Are tetraploid nuclei being produced by fusion of parental diploid nuclei in heterokaryons and are segregants resulting from loss of whole chromosomes (diploidisation?) in some sort of parasexual cycle?

Although in all work of this kind, some "recombinants" could have arisen through simple genetic changes such as mutation or mitotic crossing-over in one of the parents (Wilde and Denward identified race instabilities in parents) or even through faulty diagnoses of races by the use of poor differential lines, it seems likely that others were true somatic hybrids. These would seem to arise from the interaction of nuclear or cytoplasmic genes (or both) from each parent. Large differences in the ease with which somatic recombinants were recovered suggests that the frequency of hyphal fusions leading to hybridisation may be limited in some cases by heterokaryon incompatibility.

This is an exciting field in which we have yet only fragments of evidence for a somatic mechanism generating variation. More fragments might be gleaned by repeating and extending the type of experiments already done, with as many controls as possible, but I feel sure that progress will be slow until we understand the genetic basis of virulence and can use known genotypes, carrying a range of gene markers, as "parents".

V. The Inheritance of Mating Type (Compatibility Type)

A. *PHYTOPHTHORA* SPP.

Data on segregation of mating type have been collected whenever controlled matings were made with heterothallic species; the character is one which is clearly defined and thus easily scored. Most of the extensive literature, reviewed by Erwin *et al.* (1963), Gallegly (1970) and Webster (1974), dealt with mating type of *Phytophthora* spp. Since the genetics of mating type in this genus has been discussed more recently in some detail (Shaw, 1983) the main findings only will be summarised here to allow comparison with mating types and their inheritance in other genera.

In crosses where Mendelian patterns of inheritance are clearly shown, mating type segregations appear to be non-Mendelian (for example, in matings of *P. drechsleri*; Khaki and Shaw, 1974) suggesting that determination is polygenic or, at least, in part, non-chromosomal. All heterothallic species of *Phytophthora* examined by Sansome (1980) were found to carry reciprocal translocations in heterozygous condition; translocations did not occur in homothallic species. This discovery was an important one and led to the suggestion that mating type genes are linked on segments of chromosomes involved in the translocation. For convenience, the four chromosomes involved in the translocation are referred to as "the mating type complex" of chromosomes (Sansome, 1980) and can be clearly seen at meiotic metaphase I where they form rings or chains (multiple associations of four chromosomes). Disjunction from multiple associations is clearly not always regular which may account for some of the aberrant ratios of mating types in progenies. For example, a high proportion (up to 15%) of single oospore cultures from matings of *P. drechsleri* were self-fertile (Mortimer *et al.*, 1977) and some of these are now known to have been trisomic for one of the chromosomes belonging to the translocation complex (Sansome *et al.*, unpublished work); this provides excellent support for the suggestion that the association of four chromosomes carries mating type genes (i.e. is a "mating type complex").

The simplest hypothesis for the determination of mating type involves a single mating type gene with two alleles *a* and *A*. If A1 mating types are homozygous, *aa*, and A2 mating types are heterozygous, *Aa*, then A2 will segregate A1 mating types by mitotic crossing-over during vegetative growth, unless mitotic recombinants are eliminated. This elimination is assured by linkage of *a* and *A* to the mating type complex; products of mitotic crossing-over would be unbalanced and would not survive unless rescued by the extra chromosome of a trisomic. This explains why somatic instabilities of mating type are confined to trisomic progeny (Sansome *et al.*, unpublished work).

B. *PERONOSPORA PARASITICA* AND *ALBUGO CANDIDA*

Studies of mating have rarely been made on members of the Peronosporaceae and the Albuginaceae. Evidence of heterothallism was reported for *Peronospora parasitica* (de Bruyn, 1937; McMeekin, 1960) and more recently for *Albugo candida* (Sansome and Sansome, 1974), *Bremia lactucae* (Michelmore and Ingram, 1980, 1981) and *Sclerospora graminicola* (Sacc.) Schroeter (Michelmore *et al.*, 1982).

During cytological studies of the sexual stages of *Albugo candida* on naturally infected *Capsella bursa-pastoris*, Sansome and Sansome (1974) noticed that primary infections were sterile but became fertile when secondary infections were established. The hypothesis that isolates of *Albugo* are self-sterile, but fertile when they interact with a compatible mate, was supported by their observation that oospore production in *Albugo* was frequently correlated with a secondary infection by *Peronospora parasitica*. They suggested that the *Peronospora* provided a chemical stimulation of gametangium formation so that oospores of *Albugo* were a result of induced selfing. More extensive studies with single spore isolates should show if the stimulation across the genus boundary is due to compatibility substances *sensu* Brasier (1982) produced by each single mating type (which seems to account for interspecies stimulation in *Phytophthora* spp. (Ko, 1978) or due to more non-specific inducers (cf. the *Trichoderma* effect in *Phytophthora* (Brasier, 1971)).

C. *BREMIA LACTUCAE*

Michelmore and Ingram (1980, 1981) made a detailed study of oospore formation in a collection of 39 field isolates of *Bremia lactucae* from Britain and other European countries. Most of these isolates were self-sterile when grown singly in lettuce cotyledons but became fertile when grown in pairs in certain combinations. A heterothallic system similar to that in *Phytophthora* spp. was identified and two mating or sexual compatibility types, B_1 and B_2 were designated.

Seven of the 39 isolates were self-fertile (homothallic) in single culture but acted as B_2 mating types by promoting oospore formation more abundantly in pairings with B_1 than with B_2 tester isolates. The majority of single conidial cultures from one of these self-fertile field isolates propagated the self-fertile/B_2 phenotype; the rest were self-sterile and of B_2 or more rarely B_1 mating type. A further generation of single conidia cultures, from one of the self-fertile, single spored isolates, showed this same pattern of segregation of self-sterile B_2 and rarer B_1 mating types. It would seem that the single conidium which propagates the self-fertile phenotype is a homokaryon whose nuclei carry determinants for both mating types. Heterokaryons probably arise by segregation of each single mating type determinant and break down to stable B_1 and B_2 monokaryons which interact to produce the oospores (Michelmore and Ingram, 1982). Since conidia are multi-nucleate, the possibility that self-fertility is propagated by conidia which are balanced heterokaryons cannot be excluded, although the persistence of self-fertility through many mass conidial transfers argues that this is unlikely.

The self-fertile isolates of *Bremia* bear a striking resemblance to those of *P. drechsleri* derived from single oospores. Analysis over three generations of zoospores in *P. drechsleri* established that self-fertility is carried by a single zoospore (balanced heterokaryons can be excluded here) which may yield a stable, self-fertile monokaryon or, more usually, a genetical mosaic of fertile sectors and sterile sectors of each mating type (Mortimer *et al.*, 1977). Suspicions of aneuploidy led to a cytological study of fertile cultures and their sterile segregants (Sansome, 1980; Sansome *et al.*, unpublished work). The tertiary trisomy identified in all fertile cultures and in their sterile derivatives, explains how single nuclei can carry determinants of both mating types. Sterile segregants would seem to arise after rearrangement of mating type genes in the trisomic nuclei by mitotic crossing-over. The cytology of the self-fertile *Bremia* and its derived lines (Michelmore and Sansome, 1982) also showed clear evidence of tertiary trisomy (see Fig. 3) and thus showed that the mating type determinants of both of these fungi are borne on a "mating type complex" of four chromosomes and that aneuploidy is the basis of self-fertility and its instability. Self-fertile isolates of *Bremia lactucae* and of many heterothallic *Phytophthora* spp. (see Shaw, 1983) would seem to be successful pathogens and may be of crucial importance in populations where there is little opportunity for heterothallic mating.

D. *PYTHIUM SYLVATICUM*

Although most species of *Pythium* appear to be homothallic (Waterhouse, 1968b), Pratt and Green (1973) have made a detailed study of one species, *P. sylvaticum*, which is clearly heterothallic. The compatibility, or otherwise, of field isolates of this species depends not on mating or compatibility type, as in *Bremia lactucae* or *Phytophthora* spp., but on the expression of their sexuality.

In *P. sylvaticum*, male isolates provided antheridia in pairings with female isolates which provided oogonia, whilst pairing of two male or two female isolates did not promote mating. The sexual strengths of isolates varied continuously on a scale from strong to weak for both males and females. The fecundity of a mating was found to depend on the combining ability or "compatibility" of the particular male and female rather than on their sexual strengths. In single culture on grass leaves, all males showed their basic bisexual nature when they became self-fertile; only a minority of females were similarly self-fertile. Isolates varied in their intensity of selfing, a characteristic which was also unrelated to sexual strength.

Progenies derived from matings of a strong male with (1) a strong female and, (2) a less strong female showed segregations of approximately 3/4 males most of which were weaker than the parent. The rest of the progeny were, in (1), weak females, neutral and bisexual types, or in (2), neutral and rare weak female types. Although, in the first cross, the male parent was only weakly self-fertile and the female parent self-sterile, all male offspring and even most of the females were self-fertile to varying degrees. Over one-third of these self-fertile progeny was classed as "superhomothallic"; their self-fertility was more highly developed than that of either parent.

In heterothallic species of both *Phytophthora* and *Pythium*, single thalli can form oogonia and antheridia. Their formation in *Phytophthora* is stimulated by a mate so that reciprocal crossing and selfing is possible, whereas in *Pythium*, a mate may stimulate the formation of oogonia only (in females), antheridia only (in males) or neither (in neutrals), even although all three may be highly self-fertile under appropriate conditions in single culture.

The relative simplicity of heterothallism in *Phytophthora* spp. and in *Bremia lactucae* contrasts with the more complex system in *P. sylvaticum*. The lack of correlations between sexuality (maleness/femaleness), sexual strength, "compatibility" and self-fertility in field isolates argues for the separate genetic determination of these characters. This is borne out by the results of the two matings in which many combinations of these characters were noted. Thus it would seem that a more flexible and complex system, based on the regulation of sexuality, controls mating in *P. sylvaticum*. Perhaps the system in *Phytophthora* and *Bremia* has evolved from such a flexible system; perhaps the compatibility (mating) type differences have developed from a kind of variable "compatibility" similar to that shown by *Pythium* and have come to replace differences in expression of sex as the mechanism for promoting outcrossing.

We still have a great deal to learn. Could super-homothallic progeny of *P. sylvaticum* be trisomic? Do forms of relative sexuality noted in *Pythium* and in *Phytophthora infestans* (Galindo and Gallegly, 1960) occur in other species or genera? Are viable selfed progeny a feature of matings in all heterothallic species of this order? Is a "mating type complex" present in all heterothallic species and if so, does it always generate self-fertile aneuploids? These questions have an important bearing on all genetical work with these fungi. Many more detailed studies are needed before heterothallism in the various genera of the Peronosporales can be understood and properly compared.

VI. The Genetics of Aggressiveness and Virulence: Can We Breed Better Crop Plants?

Although naturally occurring variation in aggressiveness and virulence has been described for many host-pathogen interactions, we still have only fragmentary data on the genetic determination of these characters in the Peronosporales. With diploid hindsight we can briefly review some of these findings to gain a little more understanding.

A. THE *PHYTOPHTHORA INFESTANS*-POTATO INTERACTION

1. *The Inheritance of Virulence*

The gene-for-gene system (of single resistance genes in the potato and their complementary but still hypothetical single virulence genes in the fungus) was outlined earlier. No definite conclusions could be drawn about the determination of virulence from the analysis of the first matings of Romero and Erwin (1969) and Laviola (1968) and it it surprising to find that this important line of research has not been continued. A characteristic of these early matings was that each virulence/avirulence difference between parents segregated in the F_1. Tentative conclusions are that each virulence/avirulence character is determined by a single pair of alleles and that one or other parent is heterozygous. If virulence were dominant, the virulent parent would carry a recessive, avirulence allele whereas if virulence were recessive, the avirulent parent would be heterozygous and would carry a virulence allele. Further rigorous genetical analyses with progeny testing should identify virulence determinants unambiguously.

However, hopes that further genetical data will be more easily interpreted should not be raised too high; Sansome has found complex chromosomal rearrangements in some Mexican material (Sansome and Brasier, 1973) and polyploidy in some Welsh material (Sansome, 1977). There is now a good case for screening isolates cytologically before selecting them as parents.

That all variation in populations of *P. infestans* outside Mexico is generated by asexual mechanisms is generally agreed. But the recent finding that Welsh isolates of *P. infestans* were sexually competent when paired with *P. megakarya* Brasier & Griffin and produced selfed gametangial combinations (see Fig. 4 and Sansome, 1977) implies that rare sexual reproduction may occur in the field. It may well be relevant that Welsh isolates of the late blight fungus can be induced to form selfed oospores in the laboratory when stimulated by an A2 mating type of *P. drechsleri* isolated from a blighted potato leaf (Shaw *et al.*, unpublished observations).

2. *Asexual Variation*

Variation generated during vegetative growth and sporulation is much more accessible for study (not requiring the Mexican A2 mating type) and has been fairly fully described. The emergence of new race characters in the field and in the laboratory, referred to earlier, is thought to result from mutation followed by selection (Gallegly and Niederhauser, 1959) but the rather high frequency with which some isolates may be "trained" on resistant material suggests that mitotic crossing-over in nuclei already heterozygous for recessive virulence may be involved.

A series of reports from the University of Birmingham (U.K.) (Jeffrey *et al.*, 1962; Jinks and Grindle, 1963; Caten and Jinks, 1968; Upshall, 1969; Caten, 1970, 1971, 1974) deals with variation in growth rate, colony morphology, aspects of sporulation and aggressiveness between and within field isolates of *P. infestans* from potato. The picture which emerges is of great variation between field isolates in these characters and of similar

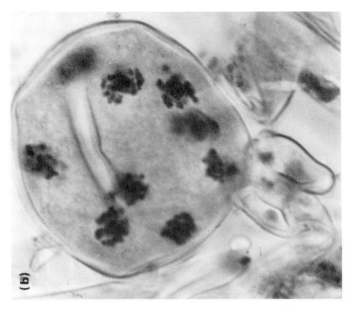

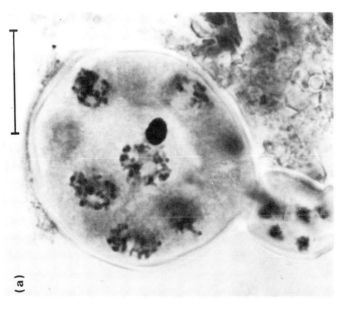

Fig. 4. Meiosis in oogonia of a British isolate of *Phytophthora infestans*. Gametangia were induced by the presence of actively mating isolates of *P. megakarya* and *P. palmivora* within the same culture. The oogonial nuclei are at diakineses (a) and metaphase I (b). Scale bar = 20 μm. By courtesy of E. R. Sansome on behalf of the late Dr. F. W. Sansome.

variation in zoospore progenies from single field isolates and even from single zoospore isolates. Selection experiments verified that this variation is genetically determined. The high frequency of somatic segregation and its reversible nature argue for the involvement of non-chromosomal genes rather than mutation and mitotic recombination of chromosomal genes. It may be relevant that virus-like particles have been found within nuclei of all field isolates of *P. infestans* examined by Corbett and Styer (1976). Intra-nuclear virus-like particles also occur in a laboratory strain of *P. drechsleri* see Fig. 5 (Roos and Shaw, unpublished observations).

Although there is evidence that isolates of *P. infestans* "grow" best on the variety from which they were isolated in the field (Jeffrey *et al.*, 1962) and

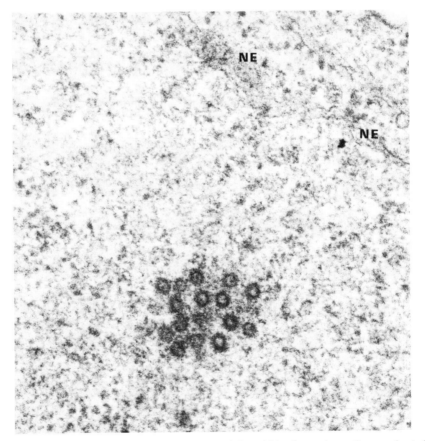

Fig. 5. Cluster of sub-spherical virus-like particles within the nucleus of a germinated zoospore cyst of *Phytophthora drechsleri*. Note nuclear envelope, NE. Magnification × 86,500. Courtesy of U-P. Roos.

that different phenotypes of a variable "atmospheric population" are selected on different cultivars (Upshall, 1969) there is little or no evidence that compatibility of an isolate and its "own" host improve by genetic change of the pathogen. This is consistent with the high stability of field resistance in commercial cultivars over many years.

3. Breeding Resistant Cultivars of Potato

The introduction of race-specific resistance into commercial cultivars has resulted in the rapid selection of complementary virulence "genes" in the pathogen accompanied by breakdown in control. The use of this major gene resistance in breeding has been all but abandoned since in many cases virulence against R genes of new breeding material is already present in the pathogen population before the resistant cultivars are introduced. Even cultivars carrying combinations of several different resistance genes are attacked by complex races of *P. infestans* which survive as effectively as races with no or few virulence "genes" (Malcolmson, 1969; Janssen, 1973; Shattock *et al.*, 1977). Experiments are now in progress (Skidmore and Shattock, personal communication) to find out if mixtures of cultivars with different R genes offer a greater challenge to the fungus population than do single cultivars. Promising results have been obtained with mixtures of cereal cultivars with different resistances against powdery mildew (Wolfe, 1978).

Most modern maincrop potato cultivars have no race specific resistance but have a high level of polygenic field resistance which is effective in limiting the growth and sporulation of all races of *P. infestans*. With possible rare exceptions (Toxopeus, 1956) this kind of resistance has not yet been eroded by genetic change in the pathogen and offers adequate control in all but the worst blight situations.

A new approach to breeding for resistance makes use of "protoclones" raised from leaf cell protoplasts. Shepard *et al.* (1980) have shown that an old blight sensitive cultivar may yield some blight-resistant plants when its protoclones are regenerated. It will be exciting if these plants turn out to have durable resistance under field conditions. The possibility that favourite old cultivars can be tailored to increase their resistance and to correct other deficiencies is a most attractive one.

Systemic fungicides have recently provided us with an alternative method of increasing the "resistance" of any cultivar. Acylalanine fungicides were initially very effective but their use as a means of protecting potato crops on a large scale has led to the rapid development of strains of *P. infestans* resistant to acylalanines (see Chapter 6). Thus it appears as if protection provided by systemic fungicides is rather like that provided by a single gene for resistance; the pathogen readily adapts to the altered chemistry of the host tissue, probably by single mutation followed by intense selection.

Crute (1981) has reviewed the various interactions found in downy mildew disease of wild and cultivated plants. Many of these downy mildew fungi, although conforming to a single morphological species, consist of several types or *formae speciales* each specialised in its ability to attack single plant taxa, or several closely related taxa. The least specialised pathogens, e.g. *Sclerophthora macrospora*, attack many genera of a single family whereas the most highly specialised exist as forms or pathotypes which attack single species and even may be confined to certain sensitive genotypes within a plant population.

1. *The Inheritance of Virulence*

Genetical studies of specificity have been confined to the *Bremia lactucae*–lettuce interaction which can result in serious disease of concern to lettuce growers throughout the world. *Bremia* is included in this review despite its unwillingness to produce zoospores; recent studies of its interaction with lettuce hosts have made this host–pathogen system the best known among downy mildews. A classical gene-for-gene system is thought to operate; races of *Bremia* can be differentiated on a series of lettuce cultivars carrying different dominant alleles (R-factors) for resistance. Ability to overcome a single resistance factor is thought to be due to the presence of a complementary single virulence allele (v-factor) in the fungus (see the review by Crute and Dixon, 1981). Parallels with the gene-for-gene system tentatively identified in potato late blight are obvious; the problems of control with resistant cultivars are similar. Although one resistance gene has protected crops on a massive scale over many years, most resistance genes are rapidly overcome when races of *Bremia* with the complementary v-factors appear. Sometimes virulence against a new resistance gene has been identified in populations of *Bremia* even before cultivars carrying the new gene were extensively grown. The similarity to late blight is again clear.

Following the working out of methods to stimulate oospore germination and to recover viable offspring, a start has been made to analyse the inheritance of virulence factors in *Bremia lactucae* (Michelmore and Ingram, 1981). Two progenies have been established from parents differing in virulence. In the first, race 1,2,5,7,8,9,10,11 was mated with another isolate of the same virulence phenotype. Progeny were uniform in their response to the differential series; all 25 inherited all virulence characters except virulence to R8 which was inherited by only three offspring. The second progeny was from a mating of race 1,2,3,4 × 1,5,6,7,8,9,10,11. All seven offspring inherited the virulence 1 character but were avirulent or showed a mixed reaction to other differentials.

Lack of a clear cut segregation of virulence phenotypes in the F1 of these crosses is consistent with a diploid somatic phase and suggests that parents were homozygous for single genes determining virulence. However, recent crosses (I. R. Crute, personal communication) have yielded segregating F_1s which, like the segregation at F_1 in *P. infestans* (see p. 95), suggests that one or other parent was heterozygous for virulence determinants. The basic avirulence of the progeny from parents of different race found by Michelmore and Ingram argues that virulence to a single R allele is determined by a recessive allele which is carried in these offspring in the heterozygous condition. It will be interesting to see if parental virulences segregate when these F_1s are sib- or back-crossed as the above hypothesis predicts. The avirulence to R8 of the first progeny is taken as evidence that v8-factor in each parent may not have been allelic (Michelmore and Ingram, 1981).

Hopefully, further crosses will soon be made which will explore virulence more fully. Improvements in percentage germination of oospores and the establishment of larger progenies would remove some of the present uncertainties. Use of more isogenic parents (both crosses involved parents from different countries) would reduce worries about the expression of virulence in different genetical backgrounds and suitably marked parents would allow the detection of induced selfing of one or other parent.

2. *Asexual Variation*

Sexual outbreeding would appear to be an important mechanism generating new combinations of virulence factors in *Bremia* (Michelmore and Ingram, 1981; Crute and Norwood, 1980) but it is likely that virulence against newly introduced R genes arises spontaneously by mutation and, when expressed, may be rapidly selected by a resistant cultivar producing a shift in the pathogen population (Crute and Dixon, 1981).

Crute and Dickinson (1976) found that a proportion of field isolates gave a mixed reaction on certain differentials and that this was due to the presence of more than one race. Propagations of such isolates from single conidia have yielded the two component races of the mixture but single conidia have also been isolated which propagate the mixed reaction. Crute and Norwood (1980) postulated that if virulence factors are single recessive alleles, the diploid vegetative nuclei of *Bremia* may become heterozygous for virulence after mating or mutation and may yield virulent homozygous segregants after mitotic recombination. The resulting heterokaryon of "virulent" and "avirulent" nuclei would segregate during vegetative growth and sporulation; thus homokaryotic conidia and some heterokaryotic conidia would be produced, the latter propagating the mixed reaction. Similar mechanisms have been suggested to explain the rather frequent race changes which can be selected in some isolates of *P. infestans* on potato differentials (Gallegly, 1968).

The finding of presumptive heterokaryons has prompted attempts to produce vegetative "hybrids" of *Bremia* with new combinations of v-factors by growing mixtures of races together on host tissue. Although many similar experiments with *P. infestans* were apparently successful, no "recombinants" have yet been reported in *Bremia* (Crute, also Michelmore, personal communications). Perhaps hyphal fusions rarely or never occur during vegetative growth or maybe heterokaryons of unrelated nuclei are unfit or very unstable. It is interesting that Michelmore was unable to re-synthesise a mating type heterokaryon from its component monokaryons (Michelmore and Ingram, 1982).

3. Breeding More Resistant Cultivars of Lettuce

According to race surveys, v-factors against R-factors 1–10 are common and widespread and do not offer much hope for effective control of downy mildew even when used in combination (see the review by Crute and Dixon, 1981). Phenotype–phenotype analysis (Wolfe *et al.*, 1976) of race survey data may help to identify combinations of v-factors which are consistently infrequent or absent from the field. Cultivars with combinations of R-factors complementary to these v-factors might have more durable resistance. R-factor 11 is a promising new gene whose complementing v-factor is rare in *Bremia* populations. There is hope that new resistant germplasm from wild lettuce species will improve control of the disease in the near future. If these new resistances prove to be race specific, as seems likely, careful management will be needed to avoid rapid erosion. Mixtures of cultivars with different new R-factors in time or in space or the combination of several new resistances in one "supercultivar" may extend the usefulness of this new resistance.

Clearly, the lettuce breeder using R genes has similar problems to his colleague using R genes in potato to control late blight. For both, field resistance or a combination of field and major gene resistance offers an alternative for stable control. The old lettuce cultivar, Iceberg, although sensitive in laboratory tests, shows slow disease development under field conditions (Crute and Dixon, 1981). Resistance of this kind may be most effective in the long run, particularly in environments in which disease is moderate or slight.

VII. In Conclusion: A Little More Speculation

Although the genetics and cytogenetics of the fungi considered in this Chapter are still in their infancy, we now have a clearer view ahead of the path we have to travel. Some, discouraged by the roughness of the road, have left to find a smoother one; others have continued, finding a fascination in the difficulties of many parts of the journey. Perhaps yet others will be enticed into the field by the exotic scenery. There are certainly many more areas waiting to be explored than explorers willing and able to tackle them. Since there are many parallels between the genetics of the more easily handled *Phytophthora* spp. and the biotrophic downy mildews, there is a good case for using the more convenient species of *Phytophthora* as model systems in pure research. A concerted effort to solve the more outstanding fundamental problems is necessary to pave the way for more applied research with the more difficult organisms.

Several new approaches are now being developed. The amino-acid analogue, *p*-fluorophenyalanine has a remarkable effect on *Phytophthora* spp. Work at Bangor and Liverpool has shown that high frequencies of drug-resistant mutants can be recovered after treatment of vegetative stages of *P. drechsleri* with the analogue (Mortimer *et al.*, unpublished observations). It is not yet clear if the substance acts as a mutagen or if recessive mutations already present are being exposed by induced loss of whole chromosomes or by induced mitotic recombination. The isolation of a stable haploid line is a possibility; it is an exciting prospect.

The genetics of heterothallic species of *Phytophthora* should forge ahead now that a simple technique is available for induced selfing of both A1 and A2 parents (Ko, 1978). Not only should this allow rapid checks to be made for heterozygosity but also should provide a way of isolating all kinds of naturally occurring (or induced) recessive mutations from heterozygotes. This method might also be exploited by those working with downy mildews.

The usefulness of electrophoresis of proteins as a method of fingerprint-ing parents in crosses is now well established (Boccas and Zentmyer, 1976; Boccas, 1981; Kaosiri and Zentmyer, 1980; Erselius and Shaw, 1982). The technique should become widely used in sexual and in asexual pairings to identify hybridisation and selfing particularly when few or no mutants are available.

A molecular study of the genomes of the Peronosporales has not yet begun but may help to solve many basic problems. Preliminary work in Bangor (Hooley *et al.*, 1982; Evola-Maltese *et al.*, unpublished observa-tions) using DNA specific fluorochromes (see Fig. 6) and fluorimetry indicates that *Phytophthora* spp. have very large nuclei and contain around five times as much DNA as fungi such as *Aspergillus* and *Neurospora*. In spite of this, the duplication cycle of nuclei is very short, e.g. divisions occur every 75 minutes in germ tubes of *P. drechsleri*. This duplication time is as short as any known for eukaryotic nuclei. Timberlake's group have started a detailed study of the genome of many fungi including *Achlya ambisexualis*. Many interesting facts are emerging; *Achlya* has more repetitive DNA than other fungi; much of this is made up of highly reiterated sequences coding for

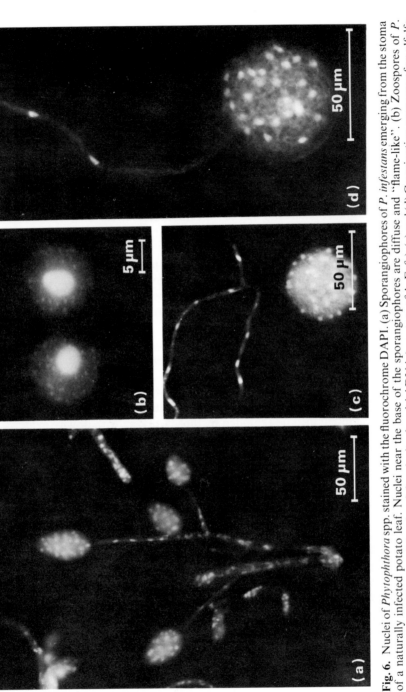

Fig. 6. Nuclei of *Phytophthora* spp. stained with the fluorochrome DAPI. (a) Sporangiophores of *P. infestans* emerging from the stoma of a naturally infected potato leaf. Nuclei near the base of the sporangiophores are diffuse and "flame-like". (b) Zoospores of *P. infestans*; note pyriform nuclei and peripheral mitochondrial DNA (courtesy of A. Fyfe). (c) and (d) Germinating oospores of an alfalfa isolate of *P. megasperma* (courtesy of H. Förster). (c) Hyphal branching in an older germling. (d) Young germling showing a single binucleate germ-tube and multinucleate oospore suggesting that rapid, repeated nuclear division takes place before germ-tube emergence.

ribosomal RNA (Hudspeth *et al.*, 1977). Techniques using DNA–RNA hybridisation and restriction endonucleases, have provided a fine structure map of the ribosomal RNA genes and their spacer sequences (Rosek and Timberlake, 1979).

Clearly, this is just the beginning!

Acknowledgements

Thanks are due to Dr. Eva Sansome, Dr. R. C. Shattock and Dr. I. R. Crute for critically reading the manuscript and to many colleagues at U.C.N.W. for helpful discussion, advice and help.

4

Some Pythiaceous Fungi

New Roles for Old Organisms

FLOYD F. HENDRIX, Jr. and WILLIAM A. CAMPBELL

*Plant Pathology Department, University of Georgia,
Athens, Georgia, USA*

I. Introduction 123
II. Mycology 124
    A. *Pythium* 124
    B. *Phytophthora* 129
III. Isolation 131
    A. Vegetable and Fruit Baits 132
    B. Selective Media 132
IV. Pathology 134
    A. Historical and New Concepts 134
    B. Algae and Fungi 134
    C. Juvenile Tissue 135
    D. Foliage, Stem, Crown and Fruit Rots 140
    E. Non-Parasitic Pathogens 142
    F. Environmental Parameters 143
    G. Interactions with other Pathogens 144
V. Distribution and Redistribution 144
VI. Ecological Niche 149
VII. Impact 152
VIII. Control 154
    A. Cultural Control 155
    B. Biological Control 156
    C. Chemical Control 158
IX. Summary 159

I. Introduction

It was long ago shown that zoospore-producing fungi could be separated into two groups, depending upon the number of flagella possessed by the zoospores. Sparrow (1943) classified those with two flagella as biflagellates in a series with four orders of which the Order Peronosporales comprised the families Pythiaceae, Albuginaceae and Peronosporaceae.

ZOOSPORIC PLANT PATHOGENS
ISBN 0 12 139180 9

The Pythiaceae, comprising the genera *Pythium* and *Phytophthora*, are of special concern to plant pathologists since species of these two genera cause important diseases of food and fibre-producing plants throughout the world. They have much in common morphologically whilst their pathogenic behaviour permits them to be classified together for such practical considerations as disease prevention and control. This chapter discusses *Pythium* and *Phytophthora* species, particularly with respect to their impact on agriculture and forestry in various parts of the world.

II. Mycology

A. *PYTHIUM*

The genus *Pythium* was established in a description by Pringsheim (1858). The name, however, goes back to 1823 when it was used by Nees but not published in an acceptable fashion. Fortunately, or unfortunately, Pringsheim's published description for the genus *Pythium* was extremely simple. The sporangium was taken for granted and the zoospore described as forming outside the opening of the sporangium from its contents. The oospore was said to be single in each oogonium.

From Pringsheim to the present a broad mycological trail exists covered with the descriptions of new *Pythium* species, speculations regarding their relationships and the value of morphological variations in the species and in related genera. Mycologists soon found bewildering variation in the size and shape of the sporangia and in the manner in which the zoospores were formed and liberated. There was less variation in the nature of the oospores, but there proved to be a great deal of variation in the number, size and structure of the attendant antheridia.

Mycologists concerned with the classification of *Pythium* species soon became themselves classified as lumpers or splitters. Hence frequent proposals were made to split the genus on the basis of morphological variations in the sporangium structure. Many of these names, as proposed, were cumbersome to spell and tongue twisters to pronounce. Fortunately, the modern concept retains the broadly based genus *Pythium* and separates its species on the basis of sporangial differences, this having the advantage of using the genus name to indicate the broad relationship and botanical position and leaving the specific differences to the specific name. An increase in the number of *Pythium* isolations recovered in the course of plant disease research showed marked variations between isolates and gave greater evidence of the variability within species.

The genus *Pythium* has been the subject of several taxonomic treatments. Those by Sideris (1931, 1932), Matthews (1931), Middleton (1943) and

Frezzi (1956) are readily accessible. Waterhouse (1968b) collected all of the figures and descriptions of the different species from the original papers and her publication is recommended as a single source for most of the *Pythium* literature up to the 1960s.

Basically, *Pythium* species can be separated on the nature of their sporangia and on the characteristics of the oogonia and attendant antheridia. Middleton's key, while oversimplified, stressed basic differences. While these basic differences work well in a key, *Pythium* species, like other botanical or biological entities, often fail to conform to a preconceived pattern. Many observed differences fail to match the published definitions, whereas observed sizes often fall between those expected.

Middleton used oogonia or oospore sizes sparingly as separation criteria, and then only if the size differences were large enough to be readily observable. Waterhouse, in her key to *Pythium* species, leaned heavily on rather small average differences in spore sizes as a means of distinguishing certain species. Attempts by us to use these criteria led to a study to determine how reliable or how constant spore size remained in culture. We found (Hendrix and Campbell, 1974) that both oogonia and oospores changed in size under standard culturing procedures and these changes could be manifest as either an increase or decrease in size over a 30-month period. It thus appears that size of reproductive structures is important in separating species only if there is no excessive overlapping in average diameter.

The genus *Pythium*, in the simplest terms, can be restricted to that group of water moulds whose asexual stage comprises sporangia of various sizes and shapes, and whose sexual stage comprises an oogonium which contains at maturity a single oospore. The asexual stage can be filamentous (but no larger than the vegetative hyphae), inflated filamentous (or with inflated lobate sections which occur singly or joined together), or in the form of globose bodies developing singly or in conidiophores. Zoospores are released through a pore or evacuation tube into a vesicle where they mature and from which they are subsequently liberated. The oogonium may have a smooth wall or be provided with spines or protuberances of various sizes and configurations. The globose oogonia are fertilised by from one to many antheridia; these take many forms and may be monoclinous, diclinous, or hypogynous.

While the above description sounds simple, in practice there is an array of puzzling variations. For example, some species lack sporangia entirely, while others lack the sexual stage. Some fungi, with all the external appearance of *Pythium* species, discharge zoospores without a detectable vesicle. Chlamydospores, sometimes indistinguishable from potential sporangia or oogonia, also occur and are used as secondary characters to delimit species. Variations in growth habit, mycelial size and host preference are common but these variations do not affect their generic placement which depends only on the sporangia and the nature of zoospore discharge and oospore condition.

Many species were described from single or relatively few isolations. Such descriptions, based on only a few isolations, represent a limited view of the variation that exists in the populations of particular fungi. However, *Pythium* species which do not conform to any described species should be published as new. Only when information is in the literature does it become available to mycologists and pathologists; today's discovery, no matter how unimportant, may help to fill out a taxonomic jig-saw puzzle confronting someone in the future. Names become important since only through names can one make use of the literature and through the literature contact those faced with similar problems.

In the past, many isolates from plants showing damping-off and early stem or root rot were referred to as *P. debaryanum* Hesse. The host index for this fungus is lengthy, but more recent studies indicate that the species has been confused with *P. irregulare* Buisman: Buisman's description of *P. debaryanum* (Buisman, 1927) stated that the oogonia often exhibited protuberances of different sizes and shapes while the oogonia of *P. debaryanum* from Hesses' publication (Hesse, 1874) showed a smooth oogonial wall. Drechsler (1930) described the development of *P. debaryanum* on wet substratum, and noted that the oogonia frequently produced up to three protuberances. Hendrix and Campbell (1970) recovered hundreds of isolates classifiable as either *P. debaryanum* or *P. irregulare* Buisman on the basis of the presence or absence of oogonial protuberances. Biesbrock and Hendrix (1967) in a study of these two same species in culture concluded that they could not be distinguished on criteria such as average spore size, host range and temperature requirements and that their separation was not valid. Hence in references here to *P. debaryanum* or *P. irregulare* both names have been retained to indicate a species complex.

Middleton (1943), citing others' and his own observations, concluded that most species of *Pythium* were homothallic. His key and that of Waterhouse recognised named isolates in which sexual reproduction was absent, even though spherical bodies which could be either sporangia or oogonia were present.

Campbell and Hendrix (1967), working with *Pythium* isolates from forest and agricultural soils, found that one frequently isolated fungus, presumably a *Pythium*, produced only globose terminal or intercalary sporangia or chlamydospores on cornmeal and hempseed agar plates. A few isolates, with all the characteristics of the above, sparingly produced functional oogonia and oospores, among mainly abortive structures. Extensive pairings of the sporangial isolates on hempseed agar demonstrated that well-formed oospores formed when the mycelium of certain isolates made contact. This heterothallic *Pythium* was named *P. sylvaticum* Hendrix & Campbell. It proved to be very common in the southern United States and stimulated a search for more heterothallic forms. This led to the discovery of *P. heterothallicum* Hendrix & Campbell (Hendrix and Campbell, 1968a) in soil from the United States and Canada. Publication of the discovery of heterothallism in *P. sylvaticum* led to the recognition of this species in Holland, and Van der Platts-Niterink (1968, 1969) demonstrated that *P. intermedium* de Bary, an established species described in *Pythium* literature as lacking a sexual stage, was actually heterothallic and would produce oospores with the proper combination of mating strains. Later Hendrix and Campbell (1969) reported on heterothallism in *P. catenulatum* Matthews and *P. splendens* Braun. These species were found to vary greatly in their capacity to produce sexual structures; some isolates were entirely lacking in functional oogonia and others produced oogonia and mature oospores only by mating with sexually active strains.

The story on heterothallic strains was never entirely concluded. After extensive soil isolations there remained a number of isolates indistinguishable from non-oospore producing strains of *P. intermedium*, *P. sylvaticum* and *P. heterothallicum* which would not mate with any other cultures with similar characteristics. Hence a considerable group of isolates may exist in culture having arisen by the germination of reproductive structures or from mycelium that gives rise only to sexually sterile organisms. Either forest soils are particularly blessed with sterile forms, or the selective media used for isolation favour the recovery or separation of them from the mass of other fungi in the soil. Heterothallic species certainly exist in the genus *Pythium*. Varying degrees of heterothallism may account for the reluctance of the sexual stage to develop in certain species. Papa *et al.* (1967), in a study on the nature of heterothallism in *P. sylvaticum*, concluded that since the antheridia and oogonia are restricted to different thalli, heterothallism in the above species can be classified as a morphological phenomenon. The nature of the reaction line which formed where mycelium of the paired cultures mingled permitted identification of the sex of the parental isolates.

The discovery of heterothallism in the genus *Pythium* and the realisation that this phenomenon made a key to species hopelessly complicated, led to a taxonomic study of the relationship between *P. sylvaticum*, the heterothallic species, and the homothallic *P. debaryanum*. Hendrix and Campbell (1974), noting that the description of *P. sylvaticum* in its heterothallic condition was comparable to that of *P. debaryanum*, suggested placing the two species in a single slot based on oogonial characters and oospore sizes. Their sporangial bodies are similar. *Pythium sylvaticum* is a widely distributed fungus and its host range may be as extensive as those of *P. irregulare* and *P. debaryanum*. Perhaps the best treatment may be to include *P. sylvaticum* in the same complex as the other two, recognising that it is isolated most frequently in a form which does not produce oospores readily on the commonly used media.

While the described *Pythium* species can be grouped according to their morphological characteristics, such groupings are purely arbitrary and may not express physiological similarities or pathogenic propensities. Krywienczyk and Dorworth (1980) approached the problem of physiological attributes by studying the serological interrelationships between nine species of *Pythium*. *Pythium aphanidermatum* (Edson) Fitz. and *P. butleri* Subram. have been considered synonymous by most mycologists and their serological reaction patterns were identical proving that these fungi are the same. *Pythium coloratum* Vaartaja and *P. dissotocum* Drechsler gave similar but not identical patterns showing that they are closely related. The serological reactions also indicated why *P. debaryanum*, *P. irregulare*, *P. sylvaticum* and *P. ultimum* can be regarded as belonging to a species complex. The investigators suggested that more intensive application of serological techniques might refine further the validity of the species complexes as stated by Hendrix and Papa (1974). Any refinement in techniques that prove relationships between species would afford greater insight into their pathogenic potential and perhaps would afford a better idea of the significance of morphological characteristics that could be used in defining species limitations.

The final judgment regarding species delimitations comes only after observing the organism in culture on several different media and determining the nature of the sporangia and the induced zoospores. Some species form sporangia readily and release zoospores if agar plates are simply flooded with water. Others require a period of continuous irrigation. Some species must grow on vegetative material such as grass or leaf sections, or even on root material, before they will form functional sporangia and release zoospores following flooding or irrigation. Temperature changes and alternating periods of light and dark are sometimes required.

Oogonia formed in or on culture media are generally obscured by mycelial growth. The origin and number of their accompanying antheridia are often difficult to trace and once antheridia have performed their function and lost their contents they remain as collapsed hyaline threads in the mass of vegetative hyphae. The search for the "typical" antheridial condition is limited to the few individual oogonia so placed in the media as to be observable under a magnification capable of revealing their structural condition. Interpretation by the observer is subject to wide variation in judgment and hence statements as to antheridial number, origin and size are often difficult to equate with published descriptions of what is presumed the same species.

There is little evidence for host specificity in plant pathogenic *Pythium* species. Some have been isolated more frequently than others from particular plants, or from the soil under these plants. For example, *P. catenulatum* seems restricted to grasses or monocotyledonous hosts. Others may appear restricted to certain plants only because they have been isolated in the course of studies confined to a particular plant species.

B. *PHYTOPHTHORA*

The famine-spawning potato blight of the 1840s in Ireland and to a less drastic extent elsewhere in Europe, provided mycologists and early pathologists with a real challenge. The fungus associated with the disease was studied intensively with the tools and techniques available. The presence of aerial conidia could be observed on the host tissue, as well as in culture, and zoospores could be induced to appear with relative ease. However, oospores were not produced in culture nor could any be located in plant tissue. De Bary (1876), as the result of his studies on the nature of the potato blight organism established it as *Phytophthora infestans* (Mont.) de Bary, creating at the same time a new genus separated from the older *Peronospora* by basic differences in the nature of the conidiophores. He noted that the swellings on the conidiophores where the conidia were produced remained after the conidia were shed. True *Peronospora* species lacked such swellings. De Bary's original article made no mention of the manner in which the zoospores were released. He did not find oospores and the mycological description of the species and genus was restricted to the naming of its host and to the form of the conidiophore branching. Conversely, Fischer (1892) stressed the manner in which the zoospores were released and direct release from the sporangium, without the formation of a vesicle, was established as the distinctive difference between *Pythium* and *Phythophthora*. Oospores of *P. infestans* escaped discovery until they were reported by Clinton in 1910. W. G. Smith illustrated what he considered resting spores of *P. infestans* in 1875 and was awarded a gold medal by the Royal Horticultural Society. However, it was later determined that the illustrated oospores were probably those of an associated *Pythium* (Carruthers, 1875).

For more than 50 years mycologists debated the advantages and disadvantages of separating *Pythium* and *Phytophthora* on the basis of the only real morphological difference: the presence or absence of an extruded vesicle into which the zoospores are released and in which maturation is completed. This point has been subject to the various interpretations that might be expected with any biological process and more than one species started its mycological history in a different genus and later became recognised as an aberrant *Pythium* or *Phytophthora* species. Fitzpatrick (1923) and others suggested that the two genera be combined. In the process of describing, or discovering new species, the familiar morphological variations came to light and so resulted in suggestions not only for new species of *Phytophthora*, but for new genera also. However, Tucker (1931) in his extensive study on the taxonomy of the genus *Phytophthora*, provided pathologists and mycologists with a clear picture of the pathogenic species, and a logical basis for their separation, both in culture and by host. Waterhouse (1956) did for *Phytophthora* what she did later for *Pythium* in collecting in one monograph the descriptions and illustrations of all the published species. This monumental work firmly established *Pythium* and *Phytophthora* in the Pythiaceae and recognized the overlapping characters of many of the species. Further, it provided the basis for species stability so necessary for research and control of these global pathogenic fungi.

Since the publications of Middleton (1943) and Waterhouse (1968b), discoveries of heterothallism in *Pythium*, and accounts of mating types in *Phytophthora* have altered somewhat the criteria for species separations. In practice, most *Phytophthora* species can be recognised from characteristics that are difficult to express in mycological language. Mycelial characters and growth habit on cultural media are relatively distinctive for many species as is the manner in which the oogonial initial grows through the antheridial cell resulting in prominent paragynous antheridia. As with *Pythium* species, variations account for classification problems, but the wide distribution of *Phytophthora* species as plant pathogens has resulted in a large body of literature in which the different species emerge as identifiable entities regardless of imperfections in their classification in a mycological sense. A surprising number of species have global distribution as the result of soil and plant movements during the long period in which pathogens were ignored or were unknown. Tucker (1931), in his extensive studies on *Phytophthora* analysed the literature up to that time and carried out comparative cultural and pathogenicity studies which provided pathologists with valuable information on how to identify individual species and on the requirements for pathogenicity and control. His section on heterothallism and hybridism in *Phytophthora* showed that heterothallism is common and is a factor in the origin of new strains. In his key he distinguished 18 species and listed those that he considered synonymous. Tucker's conclusions were based on 150 cultures from many hosts, all grown under standard conditions for comparison.

In contrast to the situation in the genus *Pythium*, the early recognition of heterothallism in *Phytophthora* species resulted in extensive investigations of the role of this phenomenon in species identification and classification. Gallegly and Galindo (1958) reported that natural mating types of *P. infestans* occur in approximately a 1:1 ratio in Mexico. They used the symbols A^1 and A^2 to distinguish the different mating strains. Galindo and Zentmyer (1964) later reported the occurrence of A^1 and A^2 mating types in *P. cinnamomi* Rands. The A^1 type is apparently rather rare in the United States where it has been isolated most frequently from *Camellia japonica* in North Carolina and Georgia. Although hundreds of soil and tissue isolations of *P. cinnamomi* were made in the course of pine littleleaf disease investigations, only the A^2 type was found. Isolates were grown routinely on agar whereby four or more cultures were commonly placed on a single plate of cornmeal or other medium. If the mating strains had been present, they had opportunity for reactions with mycelium from different sources. Haasis *et al.* (1964) worked with *P. cinnamomi* isolates from many sources but were unable to locate any local A^1 strains from hosts other than *Camellia japonica*. Ko (1978) demonstrated that oospore formation was related to diffusable hormones and still took place when the appropriate mating types were separated by polycarbonate membranes.

Sporangial production frequently does not occur in culture in the absence of special treatments. The materials in soil extracts play an important role in the formation of sporangia and the release of zoospores. In fact, certain species fail to form sporangia in culture without the addition of soil extracts or water from some source presumably containing the necessary stimulatory agents. Root extracts are known to stimulate sporangial formation in *P. cinnamomi* which explains why dormant spores of the species react to roots penetrating soil in which these spores may have been in a resting state for years. Once produced, zoospores move in a water film and are attracted to the region of elongation behind the root tip, presumably by chemotaxis (Zentmyer, 1961). In liquid cultures the swarming of motile spores and their clustering around the area of elongation can be observed. In the case of shortleaf pine, mycorrhizal activity becomes important since these structures effectively prevent infection by zoospores of *P. cinnamomi* (Marx, 1967, 1973). The protection against this fungus and other pythiaceous species in shortleaf pine (*Pinus echinata*) and presumably other trees also, depends both upon the mechanical barrier formed by the mycorrhizal covering and on the antibiotics produced by the fungal components.

III. Isolation

Isolation of pythiaceous fungi from plant material and soil is difficult, and they are often absent from old soil fungal surveys, even in situations where they are important pathogens. They are not readily recovered by conventional agar-based isolation techniques; Hodges (1962) for instance did not recover pythiaceous fungi from forest tree nursery soils in the southern United States, even though they probably account for more seedling loss there than any other agent. To recover such fungi, a totally different isolation procedure must be used and as a consequence, surveys for pythiaceous forms do not usually include other fungi since the selective isolation procedures discourage their growth.

A. VEGETABLE AND FRUIT BAITS

Owing to the hit and run tactics of pythiaceous fungi they are generally the first invaders of plant tissue where they maintain their position only by advancing ahead of saprophytic organisms. In fact, they pave the way for saprophytes and weakly parasitic fungi to continue and finish cell and tissue disintegration. Whenever *Pythium* or *Phytophthora*-invaded material is collected and brought to the laboratory for isolation, the elapsed time and prevailing temperatures usually favour the non-pythiaceous camp followers making it difficult to isolate the real pathogens, especially on media favourable for bacteria and saprophytes.

Tucker (1931) was one of the first investigators to use fruits and other plant parts with living cells to separate *Phytophthora* species from infected roots and other tissues. He found that apples, potatoes and citrus fruits provided selective mechanisms whereby one *Phytophthora* species could be separated from others in taxonomic evaluations.

Following Tucker's example, baits of various kinds were routinely used to isolate pythiaceous fungi from infected plant material. Usually these organisms, if present, would invade the living cells and keep one jump ahead of other organisms anxious to take advantage of the nutrients released by the death and disintegration of the cells.

Campbell (1949) used Tucker's apple technique in an attempt to isolate *Phytophthora* spp. from decayed small roots of littleleaf-diseased shortleaf pine. *Phytophthora cinnamomi* was recovered from only 2 % of the root pieces inserted in apples but when the cut surface of green apples was applied to moistened soil in small dishes, the fungus grew into the apples where it caused a characteristically firm rot. Later the soil isolation technique was modified such that soil was placed in a hole bored into the apple.

B. SELECTIVE MEDIA

In the 1960s many selective media were devised for the isolation of pythiaceous fungi from soil and plant tissue. One of the first, designed to isolate *P. aphanidermatum* (Edson) Fitz., was used by Hampton and Buchholtz (1959). The addition of coumarin to a non-nutrient agar limited the growth of contaminating organisms but stimulated sporulation of *P. aphanidermatum*.

Eckert and Tsao (1962) described a medium using polyene antibiotics as selective agents for *Pythium* and *Phytophthora* spp. Descriptions of differing versions of selective media using various antibiotics followed in profusion and it was discovered that at the concentrations suggested, some were inhibitory to spore germination. In other media, there were interactions between ingredients which also resulted in the inhibition of spore germination. We prefer cornmeal agar ammended with 100 p.p.m. nystatin and pentachloronitrobenzene (PCNB) to rose bengal although the primary weakness of this medium is that it must be cooled to below 45°C before adding the nystatin which at high temperatures or after storage for two weeks, breaks down and offers little selectivity. Tsao (1970), gave an excellent review of selective media for the isolation of fungi, and later Tsao and Guy (1977) devised a medium for the selective isolation of *Phytophthora* which inhibited both *Pythium* and *Mortierella*. It is worth noting nevertheless that when active lesions are the source of material for isolations, less reliance on saprophyte inhibitors is needed since the host screens out many non- or weakly pathogenic fungi.

Singh and Mitchell (1961) first developed a medium for the isolation of pythiaceous fungi from soil, utilising specific antibiotics and rose bengal. Haas (1964) added PCNB, removed the rose bengal and used additional antibiotics. Various other media have been described since that time. These media, with the exception of those of Tsao and Guy (1977) and Sneh (1972) are in general more useful for the isolation of *Pythium* spp. than *Phytophthora* spp. In considering the ingredients for selective media, it should be noted that rose bengal is absorbed in light and becomes toxic to hyphae while media containing a combination of streptomycin and rose bengal have limited application and pimaricin is generally inhibitory to spore germination at high levels.

Whilst no selective medium will effectively isolate all propagules of pythiaceous fungi from soil samples, there has been a steady increase in the reported numbers of these fungi isolated per gram of soil commensurate with improvements in selective media. Counts of over 2000 propagules per gram of soil have been recorded.

Selective media tend to be more useful to the developer than to other researchers. Whilst no valid comparisons can be made between reports based on different media, it is also difficult to compare fungal populations between reports in which the same medium is used by different workers. Flowers and Hendrix (1974), for instance, developed a gallic acid medium, which was compared with the modified Kerr's medium described earlier (Hendrix and Kuhlman, 1965). An autoclave temperature higher than that used by the developers affected the gallic acid, and compromised the isolations.

Selective media are extremely useful in comparing treatments within or between experiments after the researcher has enough data to relate his propagule counts to disease development or other variables. While it may be possible to approach a reasonable qualitative evaluation of soil populations of pythiaceous fungi, it is not yet possible to obtain strictly quantitative data. The numbers of propagules observed and recorded are dependent on the media and the skill of the researcher. Colonies that grow on selective media must be transferred to growth media for identification.

IV. Pathology

A. HISTORICAL AND NEW CONCEPTS

Pythium monospermum Pringsheim, the type species of the genus, was originally described as saprophytic on insect cadavers in water. Much of the early work dealt with these fungi as saprophytes, or weak pathogens which could rot seed or kill young seedlings. An example was the work of Ito and Tokunaga (1933) who reported that rice seedlings were almost resistant to *Pythium* spp. when 3–5 cm high. This type of research firmly established *Pythium* spp., and some *Phytophthora* spp. as major components of the complex of organisms causing damping-off.

Wager (1931) reported *Pythium* species as the cause of root rot and wilt of pawpaw, soft rot of Irish and sweet potato and rot of citrus fruits. Rands and Dopp (1938) reported on a decline disease of sugar cane caused by *P. arrhenomanes* Drechsler and detailed the environmental conditions favouring severity of the disease. Campbell and Copeland (1954) dealing with a decline disease of pine also emphasised the effect of environment on the ability of the host to recover, as well as on the disease directly. Hendrix and Powell (1970) considered the interaction of various abiotic and biotic factors on decline of peach trees in which *Pythium* species were involved.

Braun (1924) and Tompkins and Middleton (1950) reported on *Pythium* species causing stem rot diseases. *Pythium* species also have been found to cause collar rot of apples, foliar and stem blights of grass, vegetables and many ornamentals, and snow rot of winter wheat. Additionally, they infect fresh water and marine algae and many other fungi. Some species produce growth regulating substances causing the same reaction as IAA.

It is obvious that the concept of *Pythium* spp. primarily as damping-off organisms is too restricted. If this concept is expanded, however, to include other succulent tissues and other forms of juvenile tissue such as growing roots, it will encompass much of the damage now attributed to this group of fungi.

Phytophthora species caused the same types of diseases as *Pythium* species. In general, there are probably fewer diseases in which the fungi are restricted to juvenile tissue, and more where the fungi attack almost exclusively fruits and foliage. This account will be limited to those *Phytophthora* species which cause diseases similar to those caused by *Pythium* species. These phytophthoras include *P. cinnamomi*, *P. nicotianae* van Breda de Haan var. *parasitica* (Dastur) Waterhouse and *P. cactorum*.

B. ALGAE AND FUNGI

Pythium species were probably originally saprophytic rather than parasitic in aqueous environments and they evolved along with other plants to assume a terrestrial existence. Sparrow (1931) described the species occupying the aquatic habitat not only for the reason of their apparent variety, but also because phases in their morphology and life history seem little understood. Sparrow described two new species of alga-infecting *Pythium* and presented their life-histories in detail. Subsequently, Fuller *et al.* (1966) described the occurrence of a *Pythium*-caused disease on the marine alga *Porphyra* sp. with similar symptoms.

In a possibly related type of disease, Haskins (1963) described infection of fungi by *P. acanthicum* Drechsler. Some 98 species of fungi were tested as possible hosts; 69 were infected and supported the production of oospores; ten were parasitised without supporting sexual reproduction; nine were not parasitised; and ten actively inhibited *P. acanthicum*. The isolates of *Pythium* tested were also pathogenic on rape and flax and, indeed, since most *Pythium* species have a wide host range, this limits their potential applicability in biological control systems.

C. JUVENILE TISSUE

1. *Seed and Seedlings*

Reports of seed rot and damping-off caused by *Pythium* species are numerous (Hendrix and Campbell, 1973) and most species of *Pythium* affect several plant species. For example, when Middleton published his epic paper in 1943, he recorded a host range of over 150 species for *P. ultimum* and suggested that even this was not complete. Conversely, Johnson and Chambers (1973) found three *Pythium* spp. commonly causing damping-off of cotton while Halpin *et al.* (1952) described seven *Pythium* spp. affecting red clover seedlings.

Among many factors influencing damping-off are seed quality, temperature, moisture, the prevalent *Pythium* population, the cultivar concerned, and even seed colour. Two factors, however, are of prime importance in seed quality: vigour and the condition of the seed coat. Low-vigour seeds (by definition) germinate and emerge very slowly, and thus are exposed to pathogenic fungi for a longer period while in a susceptible state. Schlub and Schmitthenner (1978) found that scarifying and soaking soybean seed increased exudates from 5.3 µg glucose equivalents to 217 µg h^{-1}. Similar increases in exudation in other compounds were noted. Spermasphere effects were noted 5 mm from intact and 7.5 mm from sacrified seed. Keeling (1974) found a direct relationship between the amount of soluble carbohydrate exuded by a germinating soybean seed and seed rot caused by *Pythium* species. Perry (1973) found that faults in pea seed coats were important in enhancing susceptibility to *P. ultimum* while Gindrat (1976) found that coating sugarbeet seed with unbleached chitin furnished nutrients which enhanced germination of *P. ultimum* sporangia, and subsequent colonisation of the seed.

Wrinkled pea seed were more susceptible than smooth seed to *P. ultimum* (Short and Lacy, 1976). Furthermore, within the wrinkled seed, colour varied from yellow to green, the later being less susceptible than the former. This difference was probably due to increased exudates in the bleached seed, for Flentje (1964) demonstrated that the difference in susceptibility between smooth and wrinkled seed was due to the greater amount of exudation from the wrinkled seed. He also found that seedlings from wrinkled seed grew less vigorously.

Mitchell (1975) found infection of rye by *P. myriotylum* Drechsler increased as the number of zoospores increased, and reached 100% when the spore concentration exceeded 150 per g of soil. Similar results have been found with *P. irregulare* on soybeans (Southern *et al.*, 1976).

Species of *Pythium* have different abilities to affect the same host. For example, from 11% to 66% of shortleaf pine seedlings were killed when exposed to five species of *Pythium* (Hendrix and Campbell, 1968b). Halpin *et al.* (1952) reported similar results on red clover with seven species of *Pythium*, where from 4 to 72% of the stand was killed. Robertson (1973) found that of 16 species of *Pythium* tested against tomato, pea and *Ipomoea*, six were highly pathogenic, four moderately so, two mildly pathogenic and four nonpathogenic. Johnson and Chambers (1973) reported that more cotton seedlings became infected with *P. sylvaticum* than with *P. irregulare* or *P. ultimum*. Over two-thirds of the *P. sylvaticum* isolates were antheridial, whilst one-third were oogonial.

Many of the factors affecting damping-off are probably explained by the concept published by Leach (1947). In an extensive study in which temperature was the major variable, all combinations of host and pathogen in which the temperature was relatively less favourable for the host resulted in more pre-emergence infection. He concluded that other factors being constant, the relative growth rate of the host and pathogen determine the severity of pre-emergence infection.

2.   Feeder Root Infection and Decline Diseases

Feeder root necrosis, caused by *Pythium* spp. and *Phytophthora* spp. and associated with decline diseases has been reported on many plants, including grapevines, sugarcane, barley and wheat, pecan, Dutch iris, cotton, pine, caladium, strawberry, corn, tomato, hyacinth, safflower, hop, spruce, rape, avocado, peach, foliage plants, muskmelon and blueberry. In general, infection is restricted to the juvenile tissue at the root tips. The description by Rands and Dopp (1938) of *Pythium* root rot of sugarcane was the first complete description of this type of disease. They described the soft terminal portion of the roots as flabby and water-soaked. A noticeable deficiency of secondary roots and still finer branches or rootlets are indicative of the activity of root rot even though no flabby root ends are found at the time of observation. Roots killed by *Pythium* are invaded immediately by soil fungi and bacteria and shrivel up and disappear quickly under usual field conditions. While many pages have been published since this description, it has not been improved. Above-ground symptoms on susceptible cane varieties were described as poor growth, lack of ratooning, severe wilting under adequate moisture conditions, yellowing of the leaves and eventual death of the plants. For less susceptible varieties, the only above-ground symptoms were an unthrifty appearance, deficient and delayed suckering and closing in of the rows. More vigorous varieties were less affected by the disease and rapid root replacement apparently counteracts the effect. This aspect was reported more fully by Campbell and Copeland (1954), in their investigations of littleleaf disease. They offered the first full description of a decline disease of a perennial plant, and investigated more fully the effect of soil conditions on severity of the disease.

Littleleaf disease affects shortleaf pine and, to a much lesser extent, loblolly pine (*Pinus taeda*), on six million hectares east of the Mississippi River. On one-third of the area, the disease is severe enough to interfere significantly with forest management plans. The symptoms of littleleaf disease are those commonly associated with some form of malnutrition. Affected trees have a sickly appearance, generally decline slowly, and die prematurely. Growth is reduced and the needles are yellow and shorter than normal and develop in tufts on the ends of the branches. Owing to the decreased needle weight, the branches become more upright. Twigs and branches die throughout the crown, starting in the lower part and although a very heavy cone crop is set as the crown symptoms develop, the seeds are reduced in size and vigour. Below ground the major symptom is a lack of feeder roots.

Since the feeder roots collapse and disappear quickly following infection, Campbell and Copeland's root isolations resulted in the recovery of *Phytophthora cinnamomi* from only 2% of the roots. No other pathogenic fungi were isolated. Soil isolations were more fruitful, however, and correlated more nearly with the occurrence of the disease. Although the fungus may be found in shortleaf stands outside the littleleaf belt, the disease only occurs on soils with low porosity and poor internal drainage. Typical littleleaf sites are eroded, have extremely firm subsoil, short depths to zones of greatly reduced permeability, and strongly mottled subsoils. Under these circumstances, shortleaf pines are not able to replace killed feeder roots, although loblolly pine is more efficient in this respect and thus is less susceptible to the disease. Continual killing of the feeder roots and lack of adequate replacements result in restricted colonisation of new soil areas and when nitrogen in the limited area of colonised soil is depleted, the littleleaf symptoms appear. Zentmyer and Klotz (1947) described a similar disease on avocado in California. The symptoms that developed on young as well as older trees were associated with shallow soils having poor internal drainage and subjected to excessive water (Zentmyer and Richards, 1952).

Phytophthora cinnamomi is restricted to relatively few hosts in the south-eastern United States and, with a few exceptions, is limited to the feeder root tissues. It is more damaging in Western Australia, however, where it has devastated more than 100,000 hectares of complex forest, woodland and heath communities (Newhook and Podger, 1972). It was inadvertently spread by movement of soil from disease centres for use in road stabilisation and *Phytophthora cinnamomi* has caused severe chlorosis, stunting, decline and death of over 400 native Australian plants. Invasion of large roots occurs extensively in many species. Although the disease occurs on well-drained soils, it frequently starts on those areas with impeded drainage. It has been suggested that the pathogen was artificially introduced into Western Australia, thus accounting for its aggressive behaviour.

Hendrix and Powell (1970) further defined the role of interacting factors on decline diseases involving the Pythiaceae. Studying peach decline, they found that many factors affected the feeder root system. These included cultivation, hard pans, nematodes, fertility, pH, time of pruning and the predominant *Pythium* species (Taylor *et al.*, 1970). They concluded that unless herbicide usage was substituted for mechanical cultivation, benefits could not be reaped from controlling *Pythium* spp., nematodes and other factors. At least 75 % of the feeder roots occur in the top 25–30 cm of soil and mechanical cultivation destroyed those in the top 15–20 cm. Destruction of the surface roots alone stressed the trees too much for continued production and survival although subsoiling before replanting to improve the drainage and water availability during drought did improve the performance of the trees by about 10 %. Chemical control of nematodes or *Pythium* spp. improved performance by about 25 %, but controlling both did not further improve longevity or fruit production. Synergistic effects between *Criconemoides* and *Pythium* species were measured (Hsu and Hendrix, 1973) but the factors were mainly additive.

None of the stress factors, alone or in combination, are usually given the credit for killing trees. They probably only weaken them, leaving them susceptible to cold injury, *Pseudomonas syringae*, and other agents which do not affect nonstressed individuals. Peach decline is controlled only by removing the effects of the pathogens and the other agents causing stress.

Bumbieris (1972) described a decline of grapevines in Australia in which *Pythium* spp. were involved, but other unidentified factors were also present.

Apple replant problems also exist in many parts of the world. While nematodes may be involved in individual situations, they are not considered to be a part of the general causal complex (Savory, 1966; Sewell, 1981). Among the soil microflora. *Pythium* spp., especially *P. sylvaticum*, have been shown to lower growth. The amount of reduction was roughly comparable to the increase in growth which occurred when replant soils were fumigated with chloropicrin. Isolates of *P. sylvaticum* from apple soils are more pathogenic to apple roots than isolates from non-apple soils (Sewell, 1981). This leads to the conclusion that with fungi having wide host ranges such as *Pythium* spp., physiological races most adapted to particular hosts will evolve with long term associations.

As a result of these various studies, starting with those of Rands and Dopp, a picture of decline diseases has emerged. Feeder roots are killed by phythiaceous fungi, frequently under conditions of excessive soil moisture which are not conducive to rapid root regeneration. In many cases, other factors also kill feeder roots and this results in great stress to the plants, which are consequently killed by weak pathogens, nutrient deficiencies and adverse weather. Nonstressed trees are not affected by these factors.

Declines of annual plants are frequently more dependent on environmental factors than are declines of perennials. Roncadori and McCarter (1972) found that cotton root rot caused by *P. irregulare* was greater at lower temperatures, and the effects, as measured by root and shoot growth, occurred primarily in the first two months after planting. In further studies in which they measured yield, they found that flowering and boll set were delayed. In short season locations, this delay resulted in a yield reduction which was not observed at long season sites (Roncadori *et al.*, 1974).

Studies on the histopathology of pythiaceous fungi emphasise that zoospores are attracted to the root ends at the zone of elongation where they encyst, germinate and penetrate directly. Zoospores frequently encyst on root hairs (Nemec, 1970, 1971, 1972), but hyphae also infect through root hairs, a slight hyphal swelling occurring at the site of penetration. Cell invasion occurs below the level of secondary thickening of the endodermal cell walls and penetration occurs in two to eight hours after inoculation, colonisation of the cortex following within 24 hours (Miller *et al.*, 1966; McKenn, 1977). Sporangia and oospores are found in the invaded tissues.

Pythiaceous fungi germinate, penetrate and colonise tissues rapidly, and form both sexual and asexual reproductive bodies shortly thereafter, but they are no longer actively present when secondary organisms start to build up.

Bruehl (1953) reported that the general destruction of fine rootlets of barley and wheat by *P. arrhenomanes* resulted in a loss of vigour, discolouration of the leaves, short straw, lack of tillering, delayed maturity and poor spike exsertion. Low soil organic matter and nitrogen increased disease severity. The differential response of cotton and small grain is probably due to the ability of the small grain pathogen to parasitise at higher temperatures than the cotton pathogen.

D. FOLIAGE, STEM, CROWN AND FRUIT ROTS

Pythium species also cause stem and crown rots, usually involving succulent tissues on hosts including tobacco, poinsettia, geranium, calla, taro, carrots, peanuts, Irish potato and various vegetables such as cucumber and watermelon.

Meurs (1934) reported that a parasitic stem burn of tobacco, originally described in 1921, was caused by *P. aphanidermatum*, *P. myriotylum*, and *P. deliense* Meurs. The symptoms were similar to those of black shank, caused by *Phytophthora nicotianae* var. *parasitica*. The disease, described as rare in seedbeds, resulted in a blackened stem and root with the affected parts shrivelled and brittle, the pith hollow, and the stems commonly breaking. The disease developed shortly after transplanting and was favoured by rainy weather. Braun (1924) described a similar disease on pelargonium caused by *P. complectens* and *P. debaryanum*, and Griffin (1972) ascribed the same problem to *P. splendens* Braun. The affected stems become blackened and shrunken, this being followed by defoliation and death. The roots are usually rotted, the leading edge of the lesion is well defined and noncutinised and nonlignified tissues of cuttings are normally the infection court. The disease occurs during warm humid conditions, and is more severe under lower light intensities, such as occur during cloudy weather. On poinsettias, a similar disease of cuttings is caused by *P. debaryanum*, *P. perniciosum* and *P. aphanidermatum* (Bolton, 1978) while Gay and McCarter (1968) found *P. myriotylum* causing a stem rot of snapbean in the field in South Georgia.

Taro and calla corms, Irish potatoes, cucumbers and watermelons develop a soft, watery rot caused by several *Pythium* spp. (Tompkins, 1950; Boyd, 1972; Ooka and Yamamoto, 1979). These organisms probably enter through wounds caused by harvesting operations.

Carrots grown in muck soils throughout North America are subject to brown root and forking, caused primarily by *P. sulcatum*. Dead, brown lateral roots and forked tap roots are the typical symptoms although a rot of the tap root occasionally occurs. The disease is more severe in fields planted to carrots for several years, especially on land where damping-off is common (Mildenhall *et al.*, 1971; Sutton, 1975; Pratt and Mitchell, 1973).

Garren (1970) described a peanut pod rot caused by *P. myriotylum*. Frank (1968) described the symptoms as a dry rot with a fairly dark discolouration of the pod and with mycelial development in the pod cavity. Seeds mummify and the pods disintegrate. Later, he described an interaction with *Fusarium solani* in which this fungus predisposed the pods to *P. myriotylum* attack (Frank, 1972). Shew and Beute (1979) described an interaction with mites of the genus *Caloglyphus* in which the mites were acting as a disseminating rather than a wounding agent. The mites feed selectively on *P. myriotylum* and the fungus survived passage through the mites' alimentary canal.

Pythium species were recorded as the causes of a foliage blight of various turf grasses in the early 1930s. While they occasionally cause problems in pasture situations, they are economically damaging in highly managed turf, particularly in turf grown somewhat outside its natural range, i.e. cool season grasses grown in warm climates. Pathogenic species in such conditions include *P. aphanidermatum*, *P. catenulatum* and *P. ultimum* although, undoubtedly, many other species are capable of this type of pathogenicity. Hendrix *et al.* (1970) identified over 20 species isolated from golf greens in which a monoculture of turf had been maintained. They caused a decline by killing feeder roots but, under appropriate environmental conditions, many of these species are probably capable of causing foliar blights. *Pythium arrhenomanes* and *P. debaryanum* have been reported as foliar pathogens (Howard *et al.*, 1951) while Moore and Couch (1961) screened 45 species representing 21 genera of Gramineae for susceptibility to foliage blighting by *P. ultimum*; all were susceptible. Most, however, are not affected by the disease in the field, the exceptions being the bent grasses, Bermuda grass on golf greens and in drainage ways leading from greens, and rye grass when used for overseeding golf greens. *Pythium* diseases are also found on highly managed lawns. Muse *et al.* (1974) found that *P. graminicola* Subram. was a strong pathogen, *P. vanterpoolii* V. Kouyeas & H. Kouyeas a moderate pathogen and *P. torulosum* Coker & Patterson a weak pathogen on bent grasses, blue grasses, fescue and rye grasses, but diseases caused by these fungi were observed in the field only on a golf course of bent grass.

Yet another example of the diverse types of disease caused by these fungi is snow rot of winter wheat in Washington. Lipps and Bruehl (1978) studied this problem, first reported in Japan in 1933, and named *P. iwayamai* Ito as the primary cause. Infection and disease development occur at 0.5°C in melted water under a snow cover. *Pythium ultimum* was shown to grow on agar at this temperature as well as *P. iwayamai*, and also to invade plant tissues, but it did not cause snow rot.

Pythium ultimum causes collar rot of apples, as does *Phytophthora cactorum*, East Malling IV, VII and IX rootstocks being susceptible (Bielenin *et al.*, 1976).

Phytophthora heveae Thompson, isolated by Campbell in western North Carolina and eastern Tennessee and reported by Campbell and Gallegly (1965), has been associated with rhododendron dieback in a nursery in that area. Infection occurred on young leaf and shoot tissue and many infected plants died (Benson and Jones, 1980). *Phytophthora* species involved with similar conditions at other nurseries included *P. cactorum*, *P. citricola* Sawada and *P. nicotianae* var. *parasitica*. These fungi also infect roots and rot the crowns of the plants. Zentmyer *et al.* (1978) reported a crown rot and canker of avocado caused by *P. heveae*, whilst *Phytophthora cinnamomi* is known to cause a stem canker and crown rot on both avocado and rhododendron.

Cankers and crown rots on apple and citrus are caused by *P. cactorum, P. citricola* and *Pythium ultimum.*

E. NON-PARASITIC PATHOGENS

The ability of *Pythium irregulare* to cause pathogenesis on table beets without infection was described by Brandenberg (1948). It was not until the 1970s, however, that J. W. Hendrix and co-workers published a series of articles showing that this phenomenon occurs with both *Pythium* and *Phytophthora* species. Blok (1973) described a growth-regulating substance produced by *P. sylvaticum* which caused symptoms in the absence of viable fungal propagules while Brandenberg (1948) had found that *P. irregulare* produced a substance in culture which was transported through the plants, causing wilting and leaf necrosis of several plant species. When Blok (1973) tried to repeat the experiments using non-purified filtrates, she was not successful in duplicating symptoms using *P. irregulare,* but the culture filtrate did lower longitudinal root growth.

Csinos and Hendrix (1977a,b), using *Phytophthora cryptogea* Pethybr. & Lafferty, found that extracts of non-viable mycelium stunted tobacco plants and caused necrotic foliar symptoms. The toxin was heat stable. Kanlong and Hendrix (1977) found the abilities of eight species of *Phytophthora* and *Pythium* to kill plants and reduce plant growth were not related. Flowers and Hendrix (1974) had previously shown that increases in fungus population did not occur in the absence of parasitism. Csinos (1979) has since shown that a toxin exists in non-viable mycelium of *Pythium myriotylum*. Roots of tomato seedlings exposed to this heat stable toxin grew slowly and in severe cases roots and hypocotyls turned brown and the seedlings died.

Blok (1973) using *P. ultimum, P. irregulare, P. paraoecandrum* Drechsler and *P. sylvaticum,* found the auxin, IAA, in culture filtrates. Roots exposed to cultural filtrates of living mycelium started to swell in the region behind the root tip as a result of the lateral growth of cells and often produced large numbers of root hairs. Root growth was reduced. This phenomenon was caused by toxin diffusing from viable mycelium. Enough inhibitory substances were produced in five hours in shake culture to cause a 34 % inhibition of root growth and although IAA auxin was present in the culture filtrate, some other heat-stable substance was also involved. Inhibition of root growth by IAA alone was overcome in two days, but was not overcome when culture filtrates were used. Blok suggests that since *Pythium* spp. frequently infect through root hairs, production of the auxin-like toxin is a mechanism in disease development.

The occurrence of this phenomenon in nature is more uncertain. Csinos and Hendrix (1977b) suggested that it was a cause of poor survival of tobacco transplants or poor growth of those that survived. Moreover, when the evidence for non-parasitic pathogenicity is viewed on the basis of the findings given by Wilhelm (1965), it seems possible that such adverse plant responses occur extensively in nature. Wilhelm reported experiments designed to test the hypothesis that *P. ultimum* reduces plant growth without producing symptoms other than those expected in soils low in fertility. *Pythium ultimum* induced a cessation of plant growth, and Wilhelm had the impression that invasion of the roots did not extend beyond the root hairs. It is of interest also that Hoppe (1949) did not consider invasion of corn roots by *Pythium* spp. necessary for stunting to occur.

Some isolates of *P. ultimum* are clearly capable of penetrating and killing strawberry roots (Watanabe *et al.*, 1977), yet the isolates tested by Wilhelm did not cause this type of root damage. The isolates of *Phytophthora cryptogea* used by J. W. Hendrix and his co-workers in Kentucky were not parasitic to tobacco, but were clearly pathogenic in that they induced plant decline. There may be as much diversity within species concerning the mode of disease development as there is with response to environmental factors. This area of the pathology of pythiaceous fungi is interesting and may explain some baffling disease problems. It probably represents the next new frontier.

F. ENVIRONMENTAL PARAMETERS

The effects of environmental factors on diseases caused by pythiaceous fungi were summarised by Hendrix and Campbell (1973). With some species, particularly those readily producing zoospores, fluctuating moisture levels are important. Those not readily producing zoospores respond more to temperature. Undoubtedly, much of the plant mortality attributed to pythiaceous fungi in low-lying areas is due to drowning. Biesbrock and Hendrix (1970) found little or no effect of *Pythium vexans* de Bary or *P. irregulare* on mortality at 90% soil moisture capacity, even though both were capable of killing large numbers of roots at lesser soil moisture levels. It is relevant that Gardner and Hendrix (1973) found that *P. vexans* and *P. irregulare* could tolerate increased CO_2 levels as long as O_2 levels were not lowered too drastically. Another effect of soil moisture is on the distance that exudates diffuse through soil; at 28% moisture capacity, they diffused 10 mm in 24 hours, but only about half that far at 15.7% (Stanghellini and Hancock, 1971). This phenomenon is certainly a compounding factor in soil moisture studies, and has been only infrequently considered. More specific effects of environment have been considered in previous sections but a generalisation, admittedly with many exceptions, is that diseases caused by pythiaceous fungi occur under conditions unfavourable both to other pathogens, and to the host. Pythiaceous fungi have evolved to react rapidly and under conditions frequently not favourable for other organisms; conversely as other organisms become active, pythiaceous fungi usually are killed or become dormant.

G. INTERACTIONS WITH OTHER PATHOGENS

In nature, pathogens frequently do not occur alone. There are often additive or synergistic effects with other pathogens. In so far as pythiaceous fungi are concerned, this may be the rule rather than the exception when they are involved in decline diseases and the subject was discussed in this context earlier in the chapter. Interactions involving non-decline disease are discussed below.

Kerr (1963) measured a synergistic effect of *Fusarium oxysporum* Schlect. ex Fr. and *P. ultimum* on dry weight of peas. He suggested that *P. ultimum* caused root damage allowing the wilt organism to enter the plant. Whitney (1974) found a synergistic effect of *P. ultimum* and *Heterodera schachtii* on damping-off, but not root rot of sugar beet; it was associated with increased growth of the fungus around infection sites and not an increased number of sites. Interestingly, however, interaction between this nematode and *P. aphanidermatum* was additive not synergistic. Powell *et al.* (1971) found a synergistic effect of *Meloidogyne incognita* and *P. ultimum* in rotting tobacco roots and suggested that root-knot nematode infections predisposed roots to subsequent invasion by other organisms, many of which were not otherwise capable of invading the roots. Koike and Yang (1971) found both additive and synergistic interactions between sugarcane mosaic virus and *P. graminicola* on the growth of sugarcane while Pieczarka and Abawi (1978) observed an antagonistic interaction between *P. ultimum* and *Rhizoctonia solani* Kühn and a synergistic interaction between *P. ultimum* and *Fusarium solani* (Mort). Appel & Wollenw. f. *phaseoli* (Burkholder) Snyder & Hansen on bean roots. Antagonistic reactions with *R. solani* have also been reported for other *Pythium* spp. and *Phytophthora* spp. In one of the few instances where synergistic effects were quantified, Hsu and Hendrix (1973) found that *Criconemoides quadricornis* extended the temperature range over which *P. irregulare* damaged pecan roots and in general, synergistic interactions are probably more common when one or both of the organisms are weak pathogens in a particular situation, whether that results from host resistance or unfavourable environmental conditions.

V. Distribution and Redistribution

The literature on pythiaceous fungi enables one to trace the expanding interest in this group of fungi over the past 150 years. The first accounts of the isolation and identification of *Pythium* and *Phytophthora* species are related to diseased hosts or the infection of plant material placed in situations where such material would attract zoospores. By the mid-1900s pathologists and mycologists had collaborated on impressive host lists whereby named species could be positively related to agricultural, horticultural and forest vegetation. These lists demonstrated wide host preferences and extensive distribution for certain pythiaceous fungi and the possibility that others were less widely distributed and restricted to only a few plant species. It is possible, however, that such lists may reflect greater activity on the part of pathologists in certain regions of the world where diseases associated with these fungi have greater economic significance.

The isolation of pythiaceous fungi from soil and from plant material containing saprophytic fungi was hampered by the greater adaptability of these ubiquitous spore-formers to agar media and their rapid growth in culture. Following the development of selective media (see p. 132) pathologists were able to obtain a better picture of the relative abundance of specific groups of fungi and their geographical distribution. These studies have shown that most *Phytophthora* and *Pythium* species are truly worldwide in distribution and that the few with more limited distribution records may simply reflect a lack of interest in their hosts, or that some areas are less accessible to pathologists and mycologists. A few may be restricted by temperature and rainfall but this again may equally be a question of the restriction of the hosts and those pathologists interested in the diseases of these hosts.

For thousands of years prior to the recognition of plant diseases caused by soil-inhabiting fungi, man and animals criss-crossed continents and the world, distributing soil and plant material. Certain ancient civilisations that depended heavily on agriculture no doubt played an important part in the distribution of soil pathogens. Soil movement by wind and water also served to move resistant spores of many species over long distances. Hence it is not surprising that many identifiable species of *Pythium* and *Phytophthora* are found wherever any search is made.

The distribution history of only a few species can be traced at this late date. *Phytophthora cinnamomi* was described in 1922 causing a stripe canker of cinnamon bark in the West Coast Highlands of Sumatra (Rands, 1922). The subsequent interest in this fungus has led to a fairly complete dossier on its distribution and host range. Shortly after the species was described, Tucker (1928) reported it from avocado roots in Puerto Rico and White (1930) found it causing a disease of rhododendron in nurseries and plant beds in New Jersey. Crandall *et al.* (1945) presented evidence to show that *P. cinnamomi* had been present in the south-eastern United States for many years and had killed much of the chestnut in the lowlands and mountain valleys long before chestnut blight (caused by *Endothia parasitica* (Murrill) P. J. & H. W. Anderson) eliminated this species as an important component of eastern hardwood forests. Their contention is borne out by reports of dying American chestnut in Georgia, North and South Carolina long before the fungus was associated with littleleaf of pines or other diseases. Wager (1940) associated *P. cinnamomi* with dying avocado trees in southern California and also reported the disease from western Transvaal and Natal in South Africa in 1941. Crandall (1947) observed dying and declining avocado trees at Tinga Maria in eastern Peru in 1944.

These observations made in the short span of 20 years after the recognition of *P. cinnamomi* as an identifiable species distinct from those already described raised questions as to its origin and distribution. Conversations with Dr. Rands on this subject suggested no sudden spread of the fungus from the highlands of Sumatra but that it was already widely distributed in the tropics and needed only the activities of plant pathologists to isolate it and determine the conditions conducive for its spread and pathogenicity. The subsequent history of the fungus indicates that its place of origin is southeast Asia and that it had already reached its maximum spread before it was finally isolated and named. While it is more or less confined to tropical and warmer areas of the world, it persists in nursery and plant bed situations as far north as New Jersey and northern California and has been isolated from pines in northern Kentucky.

The recognition of mating types and their distribution suggested the possible relationship between the ratios of A^1 and A^2 types as the means of pinpointing the areas in which *P. cinnamomi* can be considered indigenous. Ko *et al.* (1978a,b) found the fungus in the soil in natural healthy forests in central Taiwan. Their isolations produced approximately a 1:1 ratio for the A^1 and A^2 types. These isolations were from an area where the fungus was causing no detectable damage to the natural plant populations, indicating a more or less stable condition relative to mating types. In most areas of the world the A^2 type is common with the A^1 restricted to a limited number of hosts. Shepard (1975) suggested, on the basis of a northward increase in the frequency of isolation of the A^1 type in Australia, that *P. cinnamomi* was probably indigenous to the New Guinea/Celebes region. If the ratio of the two types occurring in the natural plant communities is important in determining the possible centre for the spread of *P. cinnamomi* worldwide, then Taiwan would appear to be within the boundaries of the indigenous home of the fungus.

As the fungus became distributed outside its original indigenous areas the A^1 and the A^2 types became separated, perhaps because of different host susceptibility or separation of the soil matrix containing propagules of the fungus. In the United States only the A^2 type was isolated from soil under shortleaf or other pines in the south and southeastern states. The A^1 type was isolated from *Camellia japonica* in Georgia and elsewhere in the south-east by Zentmyer (1980), who found the A^1 type restricted to camellia in an extensive survey of *P. cinnamomi* on various plants in North and South Carolina. Hendrix and Campbell (1966) studied the identity and abundance of pythiaceous fungi on ornamental nursery stock in Georgia nurseries. For many years *P. cinnamomi* and various *Pythium* species have been routinely distributed throughout the State on potted and balled nursery material. Where soil and moisture conditions favoured the pathogens, such infected nursery material slowly declined or died rapidly, making a market for replacement plants which were often planted in the same locations where the others had died. Recognition of the almost universal presence of pythiaceous pathogens in nurseries, especially those of long use, has encouraged sterilisation of soil, containers and plant beds to reduce the amount of contamination.

Phytophthora cinnamomi has been isolated from forest tree nurseries, especially those producing susceptible host species. Fraser fir (*Abies fraseri*) has found favour as a Christmas tree in western North Carolina. Much of the available planting stock once came from a single nursery badly infested with *P. cinnamomi* and trees from this nursery were shipped to Christmas tree growers with little regard for the spread of the fungus (Kuhlman and Hendrix, 1963). This type of distribution and redistribution is a continuing process and has probably contributed to the wide distribution of many root pathogens.

World War II interrupted research on *Phytophthora*-caused diseases or delayed the publication of results already discovered. After the war the United States Department of Agriculture intensified its efforts to find the cause of a decline disease of shortleaf pine in the southeastern States. *Phytophthora cinnamomi* was eventually linked to the problem, not because it could be isolated from diseased roots, but because of its constant presence in the soil under and around diseased trees. The distribution of the fungus could be equated with the symptoms of littleleaf which indicated its presence in the soil in the Piedmont and adjacent physiographical areas from Virginia south and west to Mississippi. However, the disease does not occur in the shortleaf pine stands west of the Mississippi even though the fungus was later isolated from soil under a few pine stands in Texas and Arkansas. It is not surprising that *P. cinnamomi* was not detected until 1948 because the fungus kills only root tips which then usually disappear within 24 hours.

The extensive research on littleleaf disease of pines in the United States stimulated interest in *P. cinnamomi* as the cause of forest tree diseases in other parts of the world. Newhook (1959) found *P. cinnamomi* and other *Phytophthora* species associated with the dying of *Pinus radiata* and other conifers in New Zealand. Later studies by Newhook (1959) showed that *P. cinnamomi* was also present in indigenous plant communities in New Zealand indicating that it may be native to soils in a number of vegetation types.

It is apparently too late to establish with certainty whether or not *P.*

cinnamomi was native or imported to the United States. It has been present and recognised for many years in nurseries along the east coast and probably killed American chestnuts in the south as far back as 1855. Campbell and Hendrix (1967), in evaluating the distribution of *Pythium* and *Phytophthora* species in forest soils of the south-eastern United States, found *P. cinnamomi* in soils of relatively undisturbed relic forest stands in the mountains of North Carolina and Tennessee at elevations that eliminate any possibility of transport by moving water. The presence of *P. cinnamomi* in remote areas where it is not causing recognisable disease suggests that it may have been present originally as a minor component of the soil fungus flora. As the land was cleared and the habitat drastically changed, the fungus survived only in isolated areas.

Nonetheless, if *P. cinnamomi* is an introduced pathogen in the southern United States its spread and damage has been less spectacular than that in Western Australia where the effects of the invasion of the fungus in the eucalypt forests has been devastating on practically all trees, shrubs and herbaceous species.

Clear evidence of the potential danger of an introduced soil pathogen can be found in south-western Australia. *Phytophthora cinnamomi* has been present for many years in other parts of Australia where it caused moderate loss in orchards and horticultural plantings. Its introduction into western forests produced a spectacular example of the damage that a pathogen can cause in an ecological situation lacking competing organisms and providing continuous contact between susceptible roots. Its introduction into the eucalypt forest areas where it encountered highly susceptible plant species has afforded information on the possible rate of spread by a soil pathogen whose competing organisms are at a mimimum. Here *P. cinnamomi* can be traced from one or two original centres to an area in which jarrah (*Eucalyptus marginata*) and practically all the associated plants have been killed on hundreds of hectares (Podger, 1972; Pratt and Heather, 1972, 1973). Weste (1973) showed that the fungus was introduced through contaminated gravel on road verges and that it followed water channels at a downhill slope rate of as much as 400 metres per year. Soil type, depth and local drainage all influence the severity of the disease, particularly the rate of spread and rate of mortality.

The distribution and host range of *P. cinnamomi* offers an interesting record of the potential danger from a soil pathogen, either when newly introduced into an area or where changes in the agricultural or vegetation patterns provide an opportunity for the fungus to increase and to spread. This pattern may have been repeated many times before the advent of plant pathologists and demonstrates the need for greater attention to the woes of plants important in the human food chain.

VI. Ecological Niche

The standard mycological texts such as Wolf and Wolf (1947), Bessey (1950) and Alexopoulos (1962) discuss the evolutionary trends in the water moulds and go into detail on the asexual and sexual reproduction in the Peronosporales. The pythiaceae represents a family whose habitat is transitional between purely aquatic and completely terrestrial. Contrary to the condition in the more primitive uniflagellates, most species of *Pythium* and *Phytophthora* can be cultured on non-living media. In nature, most species live mainly as saprophytes but many may become parasitic on living plants, particularly on young succulent tissues. A few are parasitic on other *Pythium* species. Of the 66 species recognised by Middleton (1943), three are saprophytic and one parasitic on animal matter, nine parasitic on fresh water algae, one on marine red seaweed, nine in soil and vegetable debris where they are not known to be parasites and 40 or more are capable of parasitism in higher plants.

Various representatives of the Pythiaceae found ecological conditions particularly favourable in glasshouses and horticultural and forest tree nurseries where plant density and moisture furnished abundant host material plus ideal situations for the production and spread of zoospores. The mass of succulent roots and the associated young stem tissues, permitted outbreaks of a disease condition characterised as damping-off. Beach (1946) distinguished between pathogenic and physiogenic damping-off and noted the relationship of several *Pythium* species to the condition in reported instances.

Damping-off and stunting from root damage are particularly troublesome in forest tree nurseries. Hartley (1921) reported on the condition in forest tree nurseries in Nebraska and elsewhere and advocated control by making the nursery soils more acid and hence unfavourable for *Pythium* and other damping-off organisms. By the mid-1950s, when nursery production of conifers and other tree species reached the astonishing peak of 10^9 or more plants each year, *Pythium* and *Phytophthora* species had become well established in a favourable ecological niche in nurseries. Drastic measures to decrease losses both from outright killing of seedlings and from the reduced vigour of those surviving were required.

Tomato, peppers and other vegetable plants are grown extensively in south Georgia and shipped north for the transplant trade. Much of the land devoted to transplant species is under irrigation and damping-off and other root and stem diseases have been locally severe. This led to investigations of the irrigation water, whose source was ponds so situated as to receive run-off from the planted areas. Shokes and McCarter (1979) demonstrated the presence of 13 *Pythium* and one *Phytophthora* species from such ponds and concluded that the water was a potential source of inoculum and a source of reinfestation of fumigated plant beds. Gill (1970) had previously trapped *P. myriotylum* from irrigation pond water and suspected the water as a source for infection of tomato plants by this fungus. Thorn *et al.* (1959) had no difficulty in demonstrating the presence of zoospores of *P. cinnamomi* in canals, water storage basins and drainage areas during the wet winter of 1957–1958 in southern California. Apples placed in water holes in a dirt road on the grounds of the Forestry Sciences Laboratory near Raleigh, North Carolina consistently recovered *P. cinnamomi.* The road was in an area surrounded by shortleaf pine, many with littleleaf symptoms. The above cited reports are but a few of those in the literature demonstrating that water moulds can be transported in irrigation water.

Reproductive propagules of many *Pythium* and some *Phytophthora* species accumulate under specific crops and while, with proper management, they may cause only limited loss, they remain and may become very destructive once susceptible plants are grown in these areas. A good example of the effect of residual infective material on a crop not previously grown on land once used for cotton, melons, tomatoes and other crops has been reported by Campbell and Presley (1946). The rubber-producing plant guayule (*Parthenium argentatum*) was widely planted on agricultural land in California and the southwest during World War II as a back-up source of natural rubber. A seedling stem and root rot caused by *P. ultimum* was troublesome in the Salinas Valley of California in irrigated guayule nurseries patterned after conventional forest tree nurseries with water supplied through over-head sprinklers. The fungus is a very common pathogen under various vegetable crops in the coastal valleys of California. The amount of disease could be regulated during the summer dry period by controlling the amount of water supplied and by reducing the areas where water would stand.

A root and stem rot also developed in a number of the guayule plantings made on irrigated fields not only in the coastal areas but inland in the San Joaquin Valley. This disease was caused mainly by *Phytophthora drechsleri* Tucker with some additional involvement by *P. cinnamomi.* The disease was most pronounced at the end of irrigation runs or in depressions in the fields that remained wet for some time after water was applied but there was little or no crop history indicating severe losses from these two fungi prior to growing guayule on the land. Later studies indicated that guayule would escape infection if water was stored in the soil during the winter or during the growing season when temperatures were below 15°C. Fortunately, enough water could be stored during the dormant season to permit the development of this desert shrub during the dry summers.

A systematic search was made for *Phytophthora* species in old growth hardwood and mixed hardwood stands in forest situations in eastern Tennessee and western North Carolina (Campbell and Gallegly, 1965). The area sampled was extended to include old growth stands in northern and eastern Kentucky (Hendrix *et al.*, 1971). *Phytophthora cactorum, P. citricola, P. nicotianae* var. *parasitica* and *P. cinnamomi* were found in a number of areas sampled. A surprising discovery was the presence of *P. heveae* in several old growth stands in the mountain range forming the boundary between Tennessee and North Carolina. The soil samples were not related to host hence *P. heveae* could not be connected with any pathogenic symptoms on the forest vegetation. *Phytophthora heveae* had never been isolated before in North America and its presence in several remote old growth forest stands relegated it to the position of a mycological curiosity. However, Benson and Jones (1980) found *P. heveae* to be one of several *Phytophthora* spp. causing a crown rot of rhododendron in horticultural nurseries in North Carolina. *P. heveae* was confined to a single nursery in the western part of the state, not far from the area where the fungus had been isolated from forest situations. Native rhododendrons were common in these old growth stands and the jump from wild-growing rhododendrons to those in horticultural nurseries may offer an example of the manner in which many pathogenic fungi spread from a restrictive range to a widespread distribution.

The very abundance of the Pythiaceae in forest soils raises the question as to the part they play in stand composition and natural succession. The forest floor is an arena where seedlings of all kinds struggle for existence; this struggle is against each other for light, nutrients and moisture. The presence of *Pythium* and *Phytophthora* species, whose existence depends in many cases on the availability of root tips and succulent tissues, suggests that these fungi play a major role in determining which species suffer the least damage and maintain their place in the continuing struggle for existence. Usually the forest soils have a humus layer which supports many saprophytes. These in turn may successfully compete with soil pathogens and prevent them from causing excessive damage.

Tree species, particularly pines, once their feeder roots have become thoroughly mycorrhizal, develop a system wherein the very susceptible fine roots are protected against infection by zoospores of various lower fungi, particularly *P. cinnamomi*. Marx (1969a,b), and others, have examined the influence of mycorrhiza on the ability of zoospores to infect root tips and juvenile tissues. They concluded that the protection mechanism is both mechanical and chemical, the mass of mycelium around the roots forming an effective barrier and the mycorrhizal fungus or fungi possibly exuding antibiotics of various kinds that prevent zoospores from germinating. This latter phenomenon is common in the mycological world and many fungi maintain their niche through discouraging others by means of the anti- biotics which they exude into the surrounding biosphere.

The area of elongation in shortleaf pine roots is attractive to zoospores of *P. cinnamomi* and young growing roots suspended in water culture offer a means of studying this phenomenon. Zoospores released into the system are attracted to the region of elongation where they collect in large numbers. Zentmyer (1961) has studied this strong attraction of the active meristematic tissue for zoospores and considers it chemical in nature.

Thus, members of the Pythiaceae have evolved specialised adaptations to a terrestrial life style. Part of their success may depend on the quickness with which they respond to favourable conditions, such as water and temperature. During unfavourable conditions, they exist as resting spores of various kinds but, after only a short period of favourable moisture and temperature, zoospores are produced. These may be activated by exudates from roots that have grown into the area where the resting spores have been lying dormant and they then reach and infect the root tips ahead of any other competing fungi. This enables them to form new mycelial systems and resting structures by the time competing organisms have invaded the mined-out root sections. Experience with inducing sporulation and infection by *P. cinnamomi* in pot and liquid cultures has shown that zoospore production ends as the waterlogged soils encourage organisms with lower oxygen requirements. It is interesting to observe various paramecia and other motile organisms catch and engulf zoospores as the cultures age. These motile organisms may exert considerable influence in bringing an infection cycle to an end by eliminating most of the zoospores circulating in the system. While many scientists point to plant death in waterlogged areas as being due to pythiaceous fungi, it is more likely that it is due to drowning.

An example of the host indices for various *Phytophthora* and *Pythium* species with a history of causing plant disease reveal some information about the ecological preference of only a few species. *Pythium debaryanum/ irregulare*, *P. ultimum* and *P. sylvaticum* appear in practically all lists, often on the same hosts. They have developed no strong specialisation with respect to host or environmental conditions. Hendrix and Campbell (1970) in reporting the distribution of fungi under perennial crops found only *P. catenulatum*, *P. middletoni* Sparrow and *P. periplocum* Drechsler restricted to recoveries from single hosts, in this case grasses. *P. aphanidermatum* was common on grass but was also found in soil from the root zone of citrus and peach trees. The same distribution was noted for *P. mamillatum* Meurs. *Pythium spinosum* Sawada was isolated from soil under apple, pine, pecan, citrus, peach and grass and was especially abundant under pecans and citrus. *Pythium vexans* was most abundant under citrus trees but could be isolated in small numbers from soil under other hosts.

VII. Impact

Statistics on losses from plant diseases are difficult to find, and even harder to believe. Data on crop yields and losses are often compiled either to praise or to blame weather, politicians, economists or other interests and are used to support requests for funding research projects, market prices, aid to farmers or other humanitarian projects. Crop losses are rarely caused by a single factor. In many cases, estimates are based on yield or final stand reductions and these are broken down into estimates for each factor involved. For example, damage to soybeans by *Pythium* spp. in the south-eastern United States has been estimated to be 25–30 million dollars each year. The reduction in yield from damping-off may have been that great but it is unlikely that all the reduction was caused by *Pythium* species. There are many other organisms infecting soybeans in the field which cause damping-off, especially under conditions unfavourable for seed germination and early growth.

Pythiaceous fungi are generally present in the soil of nurseries to the extent that they must be controlled by every successful nurseryman. Hendrix and Campbell (1966) surveyed 22 ornamental nurseries in Georgia, and found pythiaceous fungi on the roots or in the soil of 24 different ornamental plants in 70 of 73 samples. Similarly, 17 of 18 forest nurseries in the south-eastern United States suffered losses from pythiaceous fungi (Hendrix and Campbell, 1968b). Similar data are available for nurseries throughout the world.

Disease impact, in addition to the direct loss, includes the cost of major changes in management practices. Often, to improve drainage, organic matter must be incorporated into the soil and, in container nurseries, growers must import components for well-draining mixes rather than use materials available locally. In many cases, preferred cultivars cannot be grown because of disease susceptibility. Frequently, the percentage of product which is marketable is reduced significantly, or the time to produce a marketable product is increased. For example, when soil was treated with methyl bromide in an azalea nursery in eastern Virginia, culls were reduced from 25 % to less than 1 % and production time from three to two years. Fine root necrosis caused by pythiaceous fungi was a problem in the nursery although additional problems probably existed, but were not identified.

The direct cost of chemicals to control these fungi in nurseries also constitutes an impact. In the many nurseries which annually grow several hundred thousand container plants, the cost of labour to apply the chemical to each container is much greater than the cost of the chemical itself. In nurseries not using containers, problems with the incorporation of the chemicals into soil are frequently encountered and these too are expensive.

A major impact of pythiaceous fungi on annual crops is their effect on emergence and stand establishment. To prevent damping-off, caused to a large extent by *Pythium* spp., seed must be treated chemically; planting dates must be changed to select times when the environment is more suitable to the host than the pathogen; and, often, non-preferred cultivars must be used.

Frequently, the seeding rate is increased which leads to irregular stands in years when damping-off occurs extensively, and to too many plants when it fails to occur.

Diseases caused by pythiaceous fungi on non-seedling annual crops occur, but are less important than on perennial crops. Examples are black shank on tobacco, cottony blight of turf, stunt of cotton and cottony leak of curcurbit fruit. The consequences, however, are somewhat greater since most of the expense necessary to produce the crop has been expended at the time the disease becomes apparent. In addition, pythiaceous fungi can cause transit rots of crops such as green bean and curcurbits and, again, the total cost of production, including harvest and packing outlays have already been invested by this time. Direct chemical control in the field is often needed, coupled with changes in cultural practices to improve drainage or other conditions making the environment more favourable for the host. Fungicides or disinfectants must be added in packing lines, and sometimes drying of the product is required while the necessity for refrigerated transport and storage facilities is, in part, due to the possibility of damage from these diseases. All of these factors add to the cost of production and the costs of the marketable product, and may reduce the amount of the product produced.

The impact of littleleaf disease on wood production in the south-eastern United States perhaps best illustrates what can happen to our efforts at disease control. This disease changed management practices on six million hectares of land. First, it was recommended that new plantings be made with loblolly pine, which was less susceptible. Nurseries thus concentrated on producing loblolly and slash pine (*Pinus elliottii*), which were infected with fusiform rust in the nursery beds. When the seedlings were transplanted, the rust was distributed throughout the south-east where the alternate oak host was plentiful. Currently, 70 % of the loblolly pine in the south-eastern United States are infected with *Cronartium quercuum* f.sp. *fusiforme* and, due to this rust, some areas are now being replanted with shortleaf pine, which will be managed on a short rotation.

With *Pythium* root rot of sugar cane, the approach has been to develop new resistant cultivars every few years. To control peach tree decline, it was necessary completely to change management practices in the orchard and in nurseries producing trees for orchard use. Root stocks were changed and chemical controls added at some locations. Thus the impact of these diseases has been both large and expensive, and has continued over a period of years.

VIII. Control

Prior to World War II and the proliferation in the number of chemicals available to control plant diseases, emphasis was placed on cultural and biological control. With the introduction of many inexpensive chemicals, these were substituted for the cultural and biological measures, rather than integrated into a total control programme using all three options. The current energy shortage has increased the cost of fungicides to the point that they must be used more efficiently while government regulations are also restricting the use of many chemicals. Additionally, there are relatively few diseases which can be controlled to an acceptable economic level with chemicals alone and the current trend is away from dependence on chemicals and toward an integrated control programme.

A. CULTURAL CONTROL

Pythium-induced damping-off and root rot of cotton is a disease of major importance throughout the cotton producing areas of the world. McCarter and Roncadori (1972) found that when seeds were chilled to 10°C for 72 hours after germination had started, *P. irregulare* reduced emergence from 93% to 28% and survival to 18%. In the absence of chilling, 61% emerged and 54% survived and therefore to lower seedling disease, growers are advised to delay planting until soil temperatures reach 20°C at 10 cm depth at 09:00h for three or four days. Under these conditions, fungicides will usually offer sufficient protection to allow an acceptable stand to emerge and survive. In dealing with parasitic stemburn of tobacco, Meurs (1934) stated that plants remaining in the seed bed should not be watered for eight days to allow them to develop a woody stem before transplanting. When these resistant plants were superficially placed in the planting holes, and the holes not completely filled with soil for 5–10 days, they survived well.

To control poinsettia root and stem rot, Tompkins (1950) removed all of the sand from propagation beds before cleaning and filling with new sand. Ferbam was incorporated into the sand, and dusted on the cuttings. These measures produced healthy plants.

Hoitink *et al.* (1976) found that composting hardwood bark infested with *Phytophthora cinnamomi* and *Pythium irregulare* killed the pathogens. They stated that many nurserymen now rely on composting of potting mixes as the control for root rots. In some cases, crop rotations utilising non-hosts have offered some benefit. This practice, which lowers inoculum levels, is frequently of greatest value with pythiaceous fungi having a restricted host range. For example, a two to three year rotation is beneficial with *Phytophthora nicotianae* var. *parasitica*, but of limited value with *Pythium ultimum* which has a wide host range.

Hartley and Merrill (1914) found that treating forest nursery seedbeds with sulphuric acid decreased damping-off caused by *Pythium* spp. without damaging seedlings. The use of this practice became widespread, but was discontinued with the advent of chemicals less irritating to the applicator.

Some interpreted the effect as due to reduced pH, while others believed it was a direct action on the fungus.

Control of littleleaf disease of shortleaf pine has been directed first at reducing economic loss. Since the onset of symptoms occurs at an average age of 35 years, short rotations and harvest of trees prior to this age is recommended while reducing the interval between harvests on littleleaf sites is also suggested. In the most severely affected areas, other pine species are preferred and in situations where it is economically feasible, pine stands are allowed to regenerate to hardwoods.

In peach decline, the various contributing stress factors are additive, and the reduction of only a few produced economic control. Substitution of a combination of mowing and herbicides for mechanical cultivation was the first step. Subsoiling to improve drainage, and the addition of lime to raise the pH value were also practised. Pruning after the food reserves had been transported from the twigs was important, as was the addition of correct soil nutrients. Two major root-damaging factors, nematodes and *Pythium* spp., were evaluated, and it was found not to be advantageous to control both. Since nematode control is in general less expensive than *Pythium* control, it was decided to control the nematodes, and tolerate the damage caused by *Pythium*. Nematicides are now used extensively in the western United States, both as pre- and postplanting treatments whereas in the southern United States, preplant treatments are common, but no postplant treatments are used, the difference in approach being related to land costs in the two areas.

Composting has been investigated and found to be a suitable substitute for chemical controls in the nursery trade. This practice will no doubt increase in the future. For some control measures, such as changing of planting or pruning dates, no cost is involved, whereas in other cases, such as subsoiling, there is some fundamental penalty. In the future, cultural practices which are economically feasible may furnish a degree of control which can be increased to the desired economic level by biological and chemical means.

B. BIOLOGICAL CONTROL

Two approaches have been used for the biological control of pythiaceous fungi: the use of resistant varieties and the use of antagonistic or competitive organisms.

Resistant varieties have been used satisfactorily for many years. The first and most successful dealt with those fungi having restricted host ranges. Currently, there is also germ plasm available for many crops with resistance to fungi such as *P. ultimum*, which have a wide host range.

Resistance to *Phytophthora nicotianae* var. *parasitica* is widely used, and is sufficiently effective to allow planting of tobacco cultivars in soils with high populations of the fungus. As with other crops, when one characteristic, such as resistance, is stressed in a breeding programme, other characteristics, such as yield or quality, are sacrificed to some extent. By combining resistance with crop rotation and chemical control, cultivars which are less resistant but have better yield and quality can be used. Research on effects of various rotations on the size of the populations of these fungi in the soil and improved detection methods are being correlated with disease, making the use of cultivars with lesser resistance practical (Flowers and Hendrix, 1972, 1974). More of this type of research, coupled with integrated control measures, will allow plant breeders to accept less resistance, and improve yield and quality.

Adegbola and Hagedorn (1970) found useful resistance to three species of *Pythium* in *Phaseolus vulgaris* while Ohh *et al.* (1978) described six pea-breeding lines with resistance to *P. ultimum*. Seed of these pea lines can be soaked in a hyphal suspension for 24 hours prior to planting and still produce good stands. Mathre and Otta (1967), on the other hand, found no resistance to the same fungus in the genus *Gossypium* and since resistance to pythiaceous fungi was not available, Bird and Presley (1965) took a unique and profitable approach to the problem of *Pythium* damping-off. Instead of breeding for disease resistance, they bred cotton lines which grew at lower temperatures, thus escaping the disease. These characteristics have been incorporated into commercial cotton cultivars which are now widely used in Texas. The recent literature contains many more such reports, and one can logically expect further examples, as less emphasis or reliance is placed on chemical control.

Much research has been published on the effect of antagonistic or competitive organisms on pythiaceous fungi. In the past, this work had little or no practical application although the observations by Marx (1967, 1973) that some ectomycorrhizal fungi protect pine roots from *P. cinnamomi* were referred to earlier. Zentmyer (1963) moreover found that incorporation of alfalfa meal into the soil at high concentrations controlled *Phytophthora* root rot of avocado. Although in the latter instance control in the field proved erratic, Zentmyer suggested that it depended on the exploding microbial population following incorporation of the meal.

Locke *et al.* (1979) were able to control *Pythium* seed rot of peas by using a seed dressing of *Trichoderma* spp. and they showed that the *Trichoderma* became established in the rhizosphere of treated peas grown in the field. Hoch and Abawi (1979) found that *Corticium* sp. controlled the damping-off of table beets induced by *P. ultimum*, and suggested the possibility of using seed dressings. There is a good probability that this type of approach can be used effectively in control programmes because application of the control agent to the seed reduces the amount required to a level which may be economically feasible.

Bolton (1978) found that the addition of *Trichoderma viride* and a

Streptomyces sp. to two commercial soil-less mixes gave a degree of protection against *Pythium splendens* on pelargonium. This type of approach also has a high probability of usefulness in that the mixes are already available commercially and it would be relatively simple to add the organisms. In a field situation where competing organisms are present it is difficult to establish new ones in high numbers because most or all of the food bases are being utilised. When a biological vacuum exists following sterilisation, however, the establishment of a desirable microflora is much simpler.

C. CHEMICAL CONTROL

Chemical control of pythiaceous fungi has been successfully practised in specialised areas which involve either seed treatments, or high value crops (see Chapter 6). It is generally prohibitively expensive to broadcast chemicals over entire fields at effective rates and Richardson and Munnecke (1964) described a linear relationship between the amount of inoculum in soil, and the amount of fungicide required for effective control. One means of improving the cost effectiveness of fungicides, however, is first to lower the inoculum level culturally or biologically and thereby reduce the amount of fungicide needed for control. Another means of increasing cost effectiveness is to place the fungicide only at the infection court, rather than throughout the soil. This is the basis of the various seedling protection practices based on seed treatment and in-furrow application. Fungicides are applied to seed at rates of less than 1 kg for each 45 kg of seed. This is feasible for most crop situations and gives effective control of pre-emergence damping-off. For those crop-pathogen-environment situations where post-emergence damping-off is a problem, seed treatments may not be sufficient. In-furrow applications, where the fungicide is applied in a band from the seed to the soil surface as the furrow is closed have been more effective, but also require more chemical, thereby reducing the number of situations where they are economically feasible.

Where feeder root necrosis is involved, the chemical must be applied to the entire soil body. With existing crops, incorporation of the chemical into the soil without destroying large numbers of feeder roots becomes a problem, although for chemicals that can be moved into the soil with water, application in conjunction with irrigation is possible. Generally, efficiency is directly related to uniformity of incorporation but the cost of this type of treatment is prohibitive except in extremely high value crops, such as ornamental nursery stock. Zentmyer (1973) was able to control *Phytopthora* root rot of avocado with monthly applications of *p*-dimethylamino-benzenediazo sodium sulphonate (Dexon) at 100 p.p.m. but he doubted the practicality of such a treatment.

Several selective fungicides for pythiaceous fungi are available or in the final testing stages. Some of these are systemic and so can be applied to foliage or in transplant water at lowered rates and later translocated throughout the plant. Two types of problem occur with these materials, however; one, as shown by Richardson (1976), is that they are so selective that they do not control all species present. Pyroxychlor controlled damping-off of peas caused by *P. aphanidermatum* and *P. ultimum*, but not by *P. irregulare* and *P. sylvaticum*. The second problem is the development of resistance to the fungicide by the fungus. It can be overcome or prevented to some extent by using mixtures of fungicides having different modes of action, or by alternating fungicides. While this will not prevent the development of resistance, it will prevent the buildup of large populations of resistant strains.

Finally, the effect of fungicides on non-target organisms should be borne in mind. This is usually considered only in terms of the killing of beneficial organisms, but should be taken a step further. When crops subject to diseases caused by pythiaceous fungi are treated with benomyl for the control of other pathogens, pythiaceous fungi, which are not affected, sometimes move into the partial vacuum left when the competing organisms are killed. These pythiaceous fungi sometimes then become more of a problem than those associated with the initial disease. This problem was documented by Warren *et al.* (1975) on bent grass affected by *Pythium* spp.

Methyl bromide and methyl bromide mixtures have been used for 30 years to control a combination of soil-borne pathogenic fungi, nematodes, insects and weeds. This is economically feasible only on high value crops, generally in situations where the cost can be partially recovered by savings on fungicides, nematicides, insecticides, herbicides and the labour of hand weeding. Munnecke *et al.* (1978) have determined the concentration and time of exposure needed to kill several pythiaceous fungi, and found them to be the most sensitive fungi tested. Approximately 3,300 hectares were treated with methyl bromide in California in 1976. Problems encountered in practical applications of this method have resulted from treatment under improper conditions and re-infestation. Soil should be pulverised, free of undecomposed organic matter, temperatures above about 15–18°C, and sufficient moisture to be in good planting condition. In field applications, the methyl bromide is usually injected and covered by a plastic tarp applied mechanically. The problem of re-infestation can be approached by applying low rates of broad spectrum fungicides during soil preparation, and by sanitation. This should be an excellent opportunity for incorporating antagonists such as *Trichoderma* sp. and *Streptomyces* sp.

IX. Summary

The genera *Pythium* and *Phytophthora* are sufficiently similar to be treated simultaneously in many respects, including much of their pathology.

Mycologically, there is little problem in separating them, and the genera should be maintained separately, rather than combining them. They are not readily isolated using standard mycological techniques, and are usually absent from fungal lists compiled using these methods. *Pythium* and *Phytophthora* species are easily isolated, however, using specialised media and baiting techniques.

While pythiaceous fungi are important primary causal factors of damping-off, this traditional role is too restrictive as a definition of their pathology. They also attack juvenile root tissue of older plants. In this role, they are components of various importance of many decline and replant diseases. While some pythiaceous fungi attack many other fungi, they have little potential as biological control agents. Recent research illustrates their importance in causing diseases without infecting the plants involved and while it is commonly stated that they are responsible for death of plants in low lying areas, or in areas of standing water in fields, this is doubtful. Most injury is probably caused at soil moisture levels favourable for plant growth.

Control of diseases caused by pythiaceous fungi is seldom satisfactory when only chemicals are used. Many excellent cultural and biological techniques exist which can be used in conjunction with chemicals for more satisfactory results. Pythiaceous fungi are essentially world wide in distribution, and will continue to cause problems. The extent of the problems will depend on man's ability to change the conditions necessary for economic levels of disease to occur.

5

Plasmodiophora

An Inter-Relationship Between Biological and Practical Problems

S. T. BUCZACKI

National Vegetable Research Station, Wellesbourne, Warwick, UK

I. Introduction 161
II. *Plasmodiophora*—What is She? 163
III. The Life History of *Plasmodiophora brassicae* 171
IV. The Technology of *Plasmodiophora* Research 186
V. *Plasmodiophora brassicae*—Whence and Whither? 189
Acknowledgements 191

I. Introduction

Plasmodiophora brassicae would be an unsung biological curiosity was the family Cruciferae not such a significant provider of crop plants. It was its impact upon the cabbages and other market garden crops of the species *Brassica oleracea* L. that stimulated the investigations of clubroot disease by the Russian, M. S. Woronin, in the 1800s; an investigation that led ultimately to a masterly description and identification of the causal organism (Woronin, 1877). Clubroot disease had long been important on other cruciferous plants however, and its devastating effects on turnips grown as animal feed were well documented in the eighteenth century (Marshall, 1787). It has probably been a problem for as long as crucifers have been cultivated but its impact has never been greater than it is today because of the very wide variety of cruciferous crops now grown. Although precise figures are impossible to obtain, the annual world crucifer crop probably exceeds 10 million ha and in several countries it has been estimated that clubroot infestation exceeds 10% of the land on which they are grown.

Since the pioneering work of Woronin, studies of clubroot and its causal organism, and the scientific publications consequent upon those studies, have multiplied at an alarming rate. Nonetheless, there was for long no commensurate increase in the effectiveness of procedures for combating the disease. With certain notable exceptions, such as may be found in the work of Nawaschin, Cook and Schwartz and a few other perceptive investigators, the truly important questions concerning the biology of the causal organism were scarcely formulated before the mid-1960s let alone answered. The past ten years or so have seen progress made in unravelling the biology of this challenging organism, that is quite out of proportion to the achievements in the preceding ninety and it is no longer a pipe dream to believe that most of the remaining major uncertainties will be resolved in the next decade. Whether this will result in the hoped for solution to clubroot as a disease is a matter that possibly only time will tell but it is one that this chapter attempts to predict.

I shall seek to show that it is the combination of our inadequate knowledge of the nature and biology of *Plasmodiophora brassicae* and some of the inherent technical problems it presents to the investigator that lies at the base of our inability to alleviate the problems that clubroot disease causes for farmers, growers and gardeners. It is appropriate, therefore, to define briefly the nature of this inability and summarise what can and cannot be done at the present time.

Host plant resistance to clubroot has often been spoken and written of as the objective that pathologists and breeders should be striving to achieve (Crute *et al.*, 1980a) although it has become fashionable to make this claim in relation to a very wide range of plant diseases, irrespective of the existence of other, satisfactory means of control. It is inappropriate to widen the scope of the present brief to a critical analysis of this general tenet, but it is pertinent to see how far clubroot specialists have progressed towards the objective. Crute *et al.* (1980a) demonstrated conclusively that such progress has been very limited and concluded their wide-ranging review with a list of twelve topics that justified priority consideration in research programmes directed towards breeding for clubroot resistance. Whilst some cultivars of turnips (*Brassica campestris* L. var. *rapifera* Metz), especially stubble turnips, show good, although by no means durable resistance in some areas, quite the reverse is true of the closely related leafy vegetables of Chinese cabbage type (*B. campestris* var. *chinenis* (L.) Makino) which consistently prove to be the most susceptible of all hosts. Limited success has been achieved in combining oligogenes in *B. napus* L. for resistance to certain physiologic populations of *P. brassicae* (Buczacki *et al.*, 1975) but cultivars derived from such crosses are of value only in areas where those pathogen populations predominate, and then probably for limited periods of time only. In the important vegetable-yielding species, *B. oleracea*, the story of resistance breeding has been one of unmitigated failure for whilst a few commercial cultivars such as the cabbages Badger Shipper (Walker and Larsen, 1960) and Resista (Weisaeth, 1977) and the kale Verheul (see Buczacki *et al.*, 1975) have displayed limited resistance, their usefulness has been short-lived.

The most recent surveys of the chemical control of clubroot have been those of White and Buczacki (1977) and Buczacki (1978). Whilst the pathogen inoculum may be greatly diminished in the soil by the application of general biocides such as methyl bromide, or nematicides such as dichloropropene, the conventional fungicides sufficiently effective to be used in commercial practice are very limited. Mercury salts, although giving good protection, are prohibited in most countries whilst quintozene and systemic fungicides related to methylbenzimidazole-2-yl-carbamate have provided some control in areas where they have been thoroughly evaluated. Among less conventional fungicidal products, calcium cyanamide (Mattusch, 1978) appears to have caught the imagination of some although its use is not widespread.

The effects of clubroot can be ameliorated by the adjustment of a number of agronomic practices including the improvement of soil drainage and the elevation of soil pH. These have been well documented by Karling (1968) and there has since been little change from the picture as he described it.

With these albeit imperfect and in many cases crude procedures in mind, the biology of *P. brassicae* can be described in the light of current research and thinking.

II. *Plasmodiophora*—What is She?

The classification of an organism has long since ceased solely to be a matter for the esoteric concern of taxonomists, for in indicating the relationships of the organism with others, its taxonomic position should encapsulate a gamut of attributes, of importance to all who seek to understand it. Thus it is with plant pathogens and if there is to be a natural and logical consequence upon the demonstration of Koch's postulates, it is for the causal organism to be identified and set in the natural order of things. Then, by using the time-honoured guidance of precedent, the pathologist may draw upon the experience of others with closely related organisms to suggest ways in which the problem may be studied further and, hopefully, controlled.

The paper by Sparrow (1958) is recognised as one of the most stimulating on the taxonomy of the lower fungi. In it he asked "But what about the Plasmodiophorales? The loose thinking and fruitless argument which have surrounded this group like a miasma are due to a variety of factors, chief of which is the undue weight given to the 'plasmodial' nature of the thallus, to promitotic division figures, and to incomplete and inaccurate observations on critical structures and stages in the life history". To be fair, Sparrow himself, later in his account, did remark on the plasmodial nature of the thallus and the "protozoon (not Myxomycete-like) nuclear division figures", in concluding that the Plasmodiophorales belonged nowhere in particular but closer to fungi than Myxomycetes and in a parallel evolutionary line to Chytridiomycetes, Hyphochytriomycetes and "Phycomycetes" (the "Saprolegniacean-Lagenidian-Peronosporacean galaxy"). The emphasis he laid nonetheless upon the zoospore, "the structure *par excellence* which lies at the very base of any natural system of the lower aquatic Phycomycetes" and Sparrow felt that the four lines came from "radically different progenitors".

In the years that have elapsed since Sparrow's paper was published, the plasmodiophorids* have continued to be surrounded by a "miasma" of argument. In other respects the situation has improved, however, and much more "complete" and "accurate" observations are now available on the life history. The extra dimensions of biochemical information can be added to the morphological data available previously and I shall suggest that the ancestry of the plasmodiophorids may not be so "radically different" from that of certain other groups. I shall also hope to show that the arguments, for so important a plant pathogen as *Plasmodiophora brassicae*, need not be "fruitless".

M. S. Woronin came close to fulfilling the criteria that we now define as Koch's postulates when he recognised the causal organism of clubroot in the 1870s (Woronin, 1877). He then commented upon the biological affinities of his new organism: "Because of the simplicity of its whole structure, *Plasmodiophora brassicae* is a true Protist (according to the definition of Häckel) and therefore stands closest to the Myxomycetes. Cornu has already pointed out that the Myxomycetes are closely related to the Chytridiaceae; and through *Plasmodiophora* the affinity stands out still more plainly. *Plasmodiophora*, in common with the Myxomycetes, possesses a plasmodium that, after a certain time, breaks up into an unbelievably large number of globular spores that later produce myxamoebae. *Plasmodiophora*, however, differs sharply from all other myxomycetal forms in the total absence of a true sporangial membrane and because of its parasitism within another living organism. In every other way, but particularly in its manner of living, *Plasmodiophora brassicae* resembles most closely the Chytridiaceae".

Subsequently, improved methods of study and increasing knowledge prompted periodic reclassification of *P. brassicae* and its allies; changes carefully chronicled by Karling (1968). Whilst there has been fairly broad agreement on which genera belong in the group, their less immediate affinities are much more vague, except insofar as they have been studied traditionally by mycologists rather than zoologists or other biological practitioners. Since most popular textbooks tend to reflect rather than shape

---

* So I shall refer to them; it is difficult enough to place these organisms among their kin without complicating the issue by attempting to assign them to a level in the taxonomic hierarchy.

scientific opinion, it is reasonable to examine recent American and British examples of the genre for an indication of current thinking. Alexopoulos and Mins (1979), in their American text, have the *Plasmodiophora*-like organisms as a Class Plasmodiophoromycetes which, together with the Classes Chytridiomycetes and Hyphochytriomycetes, form one sub-division, the Haplomastigomycotina of the Division Mastigomycota; the second sub-division, the Diplomastigomycotina, comprising the single Class Oomycetes. Webster (1980) may more closely reflect British thinking in following Ainsworth (1973) and placing the Class Plasmodiophoromycetes together with the Classes Acrasiomycetes, Hydromyxomycetes and Myxomycetes as the constituents of the Division Myxomycota. It is of interest that Webster, a mycologist, groups *Plasmodiophora* close to Myxomycetes whilst Alexopoulos, an authority on the latter, groups it with the fungi. No-one seems very willing to give these organisms a home!

In practice, these very different approaches are a reflection of nothing more than general confusion and uncertainty among the biological community and it now seems time to attempt a synthesis that encompasses all pertinent modern research. The most wide-ranging and authoritative modern expression on the phylogeny and interrelationships of zoosporic plant pathogens is that by Barr in Chapter 2 of this volume. He considered all pertinent morphological, biochemical, ultrastructural and physiological attributes in reaching a conclusion that suggested a possible protozoan origin for plasmodiophorids. Barr's evidence was convincing but, like all phylogeny, ultimately speculative. I propose to speculate a little further in the light of recent findings, to highlight certain features of pathological interest and indicate their relevance to practical considerations.

Although Waterhouse (1973) and others have commented that the plasmodiophorids are characterised by a chitinous rather than a cellulosic wall, there was no firm evidence for this in *Plasmodiophora* other than the brief statement by Maire and Tison (1911) that the spore membranes "ne donne pas les réactions de la cellulose chez les Plasmodiophoracées", and the indication by Wisselingh (1898) that a chlor-zinc-iodide test was negative. More recently, Goldie-Smith (1951, 1954, 1956) demonstrated a negative chlor-zinc-iodide test for the sporangial and resting spore membranes (*sic*) of *Woronina* and *Sorodiscus* and it seemed improbable therefore that cellulose was present. Pendergrass (1950) obtained a similar reaction for *Octomyxa*. We have recently confirmed this belief and obtained fairly complete analytical data for the resting spore wall of *P. brassicae* (Moxham and Buczacki, 1983). These data are summarised in Table 1 from which it may be seen that the wall contains approximately 25% of chitin, together with abnormally high levels of protein and lipid. Although glucose was detected in very small amounts, the absence of cellulose was confirmed by a negative reaction in the Schweizer test and a non-cellulosic β-glucan was presumably present

Table 1. Chemical composition of *P. brassicae* resting spore walls (after Moxham and Buczacki, 1983)

Component	Wall dry weight (%)
Chitin	25.1
Other carbohydrates	⩾2.5
Protein	33.6
Lipid	⩾17.5
[Ash residue	10.5]
Total	⩾88.3

therefore. Van Wisselingh and chitosan sulphate tests for chitosan were also negative.

Other pertinent information may be obtained from our analysis moreover, for Vogel (1964, 1965) first demonstrated the value of the type of lysine synthesis pathway as a taxonomic criterion and, as Vaziri-Tehrani and Dick (1980c) pointed out, high lysine:valine ratios, accompanied by generally lower but variable proline:valine ratios, seem characteristic of the DAP (diaminopimelic acid) lysine synthesis pathway whereas proline values higher than lysine are a feature of the AAA (aminoadipic acid) pathway. Our data showed that the spore walls of *P. brassicae* have a proline:lysine ratio of 1.8 and thus probably synthesise lysine by the AAA pathway.

Information is now available for *P. brassicae* on another valuable biochemical indicator of relationship. Loening (1968) was among the first to demonstrate the value of the molecular weight of ribosomal RNA in considering evolutionary trends. Subsequently, Lovett and Haselby (1971) and Porter and Smiley (1979) examined the value of this feature among the fungi and showed that it tended to confirm the groupings based on other features. Fraser and Buczacki (1983) recently compared the ribosomal RNA molecular weight of *P. brassicae* with that of other organisms. A comparison of their results for plasmodiophorids with those obtained by earlier workers for other organisms is shown in Table 2 from which it may be seen that the three genera tested, *Plasmodiophora*, *Sorosphaera* and *Spongospora* have similar values and that these are quite close to those for Chytridiomycetes, Hyphochytriomycetes and Zygomycetes but clearly different from those for Oomycetes, Ascomycetes, Basidiomycetes, Myxomycetes, Acrasiales and Trichomycetes. Barr (Chapter 2) suggested that similarities in the rRNA molecular weights between organisms could be mere chance, but that a distinct difference was a powerful indication of phylogenetic unrelatedness. By this token therefore, plasmodiophorids are distinct from Myxomycetes

Table 2. Ribosomal RNA molecular weights of plasmodiophorids and other organisms[a]

Group	No. of species	25S (mean)	18S (mean)
Protozoans[b]	5	1.40	0.77
Trichomycetes[c]	6	1.34	0.68
Myxomycetes[d]	1	1.33	0.70
Acrasiales[d]	1	1.29	0.68
Oomycetes[d]	2	1.29	0.68
Oomycetes[e]	2	1.27	0.66
Plasmodiophorids[e]	3	1.25	0.66
Hyphochytriomycetes[d]	1	1.24	0.68
Chytridiomycetes[d]	4	1.23	0.68
Zygomycetes[d]	2	1.23	0.67
Zygomycetes[e]	19	1.23	0.65
Ascomycetes[d]	2	1.19	0.67
Ascomycetes[e]	1	1.17	0.65
Basidiomycetes[d]	3	1.20	0.67
Basidiomycetes[e]	3	1.16	0.65

[a] Recalculated to uniform standard *E. coli* molecular weights of 1.009 and 0.534 × 10^6 (see Fraser and Buczacki, 1983).

Data from: [b] Loening (1968, 1969, 1973), [c] Porter and Smiley (1979), [d] Lovett and Haselby (1971), [e] Fraser and Buczacki (1983).

and from all fungal groups except Chytridiomycetes, Hyphochytriomycetes and Zygomycetes. Interestingly, however, they are not very different from some protozoans; but this could be "mere chance"!

Zoospore flagellation is a most important feature in the biology, classification and, probably, phylogeny of zoosporic organisms. Since Koch (1956) and Olson and Fuller (1968) demonstrated the existence of a non-functional or vestigial kinetosome in the uniflagellate Chytridiomycetes and Hyphochytriomycetes, it has been widely supposed that these organisms originated in a biflagellated ancestor. Barr (Chapter 2) subscribed to this view for the Hyphochytriomycetes but found it impossible to support for Chytridiomycetes although conceding that, at present, substantiating evidence for his view is lacking (Barr, personal communication). I do not have such evidence either but cannot at present bring myself to believe that the non-functional Chytridiomycete kinetosome is an embryonic rather than a vestigal feature.

Thus, a biflagellate ancestor for the Chytridiomycetes is required. Klein and Cronquist (1967), in their monumental consideration of thallophyte relationships and evolution, recognised the paucity of information on the plasmodiophorids and grouped them among the Myxomycetes; "more out of convenience than from either absolute conviction or necessity". Elsewhere in their scheme, however, they sought ancestry for the line that led subsequently to the Chytridiomycetes in euglenophyte or pyrrophyte algae. There remains a considerable gulf nonetheless, at least between the modern representatives of these groups and the modern Chytridiomycetes and is not the biflagellate ancestral group with AAA lysine synthesis and lacking cellulose that Klein and Cronquist sought, better met by the ancestors of the modern plasmodiophorids? Ancient organisms, somewhat similar to modern plasmodiophorids but perhaps differing in the possession of a saprophytic mode of existence (but see also p. 171), would fulfil many of the criteria, not only for ancestral Chytridiomycetes, but also, by inference, for any other groups that have arisen from them.

There are, however, certain other features that merit consideration: since the phenomenon was first described by Nawaschin (1899), the taxonomic and general biological significance of the so-called cruciform or promitosis of the plasmodiophorids has often been discussed (Karling, 1968). In this type of nuclear division, the nucleolus persists and is elongated at right angles to the equatorial plate so giving a cross-like appearance to the chromatic material at metaphase (Fig. 1). This type of division is often said to be peculiar to the plasmodiophorids but whilst specialists on the group, like Karling, have dwelt upon it at length, it is somewhat surprising that neither this odd nuclear behaviour nor the group at large, received a mention in the wide-ranging account by Kubai (1978) of mitosis and fungal phylogeny. Excellent electron micrographs of the stages in cruciform division have nonetheless now been published for *Polymyxa* by Keskin (1971), for *Sorosphaera* by Dylewski *et al.* (1978) and for *Plasmodiophora* by Garber and Aist (1979a), and clearly it is a unique feature that sets the plasmodiophorids apart from any other organisms, including protozoans.

Barr believes, however, that such factors as the presence in plasmodio-phorids of plasmalemma-associated rootlet microtubules of protozoan type (although conceding that the rootlet microtubule system is simpler than that of any protozoan so far studied), the dense chromatin in the nucleus and the massive openings in the nuclear envelope in addition to nuclear pores, mitigate against a suggestion of any close relationship between plasmodio-phorids and Chytridiomycetes. He also believes that the penetration stachels of *Plasmodiophora* and *Polymyxa* bear similarities to the modified trichocysts of protozoans such as *Paramecium* although in this connection it should be noted that Barron (1980) recently reported a similar structure in the Oomycete *Haptoglossa mirabilis* Barron, while Dodge (1973) has also described trichocysts in algae.

The relationship of plasmodiophorids to Chytridiomycetes and other groups is clearly not settled yet! Irrespective of this, however, few would argue that they do seem to lie either among the most primitive of fungi or to

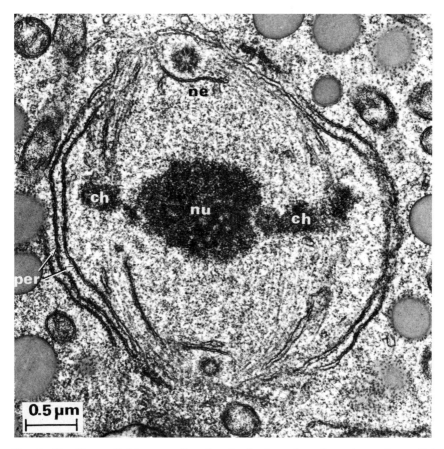

Fig. 1. Cruciform division: median longitudinal section through a plasmodial nucleus at metaphase of mitosis. Courtesy of R. C. Garber (from Garber and Aist, 1979a). Abbreviations: ne, nuclear envelope; nu, nucleolus; ch, chromatin; per, perinuclear endoplasmic reticulum.

be the modern representatives of the most advanced of fungal ancestors. A further important factor is pertinent in this context moreover. There is some evidence that the endobiotic plasmodium of *P. brassicae* ingests material from the cytoplasm of the host cell. Williams and McNabola (1967) drew attention to this possibility and speculated that such host material, contained in vacuoles in the parasite cytoplasm, was subsequently incorporated in the wall of the resting spore. While there is some doubt over the latter suggestion (see later), we have also obtained photographs that appears to show ingestion by secondary plasmodia (Fig. 2). Olive (1975), deriving his

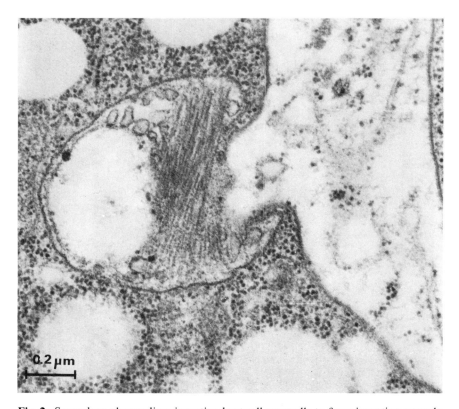

Fig. 2. Secondary plasmodium ingesting host cell organelle to form ingestion vacuole
(Buczacki and Moxham, unpublished work).

classification of lower organisms from the earlier proposals of Copeland
(1956), Olive (1969) and Whittaker (1969), included the *Plasmodiophora*-like
organisms in a sub-Phylum of the Kingdom Protista, quite separate from the
Kingdom Fungi into which the Chytridiomycetes (and other "more conventional") fungi were placed. He concluded, however, with a familiar note; that
the plasmodiophorids were so categorised "primarily as a matter of
convenience because . . . their actual relationships are obscure". Olive did
not consider them to be related to the Mycetozoa (one of the other two sub-
Phyla grouped within the Phylum Protista) but both Whittaker and Olive
laid especial emphasis on the feature of holozoic nutrition and Olive used this
as the main justification in advocating that the Phylum that contained the
Mycetozoans and their associates be transferred from Whittaker's Kingdom
Fungi.
    If holozoic nutrition is truly present in the plasmodiophorids and if it is

accepted as being absent from the fungi, then the evidence is increasingly in favour of considering the plasmodiophorids as a genuine linking group between fungi and their ancestors. Holozoic nutrition, however, would seem to be a requirement of free-living rather than endobiotic organisms and there is thus some potential for speculation on the significance of the universal occurrence of parasitism among the plasmodiophorids. It is now frequently claimed that the parasitic mode of life is not always the advanced state that was once imagined. Savile (1968) for instance, commented, "parasitism is an old, perhaps original characteristic of the fungi, rather than a belated acquisition . . . it was as parasites that the fungi left the water protected by the tissues of their hosts". It is not unreasonable therefore to suggest that *Plasmodiophora* is in the process of losing holozoic in favour of absorptive nutrition, having adopted the parasitic habit. Whether this is a recent development or a very ancient and inexorably slow process is impossible to guess in the absence of a fossil record. It is of interest that *Woronina* and the closely related genus *Octomyxa* are the plasmodiophorids in which holozoic nutrition has been most clearly observed, but these two are confined to algae and Oomycetes rather than higher plant hosts and might on this token be imagined to have been the earliest of the group to adopt parasitism (see Chapter 8). This is another characteristically plasmodiophoridean conundrum. It is finally worth noting nonetheless that in common with other, apparently obligate parasites such as the Uredinales, the plasmodiophorids have not become so specialised as to be incapable of growth on a nutrient medium: Diriwächter *et al.* (1979) reported restricted development of *Spongospora subterranea* (Walr.) Lagerh. var. *subterranea* Tomlinson on a range of culture media including water agar.

III. The Life History of *Plasmodiophora brassicae*

I have not counted the number of published versions of the life cycle of *P. brassicae*; Karling (1968) chose to illustrate three examples but there are many more. Up to the time that Karling published his account, almost all of the evidence had accrued from studies made using the light microscope, usually with fixed and stained material although with certain features drawn from the examination of living specimens. Since then the electron microscope has made major contributions although when Aist (1978) posed some challenging questions regarding the life cycle, the full impact of ultrastructural investigations had only just begun to be realised.

It is of interest to note the varying emphasis given to the diploid phase in successive published versions of the life cycle. Using the terminology of Burnett (1975), we have progressed from the inference by Prowazek (1905) of a haploid life cycle with a very short diploid phase, to the classic haploid–

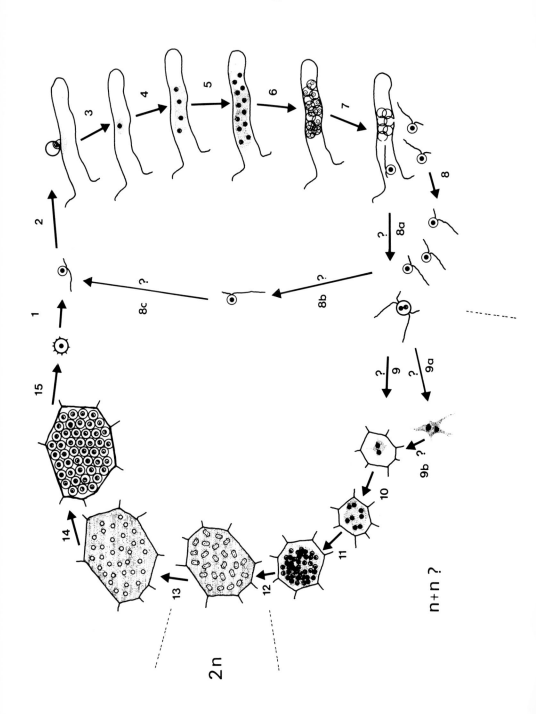

diploid life cycle of Cook and Schwartz (1930) and the aberrant one of Heim (1955), to the haploid dikaryotic cycle of Ingram and Tommerup (1972). The latter version is presently considered the most plausible and is based largely on the authors' own detailed light microscope studies but including a synthesis of other workers' observations too. This life cycle is shown in Fig. 3 and will be used as the basis for the detailed consideration that follows, for no convincing evidence has been produced in the years since it was published to refute any of the suggestions whilst considerable information has been obtained to confirm it.

That for a century after its identification, pathologists struggled to control a disease whilst the life history of its causal organism remained so imperfectly known is a bizarre state of affairs, probably without parallel in plant pathology. It is a fact that will be alluded to again in this account.

It is appropriate to commence this consideration of the life cycle with the resting body, hitherto referred to as a cyst or a resting spore, but because of possible confusion with other life cycle stages termed cysts, Karling (personal communication) has suggested that the term resting spore be adhered to. As the vehicle by which *P. brassicae* possibly achieves its considerable longevity in the absence of cruciferous crops, the resting spore has received a great deal of attention. And as it is the only phase of the life cycle of any persistence outside the host tissues, the resting spore has been the target, deliberately or incidentally, of most direct measures for clubroot control.

The ultrastructure of resting spore development from secondary plasmodia was described by Williams and McNabola (1967). One of the most

---

Fig. 3. The life cycle of *Plasmodiophora brassicae*, based essentially on the version published by Ingram and Tommerup (1972) but with possible additional features. All stages for which firm evidence is lacking are denoted "?" Not to scale. **1.** Germination of uninucleate haploid resting spores in soil to produce a single, biflagellate, uninucleate, primary zoospore. **2.** Primary zoospore encysts upon root hair or epidermal cell of host plant and injects a uninucleate plasmodium. **3–5.** Plasmodium develops within host cell and by mitotic nuclear division becomes multinucleate. **6.** Plasmodium cleaves to form a mass of zoosporangia. **7.** Uninucleate, biflagellate secondary zoospores differentiated within zoosporangia. **8.** Secondary zoospores released from zoosporangia, through root hair wall into the soil where they possibly fuse in pairs. **8a.** Secondary zoospores migrate to base of root hair, possibly to encyst and penetrate further into root tissues, possibly after fusing in pairs. **8b–8c.** Possible "microcycling" by secondary zoospores, reinfecting root hairs without meiosis or resting body formation. **9.** Formation of binucleate (dikaryotic?) secondary plasmodia in cortical cell of root either by direct penetration of binucleate zoospore fusion body or possibly via a myxamoebal stage **(9a–9b)**. **10–11.** Secondary plasmodium develops within host cell and by mitotic nuclear division becomes multinucleate (stage of root gall formation). **12.** Karyogamy in secondary plasmodium. **13.** "Akaryote" stage (meiotic prophase) in secondary plasmodium. **14.** Following meiosis, plasmodium cleaves to form resting spores. **15.** Resting spores liberated into soil as host root decays.

intriguing suggestions arising from their paper was that the layer of the spore wall comprising the spines (and to which attention had first been drawn by Yukawa (1957)) was derived from "granular aggregates" of the "inter-sporangial matrix" which they believed to comprise host plant material. Williams and McNabola were erroneous in considering this granular, spiny layer to be the outermost of the wall layers (see below) but their contention that it was host-derived merits careful consideration. They stated that during the active growing phase of the secondary plasmodium, "the plasmodial envelope appeared to invaginate, ingesting host cytoplasm and organelles within vacuoles". Whilst Williams and McNabola's illustration of this is not entirely convincing, their conclusion seems valid for we have also obtained images of this process (Fig. 2) and there seems little doubt that host mitochondria and, most tellingly, plastids, are visible within the plasmodia in Williams and McNabola's pictures. The taxonomic significance of this holozoic mode of nutrition has already been mentioned in this chapter but its biological significance in spore wall genesis is much more of an open question.

Williams and McNabola believed that groups of vacuoles, containing the ingested host organelles, aggregated around the nuclei of the plasmodia and merged with each other. The vacuolar membrane thus became the sporangial membrane, the host organelle material was deposited upon it and sub-sequently the inner part of the spore wall was formed between the sporangial membrane and the deposit of host organelle material (see Fig. 4). We have since shown (Buczacki et al., 1979) that the deposition of material from the interspore matrix is a two-stage process, a coarsely granular material being deposited first, followed later by a cobweb-like mesh of fine fibres (Fig. 5). Most of the material comprising these two outer parts of the wall is believed to be protein and lipid while the inner region is a fairly conventional, predominantly chitinous wall (Moxham and Buczacki, 1983 and Fig. 6). Whilst the protein and lipid do seem to be derived from the contents of the vacuoles as Williams and McNabola suggested, they have clearly progressed further down the digestive process than still to be regarded as "granular aggregates" of "the intersporangial matrix". It remains moreover a curious, probably unparalleled, aberration that the ingestion vesicles should function also as cleft vesicles. Very few ultrastructural studies have been made of the delimitation of resting spores of the plasmodiophorid type. If Williams and McNabola were correct in their interpretation, however, the cleavage process is quite different from that during the development of zoospores in *Plasmodiophora* (see below) or in other zoosporic organisms where the cleft vesicles arise *de novo* in the cytoplasm, some of them at least being derived from the vesicles that surrounded the developing flagella (see, for example, Lessie and Lovett, 1968). Braselton and Miller (1978) reported briefly the ultrastructure of resting spore formation in the plasmodiophorid

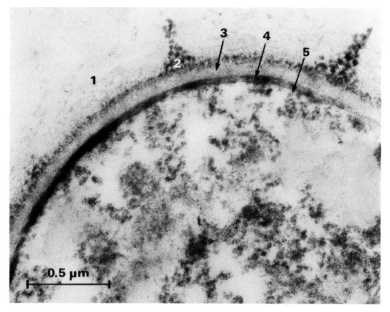

Fig. 4. Resting spore wall: transverse median section indicating the layers described by Moxham and Buczacki (1983). The composition of the layers is believed to be: **1,** proteinaceous fibres; **2,** lipoid granular material in a proteinaceous fibrous matrix; **3,** chitin; **4,** chitin, possibly with phospholipid components; **5,** spore membrane.

Sorosphaera veronicae Schroeter, in which the incipient spores were delimited in the plasmodium by sheets of plasma membrane that also seemed to arise *de novo* in the cytoplasm. Indeed the host–parasite interface of *S. veronicae* with its host *Veronica persica* Poir had earlier been shown by Braselton and Miller (1975) to comprise a single unit membrane, presumably the parasite plasma membrane. In this respect it was similar to the interface of such other non-mycelial parasites as *Olpidium brassicae* (Woronin) Dang. (Temmink and Campbell, 1968, 1969), and quite unlike the seven-layered structure that Williams and McNabola (1970) reported for *Plasmodiophora*. They described the disintegration of the outer part of the plasmodial envelope at the onset of sporogenesis and thus the ingestion vesicles were bounded by a three-layered structure that was later observed forming the sporangial membrane. It is of some interest, however, that developmental continuity between the unit membrane of developing zoospores and the similar unit membrane of the plasma lemma is well recorded (see, for example, Gay and Greenwood (1966) on *Saprolegnia*); it is the way in which the two are connected that is so peculiar about the *Plasmodiophora* resting spore.

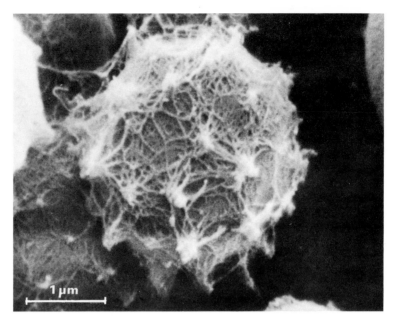

Fig. 5. Resting spore morphology: resting spore *in situ* in host cell, showing surface covered with proteinaceous fibres. From Buczacki *et al.* (1979).

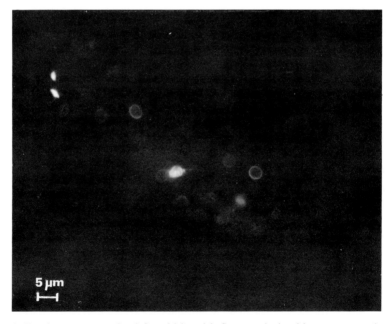

Fig. 6. Resting spores stained for chitin with fluorescein isothiocyanate conjugated wheat germ agglutinin and viewed under ultraviolet illumination to indicate localisation of chitin in spore wall. From Moxham and Buczacki (1983).

The existence of a specialised region of the wall associated with resting spore germination has been recognised for many years (see, for example, Macfarlane, 1970) but the first observations with the electron microscope of what could reasonably be termed an exit pore were reported by Kole and Gielink (1962). Subsequently a thickening of certain layers and a thinning of others in this region were shown to occur (Buczacki et al., 1980). The details of the chemistry and significance of these differences in the wall in relation to the actual release of the zoospores have not yet been elucidated.

The most complete description of the internal cytoplasmic structures of the resting spore was also the report by Williams and McNabola although they made little comment on the nuclear cycle at this stage of development, other than noting the disappearance of the nucleolus at the onset of sporogenesis, a feature observed also in the related plasmodiophorid, *Sorosphaera veronicae* (Braselton and Miller, 1973). The absence of centrioles from resting spore nuclei has been noted by Garber and Aist (1979b), and we have confirmed this observation. Their fate is not known and as the ultrastructure of resting spore germination has not yet been described in detail, it is not known when they reappear as a prelude to flagellar development (Fig. 7). Indeed, whilst

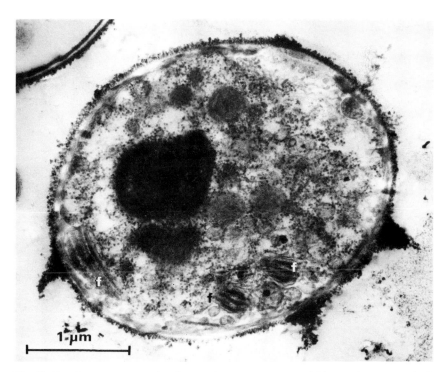

Fig. 7. Resting spore germination; primary zoospore within resting spore body shortly before germination (Buczacki and Clay, unpublished work). Abbreviation: f, flagellum.

the germination of resting spores has been observed with the light microscope (see, for example, Macfarlane, 1970), the only published electron micrographs of sections of the primary zoospores appear to be those of Aist and Williams (1971) although little detail was revealed of the zoospore in its free-swimming phase.

The encystment of the primary zoospore on a root hair and the subsequent penetration of the root hair wall are now too well known to justify repetition (Aist and Williams, 1971). It remains, nonetheless, a remarkable process,

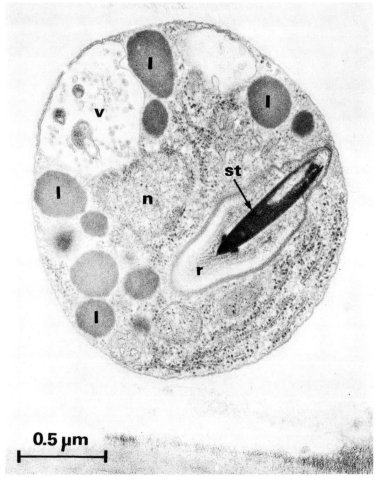

Fig. 8. Primary zoospore cyst showing location of the penetration stachel within the rohr. Courtesy of J. R. Aist (from Aist and Williams, 1971). Abbreviations: n, nucleus; l, lipid droplet; v, cyst vacuole; st, stachel; r, rohr.

matched only by the previously recorded penetration of sugar beet roots by *Polymyxa betae* (Keskin and Fuchs, 1969). The appearance of the penetration body is shown in Fig. 8 and a key stage in the process in Fig. 9.

The development of *P. brassicae* within a root hair or epidermal cell has not yet been fully described but sufficient is now known to enable a fairly complete picture to be drawn. Aist and Williams (1971) illustrated a root hair cell containing several uninucleate "amoebae" (Fig. 10), each with the same seven-layered host–parasite interface previously described for secondary

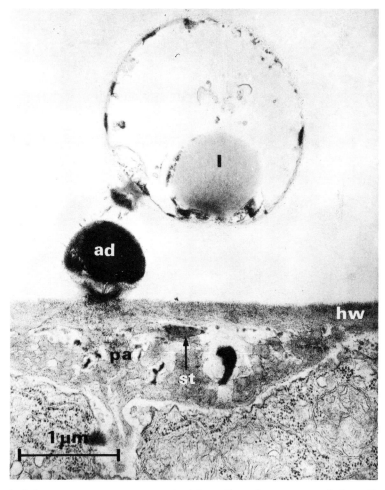

Fig. 9. Primary zoospore cyst attached by adhesorium to root hair wall after penetration of the stachel into the host cell. Courtesy of J. R. Aist (from Aist and Williams, 1971). Abbreviations: l, lipid droplet; ad, adhesorium; hw, host cell wall; st, stachel; pa, papilla.

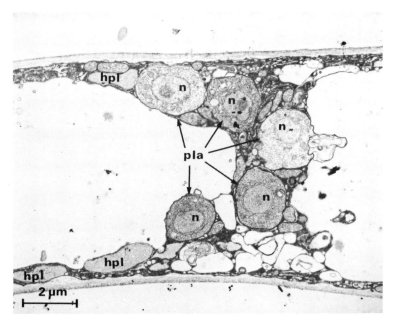

Fig. 10. Uninucleate primary plasmodia in host root hair. Courtesy of J. R. Aist (from Aist and Williams, 1971). Abbreviations: n, nucleus; hpl, host plastid; pla, plasmodium.

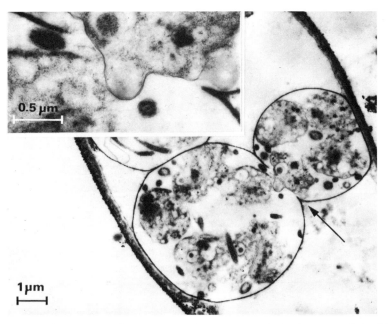

Fig. 11. Developing secondary zoospore in zoosporangia in host root hair. Note contiguity between adjacent zoosporangia (arrowed) and amoeboid movement (inset) of zoospore from one zoosporangium to the other (Buczacki and Clay, unpublished work).

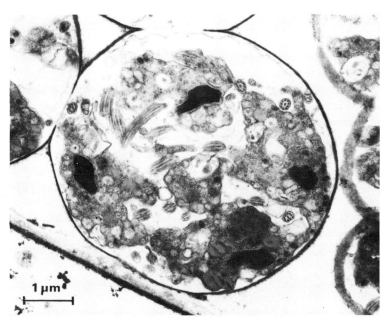

Fig. 12. Mature secondary zoospores in zoosporangia in host root hair. Note numerous longitudinal and transverse flagellar sections (Buczacki and Clay, unpublished work).

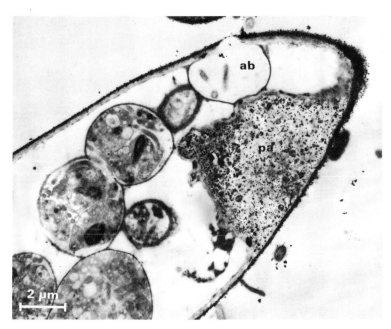

Fig. 13. Mature secondary zoospores in zoosporangia in host root hair. Note nearly median section through the aborted zoosporangium (ab) that forms the exit hole through the host wall. Note also the penetration papilla (pa) through which the primary infection took place (Buczacki and Clay, unpublished work).

plasmodia. Whereas the zoospore cyst and the very young plasmodia were anucleolate, a nucleolus was reported to develop later. Further stages in the differentiation of the plasmodium into zoosporangia and zoospores are shown in Figs. 11–13 and in Fig. 10 of Chapter 1.

The events consequent upon the maturation of the secondary zoospores represent the greatest remaining hurdle to our total appreciation of the life history of *Plasmodiophora brassicae*. Zoospores have been seen swimming freely in the lumen of the root hair by many workers with the light microscope; they have been observed in the process of liberation through the wall of the root hair cell into the external environment by both light (Samuel and Garrett, 1945) and transmission electron microscopy (Fig. 13) and they have been observed by electron microscopy encysting on the inside of the root hair cell wall in a manner analagous to that adopted by the primary zoospore (Aist and Williams, 1971). Apart from a much vaunted photograph obtained on the electron microscope from a shadowed preparation by Kole and Gielink (1961) and purporting to show a quadriflagellate object derived from the fusion of two zoospores, nothing further is known of the life cycle until the stage when small secondary plasmodia are visible in the cortex of the root tissues. This is not to ignore, however, that the literature is replete with speculative accounts of the intervening events, most of them having been discussed by Karling (1968). The most recent addition to this speculation was the contribution by Dekhuijzen (1979) who very circumspectly suggested that structures he observed on the transmission electron microscope represented a myxamoebal phase of the life cycle, this being the form in which *P. brassicae* penetrated from the outer tissues into the cortex. This is an old idea (see, for example, Fedorintschik, 1935) and Dekhuijzen's evidence was no more credible than any that had gone before; he conceded that his "myxamoeba" in a cortical cell could have been a host nucleus.

At the present time therefore there is no conclusive evidence either to refute or support the view that at least some of the secondary zoospores are liberated into the soil, fuse in pairs and thus function as gametes in establishing a dikaryon. It also seems likely that some zoospores remain in the root hair lumen, again possibly fusing in pairs, before entering the main body of the root at the hair base. The fusion in the soil of secondary zoospores of different genetic origins is indeed the only really plausible way that *P. brassicae* could maintain an outbreeding potential. The alternative would be the fusion of genetically different primary or secondary plasmodia within the host tissues, but it might be expected that firm evidence for so readily recognisable a process would have been produced by one of the army of conscientious light microscopists who have studied the problem in the past. That evidence is lacking.

It is tempting to suggest that the development and maturation of the secondary plasmodia in clubroot galls are the aspects of the life cycle most clearly understood. They are certainly the easiest to observe both with the light and electron microscopes and have been studied extensively. In broad outline, the sequence of events seems to be as Williams and McNabola (1967) described it. There are three major unexplained features, however. Firstly, what is the significance of plasmodial cleavage vis à vis plasmodial fusion (see the discussion by Dekhuijzen (1975)). Secondly, what is the means by which plasmodia "spread" through the host tissues? Is there dynamic movement of the plasmodia as many of the earlier investigators believed, or is spread solely by division of the plasmodia concurrently with the division of the host cells? (see Fig. 14). It is of interest that Yukawa (personal communication), one of the most experienced of investigators and the first to examine *P. brassicae* with the electron microscope, considered this the most taxing of the problems still awaiting a solution. Thirdly, what is the significance of the encystment of plasmodia, observed by several investigators? (Fig. 15). Very recently (Garber and Aist, 1979a,b; Buczacki and Moxham, 1980) ultrastructural evidence has been obtained to confirm the Ingram and Tommerup (1972) version of the nulcear cycle in secondary plasmodia. It has been shown that the commonest, cruciform nuclear divisions are mitotic (see Fig. 1), that karyogamy occurs at this stage (Fig. 16) and that the akaryote state (Fig. 17), when the nucleolus disappears and the chromatin becomes dispersed in the nucleoplasm, is the prophase of a first meiotic division, immediately preceding sporogenesis (Milovidov, 1931). Examination of the above mentioned papers nonetheless leaves some tantalising questions. Firstly, what is the relationship between nuclear volume and ploidy? Buczacki and Moxham (1980), in their observations of karyogamy, commented that, as a result of fusion, it might be expected that populations of nuclei would be observed with approximately twice the volume of those seen at interphase. These were not found. Moreover, Garber and Aist (1979b) stated that while interphase nuclei had a pole to pole distance of 3–4 μm, the diameter by prophase I was only 2 μm and by prophase II only 1 μm. These differences, representing at least a 30-fold volume change, emphasise that the relationships between nuclear volume, the nuclear cycle and ploidy remains elusive. Secondly, if the akaryote state in secondary plasmodia is a meiotic prophase, what of the akaryote state that Buczacki *et al.* (1980) observed in primary plasmodia? Evidence is required to show whether the intriguing suggestion, sometimes put forward, of a meiosis at the primary stage, is at all possible. Thirdly, and most intriguing of all, from whence come the nuclei that fuse in pairs in the secondary plasmodia—is there a true dikaryon? It is interesting moreover to note the comment by Garber and Aist (1979b) concerning the chromosome number of *P. brassicae*; because a "reticulate mass of chromatin" rather than discrete chromosomes is present at meiotic metaphase, they believed the many published estimates of chromosome number based on light microscopy to be of very doubtful validity. They suggested that the reconstruction of first

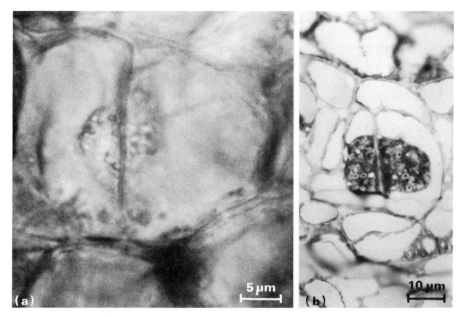

Fig. 14. Plasmodia associated with developing cell wall of host. (a) Living specimen, (b) resin-embedded thin section stained with methylene blue/azure II/basic fuchsin (Buczacki, unpublished work). Does this represent movement by a plasmodium through the cell wall or is the new wall laid down through the plasmodial body?

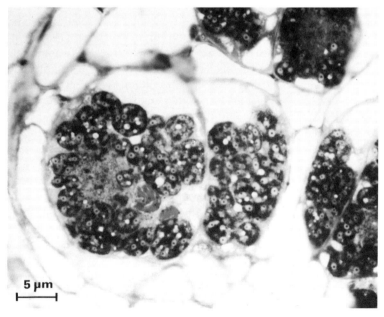

Fig. 15. Is this what has been interpreted as plasmodial encystment, or is it merely a section through a many-lobed plasmodium? Resin-embedded thin section stained methylene blue/azure II/basic fuchsin (Buczacki, unpublished work).

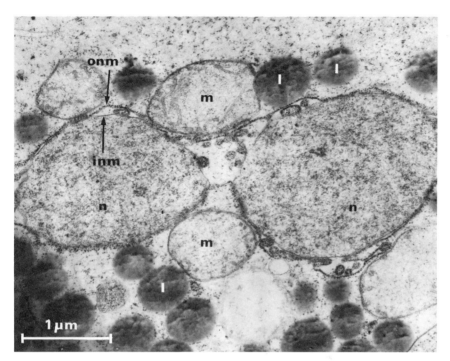

Fig. 16. Karyogamy in secondary plasmodia (from Buczacki and Moxham, 1980). Abbreviations: n, nucleus; l, lipid; m, mitochondrion; onm, outer nuclear membrane; inm, inner nuclear membrane.

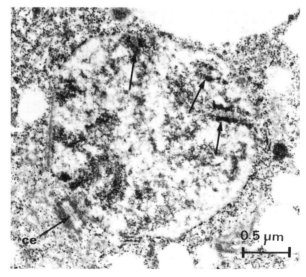

Fig. 17. "Akaryote stage"; plasmodial nucleus at prophase I of meiosis showing segments of at least three synaptonemal complexes (arrows) and paired centrioles. Courtesy of R. C. Garber. Abbreviation: ce, centriole.

meiotic prophase nuclei and the counting of synaptonemal complexes might
be a more fruitful approach. Harris *et al.* (1980) have since used this
technique very successfully to obtain a haploid chromosome count of 33 for
Sorosphaera veronicae and more recently Braselton (1982) obtained a count
of 20 synaptonemal complexes for *P. brassicae*, both numbers considerably
in excess of any of the previous estimates (Karling, 1968).

Overall, we do seem to be making slow but quite definite progress towards
the confirmation of the Ingram and Tommerup version of the life cycle in all
its broad features. Even when this is complete, however, some nagging
doubts must remain. What is the significance of the infection by *P. brassicae*
of non-crucifers? Since the first reliable report by Webb (1949), a number of
investigators have studied the phenomenon; all have observed the presence
of primary plasmodia, zoosporangia and/or zoospores in the root hairs or
epidermal cells of plants representative of a wide range of families. None
have observed any secondary stages of the life cycle. Whilst the absence of
galling is fairly readily explained by the absence from such plants of the auxin
precursors responsible for gall development in crucifers (Butcher *et al.*,
1974), the absence of secondary plasmodia from non-galled root cortex may
simply be a matter of investigators not looking carefully enough. It would
seem unlikely that having progressed satisfactorily to the stage of secondary
zoospore release, this exercise should be so much wasted effort. Are
secondary plasmodia and resting spores produced in non-crucifers in very
small numbers? Do the secondary zoospores from non-crucifers migrate to
cruciferous roots to infect them? Or do the secondary zoospores function in
the same way as primary zoospores in reinfecting root hairs or epidermal
cells and so perpetuating a micro-cycle? If the latter argument is continued to
its conclusion it is possible to consider the role of *P. brassicae* as a parasite of
wild plants. Is the infection and galling of cruciferous roots an abberant
process, occurring primarily on cultivated plants, with resting spores being
produced only because of some peculiar chemical characteristic of the roots
associated with the galling process? Is "wild" *P. brassicae* a much less
significant pathogen of the roots of a wide range of plants, progressing from
plasmodium to zoospore to plasmodium (cf. *Woronina*, Chapter 8) and only
rarely, in unfavourable conditions, forming resting bodies? More than a
century after Woronin's studies, some very challenging questions about the
life cycle remain unanswered.

IV. The Technology of *Plasmodiophora* Research

Whilst clubroot presents problems and frustrations to the crucifer grower, so
Plasmodiophora presents problems and frustrations to the research worker. I

believe that these two situations are not mere coincidence for this organism is so remarkably unco-operative in its response to many of the tests and procedures used with more conventional fungal pathogens that anyone wishing to study it and establish control measures must devise new techniques at almost every turn.

It has become a conventional preface to any description of the plasmodiophorids to state that they are unculturable, the report of the *in vitro* culture of *Plasmodiophora* by Jones (1928) usually being dismissed as erroneous and resulting from confusion with contaminating organisms. As with several obligate biotrophs, however, the dual culture of the organism with callus tissue of its host plant has proved possible with *P. brassicae*. This was first achieved by Yukawa (personal communication) in the 1950s but was not consistently successful until later (Strandberg *et al.*, 1966; Ingram, 1969a). Subsequently, Sacristan and Hoffman (1979) infected sterile stem embryo cultures of *Brassica napus* in a liquid medium without hormones, and also sterile callus derived from single cells in suspension in a similar but supplemented medium, using an inoculum of well-washed resting spores. The procedures for the dual culture of *P. brassicae* and its hosts have recently been summarised by Buczacki (1980) but despite the optimistic note sounded by Williams *et al.* (1969) in their paper, little practical use has been made of the technique. A few workers have used callus-produced resting spores as contaminant-free inoculum while Keen *et al.* (1969) and Dekhuijzen (1975) have isolated plasmodia from callus and studied aspects of their behaviour and physiology. Attempts to induce such isolated plasmodia to proliferate in a culture medium have been unsuccessful (Buczacki and Moxham, 1979). Tommerup and Ingram (1971) first observed the germination of resting spores *in situ* in callus tissue but later (Ingram and Tommerup, 1972), showed that this could occur in galled roots too. The technique has also been used to shed some light on the biochemistry of the galling process; Dekhuijzen and Overeem (1971), for instance, confirming earlier observations that infected callus, unlike healthy tissue, was independent of an exogenous supply of cytokinins and hormones for its continued growth. They were uncertain of the origin of the endogenous growth substances in infected tissues but drew attention to the observation (Ingram, 1969b) that when the parasite had been lost from a callus clone, that clone did not retain the ability to grow independently of an external supply of such substances. Recently, in an unexpected and most interesting report, Diriwächter *et al.* (1979), described the limited development in culture of *Spongospora subterranea* var. *subterranea* on a range of synthetic media, including water agar. We have recently confirmed these findings but have been unsuccessful in extending the technique to *P. brassicae*.

Whilst the practical problems of field experimentation with clubroot disease have long been recognised and known often to be associated with the commonly patchy occurrence of the problem in affected fields, little attempt has been made to overcome these difficulties. The artificial infestation of field soil to obtain a uniform inoculum is not easy and the rotavating of galled roots into the land, the inoculation of transplants by dipping the roots into a resting spore suspension and the inoculation of seeds, have all proved difficult and only partially successful. Moreover, the assessment of the degree of infestation present in field soil has also proved problematic, whether the information is required for advising farmers and growers or for research purposes. The collection of field soil samples which are subsequently sown with seeds of a test plant under more or less controlled conditions has hitherto been the most widely practised procedure (see Melville and Hawken, 1967). Too little attention has been paid, however, to the optimum method of sampling, to the incubation conditions and to the test plant used, although Chinese cabbage is now generally recognised as widely susceptible and sensitive (see Buczacki et al., 1975).

Direct methods of measuring spore numbers have been few; Buczacki and Ockendon (1978) described a procedure by which the resting spores could be extracted from soil and then counted electronically but the technique, although accurate, was dependent for its extensive use on the availability of specialised equipment. Similarly Brach and Crête (1981) devised a capacitance measuring technique to determine the numbers of spores in inoculum suspension. As an extension of such procedures, however, and as a desirable pre-requisite to the artificial inoculation of plants, a method of assessing resting spore viability is desirable. No reliable method is presently available although several procedures have been investigated, most notably those depending on the fluorescence characteristics of spores treated with various fluorochromes. Budzier (1956) adapted the acridine orange procedure of Strugger (1949) and claimed to be able to differentiate between the red-fluorescing dead spores and the green fluorescing live ones. We did not find the technique satisfactory (White and Buczacki, 1979) and have since investigated other fluorochromes including fluorescein diacetate and have found them similarly unreliable. The disadvantage of being unable to quantify inoculum viability is considerable and whilst significant effort has been made in recent years to define the optimum environmental conditions required for maximum disease expression (see, for example, Buczacki et al., 1975, 1978; Dixon, 1976; Horiuchi and Hori, 1980), those experimenting with clubroot disease must perforce accept a situation vis à vis the inoculum, that most plant pathologists would find intolerable.

At a more fundamental level, investigations of the cytology and cyto-genetics of Plasmodiophora have been far from easy. Whilst the secondary plasmodia and incipient resting spores in galled root tissue can be prepared relatively easily for the light and transmission electron microscopes, the nuclei are small and not always responsive to stains such as Feulgen, the chromosomes are apparently not discernible at all with the light microscope and on the electron microscope only "reticulate masses of chromatin" (Garber and Aist, 1979b) can be determined, thus making conventional chromosome counts impossible to achieve and hindering appreciation of the life cycle. Even when Braselton (personal communication) counted synaptonemal complexes in serially sectioned material, the observations were very much more difficult and the complexes very much less distinct than in *Sorosphaera*. Moreover the resting spores are unresponsive to attempts at stimulating their germination (Macfarlane, 1970), meaning that primary zoospores are not readily available in quantity for study.

Studies of resistance to clubroot in host plants were for a long time hampered by the lack of a generally available set of differential host plants to facilitate appraisal of the spectrum of pathogenicity in the pathogen. Whilst the European Clubroot Differential Set (Buczacki *et al.*, 1975) has to some extent improved the situation in that comparison is now more immediately available between studies made in different parts of the world, the genetics of host–parasite interaction are still imperfectly understood. Many workers have found moreover that particular test results are unrepeatable and it seems that the resistance, even in the hosts of the differential set, is not always differential! (Crute *et al.*, 1981).

V. *Plasmodiophora brassicae*—Whence and Whither?

Where has the accumulated knowledge outlined above brought us? Are we nearer now than we were a few years ago to providing the crucifer grower with an answer to clubroot? If, as I believe, the taxonomic affinities of plasmodiophorids seems to be as discussed earlier, we should no longer be surprised (Buczacki and Cadd, 1976) that fungicides with especial activity in controlling diseases caused by Oomycetes should be ineffective against *P. brassicae*. Conversely, nor should it now seem as strange, as once it did (Jacobsen and Williams, 1970), that *P. brassicae* is affected by such compounds as benomyl, known either to have activity against most fungi outside the Oomycetes or to be fairly specific for certain groups within the Ascomycete-Basidiomycete assemblage. Whilst the specificity of fungicidal activity is under highly diverse controlling mechanisms (Kaars Sijpesteijn, 1977), it is clear that of the many plasmodiophorid attributes described earlier, the presence of the chitinous cell wall may go part of the way towards explaining the relative susceptibility of resting spores to particular fungicides. Investigators should be less willing to reject as potential clubroot controlling agents, chemicals known to be limited in their effectiveness to Ascomycetes and Basidiomycetes, whilst, conversely, they should be cautious in expending too much effort in evaluating the many candidate materials that in recent years have been developed for the control of Oomycete-caused diseases. Nonetheless, as a "border-line" organism, *Plasmodiophora* may lend itself, in the short term at least, amenable to partial control by materials of more diverse biocidal activity such as insecticides, acaricides or nematicides.

The advent of the European Clubroot Differential Set (Buczacki *et al.*, 1975) has undoubtedly helped the process of screening host plants for clubroot resistance by enabling pathogen populations to be selected with particular virulence characteristics and, through field surveys (such as the multitudes reported in issues of *Clubroot Newsletter*), by providing some information on the geographical distribution of pathogenicity determinants. Nonetheless, the results obtained with *B. oleracea* do not lend themselves to very optimistic interpretation (Crute *et al.*, 1981) and for long it seemed increasingly improbable that adequate resistance to clubroot existed in that species to enable practical benefit to be derived for field crops; in Western European conditions at least. With other crop species and in other parts of the world, the future seemed a little less gloomy; in Japan, for example, where the spectrum of pathogen virulence seems less, considerable success is being achieved in transferring resistance into the highly susceptible Chinese cabbage from the closely related turnips (Ashizawa, 1977). More recently, however, detailed screening has revealed that some resistance sources in *B. oleracea* may be more useful than was previously believed (Crute *et al.*, unpublished work). I do not dismiss the possibility that a new and potent chemical control for clubroot may be just around the corner, or discount the albeit rather remote likelihood of durable and totally effective resistance being found in forms closely enough related to cultivated varieties of *Brassica oleracea* to enable it to be transferred without too much genetical gymnastics. Nonetheless it does seem increasingly that if a permanent and satisfactory solution to the problem of clubroot is to be found it will come from an area not yet investigated. Leaving aside outlandish possibilities, there remains one major stone still to be turned—an understanding of the natural factors that regulate *Plasmodiophora* in the soil, that are responsible for the germination, longevity or demise of resting and/or zoospores and that might be amenable to artificial manipulation. What is being defined therefore is a need for information that could be a prelude to biological or, at least, to integrated control.

It is scarcely credible that the spores of *P. brassicae*, exposed to the soil environment for long periods, should not be attacked, eaten, decomposed or otherwise affected by other organisms in the biosphere as the spores of so many other pathogens are. It is probable that such technical problems as the inability easily to remove spores from soil and assess their viability, have prevented this topic being investigated in the past. It is an area, however, that I believe will repay considerable expenditure of effort for such very scant information as is already available is most intriguing. Isolated observations have been made at different times (see, for example, Buczacki and Stevenson, 1981) to the effect that there is a difference between artificially infested sterile and non-sterile soil in its propensity to support infection of host plants by *P. brassicae* and the subsequent development of clubroot symptoms. The most perceptive observations were made by Smith (1955) in an admirable but largely ignored piece of research that was unfortunately never published beyond the confines of a university dissertation. Smith showed that enhanced root hair infections could be obtained when resting spores were suspended in a non-sterile rather than a sterile soil extract. He then obtained some experimental evidence to indicate that the factor responsible was a protein produced by other soil micro-organisms and he was able partially to mimic its effects by treating spores with proteolytic enzymes. Smith interpreted the increased infection to "the digestion of a protein matrix which remained attached to spores isolated from solid gall tissue". He implied that the protein matrix was probably of host origin but his results appear the more intriguing following our demonstration (Moxham and Buczacki, 1983; Moxham *et al.*, 1983) that the resting spore itself has a high concentration of protein localised in the outer part of its wall.

It is very tempting to speculate still further and relate the reported instances of clubroot being suppressed by the use of organic manures as effects mediated through the alteration of the soil microflora and fauna. Whether this proves to contain some grains of truth or whether it is to join the countless other idle speculations that *Plasmodiophora brassicae* has stimulated in the past, it is nonetheless a topic that I believe must be very fully explored.

More opinion and theories have been advanced over the biology and control of *P. brassicae* than over most other plant pathogens. A few more have been presented in these pages but whether or not a realistic and lasting control measure is forthcoming in the next few years, there can be no doubt that this most challenging and intractable of organisms will continue to fascinate and frustrate pathologists and mycologists for a great many more years to come.

Acknowledgements

I am grateful to Drs I. R. Crute and D. S. Ingram for critical appraisal of this account and to Mrs Susan Moxham and other friends and colleagues for their advice and assistance over many years of research on *Plasmodiophora brassicae*.

6

The Chemical Control of Diseases Caused by Zoosporic Fungi

A Many-Sided Problem

G. C. A. BRUIN and L. V. EDGINGTON

Department of Environmental Biology, University of Guelph, Guelph, Ontario, Canada

I. Introduction 193
   A. A Many-Sided Problem 193
   B. Historical Aspects 194
II. Chemical Control 196
   A. Downy Mildews 196
   B. Pythiums 202
   C. Phytophthora 204
   D. Aphanomyces 208
   E. White Blisters 209
   F. Plasmodiophoromycetes 209
   G. Chytridiomycetes 212
III. Fungicides 215
   A. Protectant Fungicides 215
   B. Systemic Fungicides 218
   C. Unique Characteristics of Oomycete-Selective Fungicides 224
IV. Fungicide Resistance 226
V. Future Considerations 230

I. Introduction

A. A MANY-SIDED PROBLEM

For some of the pathogens considered in this chapter, zoospores are the only infectious units whereas others, like *Phytophthora infestans* (Mont.) de Bary, produce them only under certain conditions, and some downy mildews never produce them. However, since the chemical control measures presently available for zoosporic fungi do not specifically interfere with the unique mechanisms linked with zoospore production and activity, these bodies are of far less relevance in disease control than are taxonomic and ecological considerations. For instance, the habitats of zoosporic plant pathogens may vary from leaves and fruits to soil and roots while there are many behavioural and structural differences between the various taxonomic groups that contain zoosporic plant pathogenic species (Chapter 2).

From a chemical control standpoint, the Oomycetes in particular are unique, intriguing and very important zoosporic fungi. Their cell walls contain cellulose, but not chitin, which makes them insensitive to polyoxin antibiotics; further, they do not need sterols for vegetative growth, which makes them insensitive to the potent sterol synthesis inhibiting fungicides; and they appear to have differently structured tubulin proteins and enzymes active in energy production which makes them insensitive to systemic benzimidazole, oxathiin and hydroxypyrimidine fungicides. Additionally, Oomycetes and Hyphochytriomycetes utilise enzymes with properties distinctly different from those in other fungi (LéJohn, 1971; Ragan and Chapman, 1978), including some that are active in RNA synthesis as indicated by the selective action of acylalanine-type fungicides which seem to interfere with this process (Davidse, personal communication). The unique significance of diploidy in Oomycetes moreover makes them extremely versatile and adaptable, this being reflected in rapid race evolution and quick changes in pathogenic characteristics (see Chapter 4). Their adaptability is furthermore enhanced by possible heterokaryotic conditions in the coenocytic mycelium, permitting these fungi to carry a vast arsenal of genetic information (Long and Keen, 1977; Bruin and Edgington, 1981).

Particular habitats and life cycles of both pathogens and hosts therefore dictate to a large extent the approach to the chemical control of zoosporic fungi, which is, indeed, a many-sided problem.

B. HISTORICAL ASPECTS

The wide-scale use of fungicides began with the combat of downy mildew of grape, caused by *Plasmopara viticola*, (Berk. & Curt.) Berl. & de Toni in France, 100 years ago. This pathogen was introduced to France on *Phylloxera*-resistant root stock from the United States and threatened the entire grape culture there. In 1882 Millardet, who first introduced the contaminated nursery stock, noted that vines along the roadways remained green and healthy, while others were defoliated. These vines had been daubed with a poisonous-looking mixture of lime and copper sulphate to discourage pilfering. The significance of this observation was quickly realised and Millardet, assisted by the chemist Gayon, started experimenting with mixtures of copper sulphate and lime. Three years later a satisfactory formula was announced (Millardet and Gayon, 1885; Millardet, 1885a,b) and Bouille Bordelaise, or Bordeaux mixture as it became called in 1886 (Johnson, 1935) was formally introduced. Much has been done on improving the formulation of this fungicide as well as on developing spray methods for its application. Later, Burgundy mixture (in which sodium carbonate was substituted for lime) and many other "fixed copper" fungicides were developed, in all of which, the important qualification was to have a very low water solubility to minimise the available cupric ion. The tenacity on leaves was also important, an attribute in which Bordeaux mixture excelled. The fixed coppers have continued in use, and even though the copper ion is biocidal and somewhat phytotoxic, its low solubility, limited penetration of the plant cuticle, bactericidal activity and low cost are factors which keep it in use. For example, in 1978, U.S. registrations include 34, 38 and 5 uses in vegetables and fruit, ornamentals and turf, and field crops, respectively. The use of copper for diseases caused by zoosporic fungi has decreased, but estimates of total world usage for 1980 are 215 million U.S. dollars, or about 15% of all fungicide costs (Anonymous, 1977).

The second period of organic protectants arose in the early 1950s with the dithiocarbamates which were efficacious for many fungal diseases, and estimates of their total world usage for 1980 was 570 million U.S. dollars, or 40% of total fungicide costs. The dialkyldithiocarbamates, thiram, ferbam and ziram, differ from the monoalkyl or ethylenebisdithiocarbamates (EBDCs) in mode of action and uses. Thiram remains an important seed treatment fungicide, moderately effective against zoosporic fungi as well as many others, whereas ziram, initially developed for downy mildew of vegetables, has been replaced by EBDCs. Growers often tank-mixed nabam with zinc sulphate to make zineb, the zinc ion having a greater affinity for the EBDC ion and displacing sodium, and zineb conquered the large potato and tomato markets in the 1950s. Maneb later maintained some of this market and then, gradually, mancozeb and metiram, which are similar products containing both manganese and zinc ions chelated with the EBDC ion became dominant foliar sprays for downy mildews. In the 1960s a completely unrelated product, chlorothalonil also proved efficacious for many foliar diseases, including downy mildews. All of these fungicides were of low water solubility and were moderately tenacious protectants.

Within the last few years, a third group, systemic fungicides, suddenly and dramatically shifted the prospects for control of zoosporic fungi. A few slow moving systemics like chloroneb, hymexazol and etridiazole had evolved earlier in the 1970s and enjoyed a modicum of success as seed treatments. However, the discovery of fosetyl-Al, prothiocarb, propamocarb and the acylalanines, metalaxyl, furalaxyl and milfuram has opened up a whole new area of disease control for their fungitoxicity extends to many previously uncontrollable soil-borne zoosporic pathogens.

II. Chemical Control

Because zoosporic plant pathogenic fungi are so diverse, it is impossible to generalise when discussing chemical control measures. Not only are the taxonomic differences, based on morphological, biochemical and physiological features very great, but differences in etiology, habitat and infection characteristics render it necessary for chemical control to be discussed for only a small number of pathogenic species representing the most important taxonomic and ecological groups. This chapter will not, therefore, present a complete overview, but will highlight the general trends and new ideas.

A. DOWNY MILDEWS

1. *Foliar Infections*

Downy mildew resulting from airborne inoculum can be more or less effectively controlled by sprays of protectant fungicides such as dithiocarbamates, chlorothalonil, folpet, captan or copper-containing compounds like Bordeaux mixture, copper oxychloride and copper ammonium carbonate. The choice of material depends on economic considerations, phytotoxic effects, availability of the products and on tolerable residue levels. For instance, on corn, daily spray application of maneb for 4 weeks after emergence appeared highly effective in controlling sugar cane downy mildew, caused by *Peronosclerospora sacchari* (T. Miyake) C. G. Shaw. Such a spray schedule, while feasible for research purposes, however, was far too laborious and expensive for East-Asian farmers (Exconde, 1970b) and, therefore, not practicable.

Residual deposits remaining after foliar application are of particular concern, especially on leafy vegetables like lettuce. Tenacious fungicides such as Bordeaux mixture should not therefore, be recommended for use on these crops and have been replaced by ethylenebisdithiocarbamates such as maneb and zineb. Even with the latter fungicides a safety period of at least eight to ten days is practised.

Every downy mildew-causing organism has its own specific requirements for growth and pathogenesis. All, however, have in common a preference for wet weather conditions, and most downy mildews in temperate climates are favoured by cool and humid seasons. Hot and dry periods usually arrest disease progress, but seldom kill the pathogens. Spore production depends on high relative humidity, and free water is necessary for release, motility and germination of zoospores. Infection periods, therefore, can be predicted accurately when weather conditions, leaf wetness and inoculum are carefully monitored. This monitoring should be the basis for any chemical control measure to avoid excessive and unnecessary applications. However, since the effects of environmental factors on pathogenesis and spread are not known for all foliar pathogens in all geographic areas, farmers still often tend to rely on fixed protective application schedules. The advantage of such schedules is that other pathogens, that may be influenced differently by certain environmental conditions, are also controlled by the same protectants. For instance, a weather-timed spray schedule for onion downy mildew (*Peronospora destructor* (Berk.) Casp.) decreased the number of sprays of maneb required, thus permitting the leaf blight fungus *Botrytis squamosa* Walker to infect more heavily than when sprays were applied on a seven- to ten-day interval basis (Hildebrand, personal communication).

Proper timing of spray and dust applications is not only important as far as weather conditions and infection periods are concerned, but also in order to affect the pathogen at its most vulnerable stage and at the moment when it can do most damage. For instance, corn is most susceptible to Philippine downy mildew, (*Peronosclerospora philippinensis* (Weston) C. G. Shaw), between the three-leaf and one month-old stages when fungicides should therefore be applied. Exconde (1970a), indeed, attributed the failure of 71 fungicides to control this disease to improper timing of applications.

In the following account, the problems associated with the chemical control of five major groups of foliage-infecting downy mildews will be considered in detail.

(a) *Bremia lactucae* Regel. This is the causal agent of downy mildew of lettuce and is mainly controlled by cultural practices and the use of fungicides. Breeding for resistance has been successful in that major genes for resistance have been incorporated into commercial cultivars, but the fungus has responded quickly in forming new races to overcome the resistance. Moreover, the sources of major resistance genes are limited and they are not always easy to incorporate into commercially acceptable types. Pyramiding of several such genes in single cultivars, coordination of breeding programmes in different regions and the tedious breeding for horizontal resistance may provide long-term control but it seems that fungicides will be indispensable for many years to come.

Since the early 1950s, spraying, dusting and drenching with zineb or maneb have been practised with variable degrees of success, the level of control being heavily dependent on weather conditions and inoculum load. Although ethylenebisdithiocarbamate fungicides are reasonably sensitive to weathering, residues on lettuce are constant considerations and a safety margin of seven to ten days between treatment and harvest allows residue deposits to drop to acceptable levels. The search for more active and less contaminating fungicides is, therefore, a first priority. Recently, the acylalanine-type systemic compounds with metalaxyl as the most active member have shown great potential (Urech *et al.*, 1977; Smith *et al.*, 1977). Crute and Norwood (1978) obtained almost complete control of *B. lactucae* with a soil drench of furalaxyl, followed by a fortnightly spray schedule, while incorporation into soil of a granular formulation of 3 kg a.i. ha^{-1} at seeding time gave almost complete control for nine weeks. The related compound, metalaxyl was even more active (Crute and Jagger, 1979). Bruin and Edgington (1980a) obtained complete control of lettuce downy mildew through the season, with a single granular application of 1 kg ha^{-1} metalaxyl at seeding time. Four bi-weekly foliar sprays of 0.4 kg metalaxyl or RE 26745 ha^{-1} also gave excellent control.

These systemic compounds are extremely active against members of the Peronosporales at concentrations much lower than those required with conventional fungicides, and the amount of material necessary per season as a result of lower dosages and longer spray intervals is about one-tenth. These fungicides also have limited curative action, and can, therefore, be applied after infection has taken place. The therapeutic potential of furalaxyl for instance was demonstrated in both laboratory and greenhouse tests where satisfactory control was achieved by applications up to five days after inoculation (Crute and Jagger, 1979), while in laboratory tests, metalaxyl was found to have a curative effect up to six days after inoculation (Crute *et al.*, 1980b). Crute and Gordon (1980b) observed no difference in effectiveness with metalaxyl foliar sprays (200 mg a.i. l^{-1}) whether they were applied on a regular fortnightly schedule, on a three-weekly schedule, or after the first disease symptoms appeared in the field. The only complication with the curative spray appears in the case of a late infection, for the 14-day safety margin before harvest precludes any treatment being applied. As an alternative to routine fungicide applications, Crute and Gordon proposed the following strategy: either (1) in the absence of downy mildew, a single foliar spray applied three weeks before harvest as a precaution against late infections, or (2) when mildew infection is first noticed, a curative spray followed by regular applications.

The margin between concentrations that provide good control and concentrations that are phytotoxic, however, is not very great. Phytotoxic symptoms were observed at levels of 0.01 g a.i. per plant, while the recommended rate for incorporation into peat blocks is 0.007 g a.i. per plant (Crute and Jagger, 1979; Crute *et al.*, 1980b).

The systemic fungicides, prothiocarb and propamocarb have also been evaluated but although these compounds had systemic action they gave poor control even at high concentrations. Fosetyl-Al was more effective with a curative action period of 24 hours, but was still inferior to the acylalanine-type fungicides (Crute and Jagger, 1979; Crute, 1980; Bruin and Edgington, 1980a).

(b) *Peronospora destructor*. This is the causal agent of onion downy mildew and has been controlled for many years with dithiocarbamate fungicides and more recently also with chlorothalonil but fixed copper fungicides have never proved satisfactory. Fungicides do not adhere easily to onion leaves because of the very waxy cuticle and vertical growth habit. Therefore, dusts are often used or extra sticker compounds are added to the fungicide formulation. Nonetheless, since the leaves of onions that are grown for bulb production are not usually used for consumption, fungicide residues are of minor importance and, in fact, application of a chemical which moved symplastically would have the undesirable effect of it accumulating in the bulb.

The systemic fungicides metalaxyl and furalaxyl have proved very effective against onion downy mildew (Paulus *et al.*, 1977), except when inoculum pressure was very high and, since metalaxyl allows spores to germinate and establish initial infections, it is suspected that a heavy infection may cause sufficient injuries to be damaging (Hildebrand, personal communication).

In view of the possibility of resistance developing (see Part IV), the very selective action of metalaxyl (*Botrytis* leaf blight is not affected), and the absence of a specific need for a systemic compound, it would be advisable to continue using broad spectrum protectants for onion downy mildew control and reserve the use of acylalanine-type compounds to one or two applications per season for emergency situations.

(c) *Plasmopara viticola.* This incitant of grape downy mildew occurs in most parts of the world where grapes are grown under humid conditions. In seasons when the weather is favourable and no protection is provided, it can easily destroy 50 to 75% of the crop (Agrios, 1978).

Bordeaux mixture is still used for grape downy mildew control although ferbam, zineb and captan have largely replaced it and, recently, a number of systemic compounds have been reported effective; besides metalaxyl (Urech *et al.*, 1977), furalaxyl (Schwinn *et al.*, 1977), milfuram, fosetyl-Al (Lafon *et al.*, 1977; Williams *et al.*, 1977) and cymoxanil (Douchet *et al.*, 1977) were also found to give excellent control. Since at least three systemic compounds with different biological properties and high activity against *P. viticola* are potentially available, therefore, a wise and balanced use of these should delay occurence of resistant strains of *P. viticola* for a long time.

(d) *Peronospora tabacina* Adam, causal organism of tobacco blue mould, does not infect by zoospores, but is taxonomically close to zoosporic downy mildews.

The disease can strike as soon as the plants emerge and it has commonly been considered essentially as a seedbed problem. The most dependable and practicable method of control in Europe and the U.S. is with dusts or sprays containing one of the dithiocarbamates (McGrath and Miller, 1958) although Clayton *et al.* (1942) earlier found copper oxychloride to be effective, especially when amended with cottonseed oil. Good control was also obtained by treating seedlings with streptomycin (Grosso, 1954; Todd, 1955) which has curative properties and thus can be applied shortly after infection. In Australia, good control has been achieved by using night-time application of benzol vapour as an eradicant fungicide (Angell *et al.*, 1935), while *p*-dichlorobenzene is another volatile material that has been used in seedbeds (Clayton *et al.*, 1942). However, both fungicides have the disadvantages of being phytotoxic at high concentrations, are flammable, toxic to man and laborious to apply and therefore their use has never become widespread.

Until a decade ago, field applications of fungicides for tobacco blue mould control in the U.S. were considered to be too tedious and expensive, especially since warm and dry weather inhibited development of the disease in most years. Furthermore, fungicide residues were a major concern (Lucas, 1965). In Australia, however, control of blue mould in the field was more important, especially as irrigation became more common practice in the 1940s (Pont, 1959). Sprays with dithiocarbamate fungicides have been used since 1963 and generally provided good protection if the weather was not too wet and if the inoculum pressure was not too high (Johnson *et al.*, 1979).

The systemic fungicide, metalaxyl, has proved very effective (Urech *et al.*, 1977; O'Brien, 1978; Johnson *et al.*, 1979; Tsakiridis *et al.*, 1979) and has been registered for blue mould control for a number of years in Australia. A temporary permit was also obtained in Canada and the U.S. in 1979.

Although no resistance to metalaxyl in *P. tabacina* has been found, there is evidence that considerable variability in sensitivity exists among field isolates (Bruck *et al.*, 1981), and an intense prolonged selection pressure by the fungicide might give rise to resistance problems. Such a development would be disastrous, for blue mould appears to be becoming more prevalent in the field, as indicated by recent field epidemics in Greece, Cuba, Central and North America. A significant decrease in the effectiveness of blue mould-resistant varieties (Tsakiridis *et al.*, 1979), a change in pathogen virulence, increased resistance of the conidia to adverse conditions or an overall change in the weather during the growing season may all be significant factors in this trend. The systemic fungicides propamocarb, cymoxanil and fosetyl-Al are ineffective against tobacco blue mould (Bertrand *et al.*, 1977; O'Brien, 1978; Tsakiridis *et al.*, 1979), leaving only the protective dithiocarbamates and the systemic acylalanine-type fungicides and thus giving little room to avoid the development of resistance in the long-term, unless both types of compound are applied wisely and carefully.

(e) *Downy mildews of maize and sorghum.* Maize is attacked by at least nine species of the genera *Peronosclerospora*, *Sclerophthora* and *Sclerospora* some at least of which seem to have been misidentified in the past. Some infect by zoospores, others by direct germination of conidia and most are also pathogenic on sorghum and various other graminaceous plants. Together, they may cause yield losses in maize of 50 % or more in south-east Asia, India, and Africa (Shurtleff, 1980). In spite of these heavy losses, chemical control has only occasionally been economically feasible (Exconde, 1970b, 1975; Frederiksen and Renfro, 1977). The dithiocarbamate fungicides as well as triphenyltin compounds have been superior protectants, but control is often difficult to achieve under conditions of high inoculum levels or on fully susceptible varieties (Frederiksen and Renfro, 1977). Protectant fungicides, however, are not effective against established systemic infections. Promising control of sorghum downy mildew (*P. sorghi* Weston and Uppal) C. G. Shaw in sorghum was obtained by Matocha *et al.* (1974) with soil incorporation of about 1 kg potassium azide ha^{-1}.

The systemic fungicide chloroneb was reported to be effective against downy mildew diseases incited by *Peronosclerospora* spp. when used either as an in-furrow application or seed treatment, followed by six to eight applications of a dithiocarbamate, beginning about two days after plant emergence and continuing at twice-weekly intervals. Recently, chemical control has been successfully integrated with resistant varieties (Frederiksen and Renfro, 1977) while Lal *et al.* (1977) obtained good control of *P. sacchari* on maize by chloroneb seed treatment combined with rogueing of infected plants 20 days after planting and one foliar spray of neem oil. Kajiwara *et al.* (1979) reported good control of Java downy mildew (*P. maydis* (Racib.) C. G. Shaw) on corn with a soil drench of 0.6 g etridiazole m^{-2} five days after seeding, followed by a similar drench eight days later.

Metalaxyl effectively controlled sorghum downy mildew (*P. sorghi*) on sorghum and maize with seed treatments of as little as 0.1 g kg^{-1} sorghum seed (Frederiksen and Odvodi, 1979) and excellent control of many other downy mildew diseases of graminaceous crops with this product is being reported from all over the world. The relative ease and success of seed treatments and single sprays with small amounts of such acylalanine-type compounds will make chemical control of cereal downy mildew diseases more practicable and economically feasible, although their availability and cost in developing countries might be a limiting factor.

Selective breeding has resulted in maize and sorghum hybrids with high degrees of resistance to various downy mildew diseases, but new races of the pathogens are beginning to overcome this resistance. Acylalanine-type compounds, if used judiciously and properly therefore, can help to conserve the resistance now present in commercial hybrids (Odvodi, personal communication), although their widespread and continuous use might lead to the development of fungicide resistant pathogen strains. It is highly important therefore to use them in combination with other effective fungicides.

2. *Systemic Infections*

A number of downy mildew-causing pathogens are systemic in plants, and sometimes, downy mildews that usually infect locally may become systemic under certain environmental conditions.

Downy mildew of hops, caused by *Pseudoperonospora humuli* (Myabe and Tak.) G. W. Wilson is endemic in all North American and European hop growing districts. It overwinters systemically in the perennial parts of the plant and grows into the young shoots in spring (Ware, 1926). Initial infection by air-borne spores can be prevented by protectant compounds like Bordeaux mixture, other copper fungicides or dithiocarbamates, but as soon as an infection becomes established systemically, protectants are ineffective. The disease wiped out the entire hop culture in Ontario, and constitutes an annual major risk to the crop elsewhere (Burgess, 1964). The antibiotic streptomycin has been used successfully to control systemic infections (Horner, 1963), but infections in root and crown are not efficiently eradicated. Acylalanine-type fungicides have been found, however, to have superior eradicant properties and Smith *et al.* (1977) and Schwinn *et al.* (1977) reported excellent control of systemic infections with soil drenches or foliar sprays of furalaxyl. Similar results were reported by Urech *et al.* (1977) with metalaxyl and the development of these compounds may be considered as a major breakthrough in the culture of hops.

Seed-borne infections by downy mildew pathogens are forms of systemic infection that are virtually uncontrollable directly by protectant fungicides. The main control measures, therefore, are the use of healthy seed and the use of indirect spray or dust applications before and during seed development to prevent air-borne spores infecting the parent plant. Examples of such seed-borne downy mildews are sunflower downy mildew (*Plasmopara halstedii* (Farl.) Berl & de Toni) (Young and Morris, 1927; Leppik, 1962) and sorghum downy mildew (*Peronosclerospora sorghi*) (Klaveriappa and Safeeulla, 1978). Although no data on the efficacy of acylalanine-type fungicides in controlling seed-borne downy mildew diseases are yet available, they can be expected to combat such infections successfully.

B. PYTHIUMS

Pythium species are all soil inhabiting fungi that may cause seed rot, damping-off and root rot in many vegetables, flowers, cereals, fruit and forest trees (see Chapter 4). All cause the greatest damage to the seed and the seedling either during germination and before or after emergence (Agrios, 1978). Some species infect via zoospores, others penetrate plant tissue directly with hyphae but because of the nature of the organisms and the earliness of attack, foliar fungicide applications are usually ineffective.

Diseases caused by *Pythium* spp. in the glasshouse or seedbed can be controlled by sterilising the soil with steam, dry heat or with general biocidal chemicals, like chloropicrin, methyl bromide, metham-sodium or D–D mixture. In the field, however, where soil sterilisation is difficult and expensive, seed or bulb treatment is the most important control measure and the most commonly used materials include thiram, captan, quintozene, ferbam, calomel, chloranil, dichlone, drazoxolon, zinc oxide, formaldehyde, fenaminosulf and etridiazole. Besides being used to treat seed, these compounds can also be applied to the soil in the vicinity of the seeds to suppress fungal growth. Application techniques include the incorporation of dust, granule and liquid formulations directly with the seed in the seed furrow, incorporation in peat blocks for transplants, and soil drenches.

With the development of the systemic fungicides selective to Oomycetes, a new dimension in seedling control was provided. Since these compounds are taken up by the seeds and the young seedlings, they can actively protect the coleoptile from invasion by soil-borne fungi during the most susceptible growth stages. Good to excellent control of *Pythium* diseases has been obtained with etridiazole by various researchers and this compound was listed by Martin and Worthing (1976) as very promising against *Pythium* spp. The systemic fungicide prothiocarb has also been found to be active against *Pythium*, providing variable degrees of control. Papavizas *et al.* (1977) reported good control of *Pythium* blight of bean with both etridiazole and prothiocarb either applied directly to the seed or by using a seed infusion technique. Wiertsema and Wissink (1977) obtained some control of *Pythium* on ornamentals with prothiocarb, but found furalaxyl used in low dosages as soil treatments to be far superior. Similar results were reported by Schwinn *et al.* (1977) who controlled *P. debaryanum* Hesse in sugar beet with furalaxyl whereas Paulus *et al.* (1977) found metalaxyl slightly more effective than standard treatments with fenaminosulf, captan or thiram against *Pythium* damping off of pea.

Beside causing seedling diseases, *Pythium* species also appear to be pathogenic to established root systems of fruit trees and may play a major role in apple replant disease (Mulder, 1969; Sewell, 1981). Satisfactory selective fungicide treatments have not been worked out for these diseases but possibly prothiocarb, etridiazole or acylalanine-type compounds will prove useful in this regard.

A new area where *Pythium* species are of potential importance is in hydroponic cultures and they can almost be considered as aquatic fungi in such systems which maintain a continuous flow of recirculated water-nutrient solution. Spread by zoospores of pathogenic *Pythium* species may cause havoc therefore, although Price (1981) controlled *Pythium* in nutrient-film tomato cultures by adding 30 µg etridiazole and 6 µg copper (as oxine-copper) to each litre of nutrient solution every three weeks.

C. PHYTOPHTHORA

1. *Foliar Blights*

The chemical control of *Phytophthora* blights is practised generally along the same lines as control of downy mildews; *P. infestans*, causal agent of late blight of potato and tomato for instance, has many features in common with most downy mildew-causing organisms. By far the most important means of chemical control of *Phytophthora* late blights is the use of broad spectrum protectant fungicides. Most widely used are ethylenebisdithiocarbamates (EBDCs), Bordeaux mixture, other fixed copper compounds, captafol, organotin compounds and chlorothalonil. Although the copper-containing compounds must be applied with care to avoid phytotoxic effects, residue levels are usually not of concern (except sometimes on tomato fruits) since potato and tomato foliage are not used for consumption. EBDCs are known, however, to form the carcinogenic degradation product ethylene-thiourea (Ulland *et al.*, 1972), especially when subjected to heat. Therefore, tomato fruits used for processing should be free or freed from EBDC residues and this can be done with either a 0.1 % sodium hypochlorite wash or with a hot acid blanch (Marshall and Jarvis, 1979).

The new systemic compounds fosetyl-Al, prothiocarb and propamocarb, which are very effective in controlling various diseases caused by members of the Peronosporales, have little or no efficacy against *P. infestans*. On the other hand, cymoxanil, which has local systemic and eradicant properties, and is used in combination with reduced rates of one of the protectants, mancozeb, folpet or fixed coppers, has been shown to have good fungicidal action against *P. infestans* (Douchet *et al.*, 1977). The new acylalanine-type fungicides were shown to be extremely effective against *P. infestans*, needing only about one-tenth the dosage per season of older compounds (Urech *et al.*, 1977; Schwinn, 1979), and their potential as late blight fungicides looked excellent. However, in the summer of 1980, after half a season of wide-scale use of metalaxyl in European and African countries, metalaxyl-resistant strains were detected in Northern Ireland, the Irish Republic, the Netherlands, Switzerland and Zimbabwe. This development caused major losses to local farmers but, interestingly, no resistance was found in 100,000 ha of potatoes that were sprayed with Fubol, a mixture of metalaxyl and mancozeb (Anon, 1980). In fact, addition of low rates of a protectant fungicide in metalaxyl-containing formulations increased efficacy, especially later in the season (Smith, 1979).

2. *Fruit Rots*

At least 13 species of *Phytophthora* cause fruit and pod rots in a large variety of plants (Ribeiro, 1978). An extensive review of control methods against *P.*

palmivora (Butler) Butler, causal agent of fruit rot of cacao, coconut and papaya, was presented by Gorenz (1974). Fruits usually become infected by air-borne spores or by soil-borne inoculum that contacts the fruit when it touches the soil, or is carried in splashing water drops. The best available chemical control is the protection of the fruits and buds by high or low volume sprays of protectant fungicides like Bordeaux mixture, other fixed copper compounds, EBDCs, or folpet. Little is known of the efficacy of systemic fungicides to control fruit rots, but since most of these compounds are translocated with the transpiration stream, and as fruits do not transpire significantly, soil or foliar applications will be unlikely to provide effective control. Nor, by the nature of fruit tissue, should we expect efficient deep penetration of systemic materials and so their eradicant action against established infections will probably be very limited. Systemic fungicides will, however, probably be as effective as protectants if used in prophylactic applications.

A relatively new concept in fruit rot control is chemical treatment of the orchard or plantation soil to diminish the amount of viable inoculum. Although Gorenz (1974) did not consider such treatments likely to become of importance against *P. palmivora*, Harris (1979) obtained promising results against *P. cactorum* (Leb. & Cohn) Schroet. (see Postharvest Diseases p. 207).

3. *Stem and Trunk Cankers*

A number of *Phytophthora* species cause foot rot or stem and trunk cankers of woody plants. The source of inoculum and the particular infection route dictate whether chemical control is effective and how and when applications should be made. *Phytophthora nicotianae* van Breda de Haan var. *parasitica* (Dastur) Waterhouse, causal agent of foot rot of citrus, for example, mainly infects young trees just above the bud union, especially after frost damage. Infection can be prevented by fungicidal trunk paints and Timmer (1977) found captafol (60 mg ml^{-1}) to be the best material, followed by fixed copper compounds and captan.

Cocoa canker, however, develops as a result of mycelial spread of *P. palmivora* from infected pods along the peduncle via the flower cushion into the stem, although wound infection may also sometimes occur. The control of black pod (*Phytophthora* fruit rot) is therefore of first importance in prevention of canker, along with removal of infected pods (Firman, 1974).

4. *Root Rots*

At least 22 species of *Phytophthora* cause damping-off, root, foot and crown rot diseases in a wide variety of plant species (Ribeiro, 1978). They are all soil-borne and have a long survival mechanism through thick-walled oospores and chlamydospores. They are favoured by cool, wet, waterlogged soil conditions and the main key to control, therefore, is adequate drainage. Soil sterilants and fungicidal seed treatments, root dips, or soil applications as described under *Pythium* damping-off diseases can be used against soil inhabiting *Phytophthora* species also and a fairly similar range of compounds is effective. A special form of seed treatment is the solvent infusion technique, in which seeds are soaked in water or an organic solvent containing a fungicide in solution. Papavizas and Lewis (1976) obtained excellent control of *Phytophthora* root rot of soybean, caused by *P. megasperma* Drechsler var. *sojae* Hildebrand, with pyroxychlor-acetone seed soaks and found this technique more efficient in controlling soil-borne diseases than conventional water-slurry treatments. Unfortunately, the fungicide pyroxychlor was never registered in the United States, but the seed infusion technique was highly successful with prothiocarb against *Pythium* blight on bean (Papavizas *et al.*, 1977) and with etridiazole, propamocarb and metalaxyl against *P. megasperma* var. *sojae* (Papavizas *et al.*, 1979). In later tests, etridiazole applied by a one to three hour solvent infusion technique, using 10% a.i. and acetone or dichloromethane as a solvent provided excellent control while a standard seed treatment of 0.9 g a.i. etridiazole kg^{-1} gave no control. Control by propamocarb at low rates was also enhanced by solvent infusion, while metalaxyl (0.1 g kg^{-1}) proved to be effective regardless of the method of application. The organic solvents were not phytotoxic and had the advantage over water in that they did not induce germination. However, they have only been tested on a limited scale experimentally and no commercial applications are approved or in use to our knowledge.

    Soil drenches, broadcast or in-furrow applications of fungicides can also be very useful. Zentmyer (1973) reported good control of avocado root rot (*P. cinnamomi* Rands) with repeated soil drenches of fenaminosulf, while drenches of metalaxyl, furalaxyl or, to a lesser extent, etridiazole were effective against the same pathogen causing root rot of *Azalea* (Benson, 1979) and *Chamaecyparis* (Smith, 1978). Bastiaansen *et al.* (1974) and Upstone (1976) achieved control of red stele of strawberry (*P. fragariae* Hickman) using post-planting soil drenches with captafol or prothiocarb. Soil drenches of fenaminosulf, cyclohexamide and sodium ethyl phosphonate were found effective against *P. fragariae* by Montgomerie and Kennedy (1977) while leathery rot and wilt of strawberries, caused by *P. cactorum* was successfully controlled by soil drenches of prothiocarb (Aerts, 1975). Thus far, only the fungicides fosetyl-Al and sodium ethyl phosphonate have shown indication of basipetal, symplastic movement in plants and foliar sprays of these compounds were reported to control *P. fragariae* in strawberries (Montgomerie and Kennedy, 1977), *Phytophthora* root and heart rot of pineapple (Beach *et al.*, 1979), *P. cinnamomi* root rot of avocado (Zentmyer and Ohr, 1978) and *P. nicotianae* var. *parasitica* Waterhouse root rot of citrus (Frossard *et al.*, 1977; Laville, 1979). Dipping transplant roots in a fungicide-containing solution or suspension is also mentioned by various authors as an effective disease control method.

5. *Postharvest Diseases*

The initiation of some postharvest diseases caused by *Phytophthora* spp. can be traced to infection by the pathogen in the field—days, weeks, or even months after harvest. Brown rot of citrus fruits caused by *P. citrophthora* (R. E. Smith & E. H. Smith) Leonian, *P. hibernalis* Carne and *P. nicotianae* var. *parasitica* may be initiated in the field as the crop approaches maturity; therefore, the key to successful control lies in the application of protectant or eradicant sprays in the field (Eckert, 1967). This disease is routinely prevented by spraying the lower fruit (within 1 m of the ground) with a copper fungicide before the rainy season (Klotz *et al.*, 1972). An application of 1–2 kg Bordeaux mixture 1000 m^{-2}, applied as spray or via irrigation, proved to be more effective than copper oxychloride in preventing infection originating from splashing soil particles (Oren and Solel, 1978). After harvest, infected fruits can be much easier cured by heat therapy than by protectant fungicides (Eckert, 1967).

Fruit rot of apples, caused by the soil-borne fungus *P. syringae* (Kleb.) Kleb., has become prevalent in commercial apple crops in all the main apple growing areas of the United Kingdom and is regarded as the main cause of fruit wastage during storage (Upstone, 1977). Splashing of infested soil during heavy rain is the main form of primary infection and preharvest protective sprays could, therefore, reduce disease during storage. Edney (1976) found that captan was the most effective of several fungicides in reducing rotting if it was allowed to dry on the fruit before artificial application of a soil slurry to initiate disease. Preharvest sprays of captan or mancozeb with zineb gave slight control when applied two to eight days before harvest (Upstone, 1977). Theoretically, soil treatments that reduce splashing or that lower inoculum levels in the soil are of potential value. The soil stabilisers Vinamul 8170 and Vinamul 3270 (water-based polyvinyl acetate emulsions) were applied at rates of 168 to 600 l ha^{-1} under fruiting apple trees in an attempt to reduce soil splash onto fruit and, thereby, initial infection by *P. syringae* (Upstone, 1977) and successfully lowered incidence of postharvest disease. More research in this area is clearly needed, however. When sprayed onto orchard soils, furalaxyl effectively suppressed *P. syringae* inoculum in the upper 10 cm of the soil for at least one month although etridiazole, copper sulphate and fosetyl-Al were less effective (Harris, 1979). Furalaxyl and metalaxyl at rates of 0.4 and 0.15 g m^{-1} applied to orchard soils were found equally effective in suppressing disease during storage (Harris and Gibbs, 1980).

Postharvest dip treatments using captan, etridiazole and to a lesser extent captafol and chloranil reduced secondary spread of *Phytophthora* rot of apples, but no chemical was fully effective (Upstone, 1977). Cohen and Schiffmann-Nadel (1978a,b) found suspensions containing 0.2% benomyl or the bactericide ME 135 to be ineffective in preventing mycelial growth of *P. citrophthora* on citrus fruit surfaces when used in fruit dip treatments, but these chemicals and captafol were effective in preventing spread of the fungus between decayed and sound fruits when impregnated in paper used for fruit wrapping. Perhaps the release from benomyl of the butylcarbamoyl moiety which is easily converted into the volatile compound butylisocyanate may account for this toxicity. This latter compound is a non-selective fungicide that also affects Oomycetes. Dips of apples in solutions containing 0.01 to 0.05% metalaxyl or furalaxyl completely controlled fruit rot caused by *P. cactorum*, even if treatment was delayed 4, or, in one experiment 11 days after inoculation with zoospores (Edney and Chambers, 1981). Fosetyl-Al and etridiazole were not effective (Edney *et al.*, 1979).

D. APHANOMYCES

Control measures against *Aphanomyces euteiches*, Drechsler, *A. cochlioides* Drechsler and *A. raphani* Kendrick, causal organisms of pea root rot, damping-off and tip rot of sugar beet, and black rot of radish, respectively, have traditionally consisted of employing long rotation schedules and well-drained fertile soils. Papavizas and Ayers (1974) reviewed control measures, including the use of organic and inorganic soil amendments. The favourable effect of fertilisers was described by various authors (e.g. Haenseler, 1929; Smith and Walker, 1941) while Davey and Papavizas (1961) found that addition of organic and mineral salts to the soil decreased incidence of *Aphanomyces euteiches* in peas. Later papers presented reports of effective control by soil drenches of solutions containing amino acids (Papavizas and Davey, 1962a,b, 1963), and other amino compounds (Papavizas, 1964), but under field conditions these amendments might not be effective (Mitchell and Hagedorn, 1966).

Fungicides used as seed or soil treatments have never been very effective in controlling *Aphanomyces* root rot of pea and sugarbeet although Kotova and Tsvetkova (1979) reported good control of *A. euteiches* using a pea seed treatment with hymexazol. Results with the fungicide fenaminosulf moreover, were variable, but often encouraging and good control of *Aphanomyces* diseases was obtained by several researchers (Leach *et al.*, 1960; Coulombe, 1974; Byford, 1976; Mitchell and Hagedorn, 1966, 1969). Fenaminosulf was also found to have a carry-over effect into the next season when applied in amounts of 30 to 60 kg ha^{-1} (Mitchell and Hagedorn, 1966, 1969).

Of the new systemic Oomycete fungicides, neither the acylalanine-type compounds nor the ethyl phosphonates or cymoxanil possess any activity against *Aphanomyces* spp. and only prothiocarb and propamocarb are toxic to these fungi *in vitro* and have good potential *in vivo*. Since root rot diseases are often caused by a complex of various pathogens, the efficacy of narrow-spectrum fungicides like fenaminosulf, hymexazol, prothiocarb and propamocarb is greatly enhanced by the additional use of compounds like quintozene.

Recently, dinitroaniline herbicides were found to suppress the incidence of diseases caused by *Aphanomyces* spp. (Harvey *et al.*, 1975; Grau and Reiling, 1977), and may be considered valuable additional tools in combating these problems. Teasdale *et al.* (1979) believed that the mechanism of suppression of pea root rot by dinitroaniline herbicides was based on the inhibition of motile zoospore production by *A. euteiches*.

E. WHITE BLISTERS

White blister diseases are found on various plant species and are caused by *Albugo* spp., among the most important of which are *A. candida* (Pers.) Kuntze on crucifers and *A. impomoeae:panduratae* (Schw.) Swing on sweet potato. Chemical control of local infections can be achieved with preventive sprays of Bordeaux mixture, other fixed copper compounds, chlorothalonil, captafol, captan, metiram, maneb or zineb. Chambers *et al.* (1974) evaluated several fungicides for the control of white blister of spinach and found dodine, mancozeb and chlorothalonil effective when used in a regular five-day spray schedule. Benomyl was completely ineffective. Such applications, however, do not control systemic infections but Dueck and Stone (1979) compared the efficacy of a number of acylalanine analogues and found metalaxyl and CGA 29212 to have excellent eradicant activity against foliar infections in growth room tests. Foliar sprays with these compounds moreover significantly lowered the number of stagheads (systemic infections) although furalaxyl showed no activity at all against *A. candida* (Dueck and Stone, 1979; Berkenkamp, 1980). Seed treatments of metalaxyl were not effective against stagheads nor were three foliar sprays of 18 kg ha^{-1} fosetyl-Al.

F. PLASMODIOPHOROMYCETES

The class Plasmodiophoromycetes contains nine genera, of which *Plasmodiophora*, *Spongospora* and *Polymyxa* contain important plant pathogenic species. The species that are pathogenic on higher plants are all soil-borne.

(a) *Plasmodiophora brassicae* Woronin. This is the causal organism of clubroot of cruciferous plants, which is favoured by acidic soil conditions; liming is, therefore, traditionally the most common chemical control. Dipping of roots of transplants in suspensions containing calomel, benomyl, carbendazim or thiophanate-methyl is also widely recommended. Peculiarly, *Plasmodiophora*, *Olpidium* and *Physoderma* are the only zoosporic plant pathogenic fungal genera that are sensitive to benzimidazole fungicides, a fact considered in Chapter 5. However, control of *P. brassicae* by benzimidazoles is variable, never complete and depends on seasonal rainfall, temperature and level of soil infestation (Jacobsen and Williams, 1970; Finlayson and Campbell, 1971; Buczacki, 1973a; Reyes *et al.*, 1974; Il'ina, 1979). In contrast the fungicide thiabendazole was reported ineffective against *P. brassicae*, but, unlike the other benzimidazoles, thiabendazole is not a precursor of the compound carbendazim, and it is likely that this acts as the fungitoxic agent (Buczacki, 1973a). Many fungicides, including zineb, chloroneb, chlorothalonil, dazomet, dichlozolin, calomel, mercuric chloride, fenaminosulf, formaldehyde, quintozene, cyanamides, boric acid, sulphur, zinc sulphate and captan have been used at various rates and concentrations either in transplant water, soil drenches, soil treatments and irrigation water, with variable, but usually little success (Colhoun, 1958; Karling, 1968; Reyes *et al.*, 1974; Buczacki, 1978). The effectiveness of ten partial soil sterilant compounds was reviewed by White and Buczacki (1977) and it was concluded that only chloropicrin, dazomet and methyl bromide, when used at appropriate rates, gave consistently good control of clubroot in both field and small-scale tests. The efficacy of methyl bromide is largely dependent upon the use of soil covers because of its volatility. Results of experiments with formaldehyde, metham sodium and methyl isothiocyanate have been inconsistent and the usefulness of several other compounds for clubroot control has not been sufficiently evaluated. Recent results indicated that dazomet applied to soil as a prill at 400 kg ha^{-1} or 1,3-dichloropropene liquid at 900 l ha^{-1} could be effectively used against clubroot (Buczacki and White, 1979) but in spite of the excellent control achieved by applications of chemical soil sterilants, economic considerations will ultimately govern whether such applications are feasible or even possible although they are most likely to be used in seedbeds.

In glasshouse tests, the herbicide trifluralin lowered the incidence of clubroot of cabbage when mixed into the soil before sowing, as did the related compounds benfluralin and isopropalin at non-phytotoxic dosages. Nitralin and dinitramine had no effect (Buczacki, 1973b) but trifluralin and napropamid were later found most effective when applied in combination with carbendazim (Robak and Dobrzanski, 1978).

In summary, it may be said that many chemical compounds affect clubroot disease to some extent, but none is consistently satisfactory.

(b) *Spongospora subterranea* (Wallr.) Lagerh. var. *subterranea* Tomlinson. This causes powdery scab of potato and is known to transmit potato mop top virus. The spread of the fungus, and concommitantly, the mop top virus, is greatly influenced by the soil pH: acid conditions inhibit spread, and the fungus thrives in alkaline soils. The addition of lime to the soil may, therefore, increase disease incidence (Melhus *et al.*, 1916; Reichard, 1976; El Fahl and Calvert, 1976) while application of sulphur can reduce it (Calvert, 1976; Cooper *et al.*, 1976; El Fahl and Calvert, 1976). The relation between soil pH and disease incidence displays an S-shaped curve with a sharp increase in disease between pH 5.2 and 5.5 (Cooper *et al.*, 1976). Therefore, the effect of adding lime to a soil with a pH above 5.5, or sulphur to a soil with a pH below 5.2, will have little effect on disease incidence. Chemical control of this disease, and especially of the soil-borne phase, is difficult although spread of *S. subterranea* in infected tubers can be minimised by tuber treatment with fungicidal compounds. However, since the fungus can be established deep in the tuber tissue, these treatments are only partially effective. Good results have been obtained with tuber dips for 15 minutes in 2-methoxyethylmercury chloride solution or 15 minute steeping in 0.3% formaldehyde. Incorporation of 1 g of 4% mercurous chloride for each 1 of soil was also very effective, but quintozene, captan, dazomet, fenaminosulf or copper oxide incorporation in soil was less effective (Cooper *et al.*, 1976). Soil application of a granular formulation of 0.4 kg thiabendazole ha^{-1} was not effective in controlling powdery scab (Copeland and Logan, 1977), and neither was a preplant application of 124 kg benomyl ha^{-1} broadcast on ridges (Cooper *et al.*, 1976). On the other hand, partial control was achieved by the latter authors with high doses of zinc frit or zinc sulphate, an idea adopted from Tomlinson (1958) who achieved a great decrease in infection by *S. subterranea* var. *nasturtii* Tomlinson, causal agent of the crook root disease of watercress, by applying small amounts of zinc frit to watercress beds. In general, however, the use of resistant varieties and quarantine measures to limit the spread of the pathogen, are better weapons in powdery scab control than the chemicals presently available.

(c) *Polymyxa* spp. These cause root diseases of cereals and grasses and are vectors of soil-borne cereal mosaic and other viruses (see Chapter 7). Consequently, most literature on the pathogen focuses merely on the indirect control of the virus rather than on that of the fungus itself. McKinney *et al.* (1957) observed a decrease of virus symptoms after soil treatments with formaldehyde, chloropicrin, carbon disulphide, D–D, or ethyl alcohol. Good control of wheat spindle streak mosaic was obtained by Slykhuis (1970) with soil drenches containing mercuric chloride, captan, 95% ethyl alcohol, 40% formaldehyde or 2,4,5-trichlorophenol, and by fumigation with preparations containing D–D, metham sodium or methyl bromide. Fertilisers used at normal rates did not reduce disease development, but soil amendment with urea, uric acid, or ammonium nitrate at the rate of 700 kg ha^{-1}, or chicken or turkey manure at *c.* 70 tonnes ha^{-1} greatly reduced disease. Gates (1975) found 250 kg urea, 27 kg benomyl, 34 kg chloroneb or 560 l Vorlex (methyl isothiocyanate) ha^{-1} incorporated in soil completely ineffective whereas the systemic fungicide prothiocarb provided substantial control of *Polymyxa betae* when incorporated into the coating of sugarbeet seed pills at the rate of 1 mg per seed (Horak and Schlösser, 1978).

Although soil fumigants were reported effective in controlling diseases caused or transmitted by *Polymyxa* spp., the expense of such control methods is often prohibitive for field application, and reliance must, therefore, be placed on resistant or tolerant varieties.

G. CHYTRIDIOMYCETES

The species belong to this class are mostly obligate parasites with little or no mycelium. The thallus at maturity acts as a single sporangium, or divides to become a sorus of sporangia. Only the genera *Olpidium*, *Synchytrium* and *Physoderma* are of importance as plant parasites.

(a) *Olpidium brassicae* (Woronin) Dang. This causes damping-off of crucifer seedlings and penetrates the roots of several other plants, most notably lettuce. However, crucifer-infecting strains generally do not infect lettuce, or vice versa (Sahtiyanci, 1962; Lin, 1979) and its importance as a virus vector is more significant than its direct action as a plant pathogen. As a result, most attempts at chemical control have been concerned primarily with diseases such as lettuce big vein, than with the *Olpidium* vector. The effect of soil sterilants on incidence of lettuce big vein was mentioned as early as 1940 (Jagger, 1940), but it was not until 1958 that this disease was associated with the presence of the root infecting fungus *O. brassicae* (Grogan *et al.*, 1958). These reports, together with those of Allen (1948), Rich (1950), Marlatt and McKittrick (1963), Campbell *et al.* (1980) and White (1980), showed that the soil partial sterilants chloropicrin, D–D, ethylene dibromide, methyl bromide, 1,2-dibromo-3-chloropropane or carbon disulphide were partially or completely effective when injected into the soil, although some findings were inconclusive or contradictory. Dipping the roots of diseased lettuce plants for 10 min into solutions containing 0.5 % sodium hypochlorite, 0.1 % mercuric chloride or 0.025 % hydroxymercurichlorophenol (Semesan) completely prevented spread of the disease after the plants were placed in sterile soil (Grogan *et al.*, 1958). Rich (1960) reported good control of *O. brassicae* with 50 p.p.m. quintozene in sand and Marlatt and McKittrick (1963) found that 78 kg quintozene ha^{-1}, rototilled 15 cm into the soil, suppressed incidence of big vein disease for several years.

Olpidium is only slightly affected by the fungicides fenaminosulf, pyroxychlor, etridiazole and metalaxyl, which are usually effective against soil-borne Oomycetes. However, *Olpidium*, like *Plasmodiophora*, is inhibited by the benzimidazole fungicides once thought to be toxic only to groups of higher fungi (Bollen and Fuchs, 1970; Edgington *et al.*, 1971). Tomlinson and Faithfull (1979a,b,c) evaluated the fungicidal action of carbendazim on zoospores of *O. brassicae* and observed that when formulated as Bavistin 50 WP, it killed zoospores within one hour. Additional tests showed that zoospores were also killed by similar concentrations of a blank formulation containing no carbendazim, and it was concluded that surfactants were responsible for the toxic effects. Solutions containing 1 p.p.m. dodecyldimethyl-ammonium bromide, cetyl trimethyl ammonium bromide, Agral (90% alkyl phenol ethylene oxide condensate), or sodium dodecyl sulphate killed zoospores within 30 to 60 minutes. Several other surfactants were also found effective at slightly higher concentrations although Tween 20 at 100 p.p.m. was not toxic. Tests with other chemicals demonstrated that zoospores were killed after 60 minutes in solutions containing 10 p.p.m. zinc or 4 p.p.m. copper, but were unaffected by 500 p.p.m. metalaxyl or by 100 p.p.m. of the antibiotic Bacitracin. Although Campbell *et al.* (1980) doubted the significance of these findings, results of preliminary commercial trials indicate that Agral, added to a circulating nutrient solution of 20 p.p.m. in glasshouses can give effective control of big vein disease while bavistin incorporated in peat blocks (0.025 g per block of 4.3 cm^3) was also effective (Tomlinson and Faithfull, 1979c). Campbell *et al.* (1980) tested a number of fungicides for control of *O. brassicae* and found fenaminosulf, etridiazole and metalaxyl to be ineffective while benomyl or triadimefon did not inhibit zoospore motility, but did prevent infection and reproduction of the fungus.

Encouraging data were obtained by Horak and Schlösser (1978) who achieved substantial protection of sugar beets against *O. brassicae* after incorporation of prothiocarb into the coating of seed pills at the rate of 1 mg seed^{-1}. Soil application of solutions containing 100 and 250 μg ml^{-1} prothiocarb reduced considerably the number of resting spores.

(b) *Synchytrium endobioticum* (Schilberszky) Percival. This is the causal organism of potato wart and is apparently present throughout the world, although the disease seems to be most severe in Europe (Agrios, 1978) where it is favoured by cool, wet conditions. The disease is present in a few locations in the U.S. and Canada and spread is successfully limited by the use of resistant varieties on infested soil and prohibition of the export of potatoes from areas where the disease is known to exist. *S. endobioticum*, in addition to causing wart disease, has been claimed to be a vector of potato virus X. The organism is remarkably insensitive to fungicides; more than 120 inorganic and organic chemicals, singly or in combination have been assayed (Hunt *et al.*, 1925; *c*. Hampson, 1977), but the only successful treatments were either phytotoxic or biocidal soil sterilants. Olsen (1966) found sodium dimethyldithiocarbamate the most effective among 12 compounds tested, but it did not provide complete control even at the rate of 450 kg ha^{-1} and the herbicide dinoseb appeared to be equally effective at rates of 5 to 45 kg ha^{-1}, either as a pre-emergence spray or used as preplant soil incorporation. Hampson (1977) evalutated nine systemic fungicides as seed tuber dip suspensions in pot and field tests. These included benomyl, thiabendazole, thiophanate-methyl, triforine and carbathiin, but he obtained no consistently adequate control and, in field experiments, higher levels of infection were found in treated than in untreated rows. He concluded that these systemic fungicides were of little value in controlling wart and hypothesised that they might interfere with natural antagonistic organisms.

There is probably nowhere where wart disease has been tackled with more vigour and vision than the U.S.S.R. Hampson (1979) reviewed work published there from 1955 to 1977 and reported that over 200 compounds had been tried as control agents. Of these, chloropicrin was rejected as too expensive, but Nitraphen (a complex mixture of sodium salts, nitration products of alkylphenols, water and a wetting agent) was selected for use as a solution applied to fallow ground in May–June. Calcium cyanamide was found useful for both commercial and home garden plots at the rate of 150 g m^{-2} while carbamide at a rate of 1.5 kg m^{-2} was also used and was reported to completely clear the disease in three years.

Spores may remain viable in the soil for 20 or more years in the absence of host plants. Nonetheless, strict quarantine measures together with the use of immune potato varieties and an active wart eradication procedure using 2 to 10 tonnes of ammonium thiocyanate, copper sulphate or 70,000 l 40% formaldehyde solution ha^{-1}, have been successful in eradicating the disease on a local scale (Hartman, 1955). As a result, the state of West Virginia was declared freed of the disease in 1974 (Brooks *et al.*, 1974).

(c) *Physoderma maydis* Shaw. This causes brown spot of maize and is the only air-borne plant parasitic member of the Chytridiomycetes. The disease is limited primarily to areas of abundant rainfall and high mean temperatures and is usually not severe enough to warrant a wide scale use of fungicides. *P. maydis* zoospores can infect host tissue only during certain periods of the day and within a few hours of their development, infection usually occurring within the leaf whorl. Twice weekly sprays for four weeks before silking with the protectants captan or ferbam provided good control (Broyles, 1956). Lal and Chakravarti (1977) obtained good control with post-infection sprays of carbendazim, benomyl and oxycarboxin whereas carboxin and kitazin were less effective and captan gave no control. Carbendazim or benomyl as soil drenches (0.1 to 0.2 g for each plant) on young plants were also very effective, while kitazin and oxycarboxin gave just over 50% control.

III. Fungicides

Until recently, protectant fungicides and soil partial sterilants were the only chemicals available to combat zoosporic and other plant pathogens. These compounds are usually general biocides affecting many vital cell processes and their selectivity in killing fungi and not plants is based on them not entering plant tissue. To be effective they must be applied to susceptible plant tissue before infection occurs, and they must remain present as long as the tissue is susceptible and the pathogen infective. Application must be repeated regularly in relatively high amounts; the term "blunderbuss weapons" used by Schwinn (1979) is, therefore, appropriate. The limitations of protectants are obvious: they do not affect established local or systemic infections but, in many cases, are excellent in preventing disease development.

Since protectant fungicides act on many sites in the living cell, it is almost impossible for fungi to overcome this toxicity with mutations. Resistance to protectants is, therefore, a rare phenomenon, and mostly due to decreased membrane permeability or detoxification (Dekker, 1969). Conversely, systemic fungicides are taken up and translocated in plants, and must therefore be selectively toxic to fungi. As a result, they usually interfere with a single enzyme or structure which is unique to fungi or has unique characteristics in fungal cells. Development of resistance to these compounds is possible with a single mutation and is not uncommon (see Section IV).

Almost all systemic fungicides are taken up by foliage or roots and are passively translocated in the apoplast with the transpiration stream. They usually accumulate at the terminal point of this pathway, the leaf edges. Since uptake through roots cannot occur without them passing through the living endodermis cells, and since antifungal action sometimes occurs within plant protoplasts, we cannot regard these fungicides as entirely apoplastic and thus consider them to be pseudoapoplastic. In fact, some downward symplastic movement has been observed, but if chemicals are not actively loaded into the phloem or retained there by slow diffusion, the overwhelmingly greater and faster transpiration stream will carry by far the greater part of the compound to the plant parts that transpire most. As a result, little or no fungicide usually ends up in flowers or fruits (Peterson and Edgington, 1971; Edgington and Peterson, 1977).

A. PROTECTANT FUNGICIDES

The most important protectant fungicides used against zoosporic plant pathogenic fungi are only listed briefly below but their biochemical modes of action and chemical and biological properties are extensively described elsewhere (e.g. Horsfall, 1945; Torgeson, 1967; Evans, 1968; Lukens, 1971; Siegel and Sisler, 1977). Mercury-containing compounds are omitted as their use is banned in an increasing number of countries because of undesirable biocidal non-target effects.

1. *Fixed Copper Fungicides*

Bordeaux mixture was discovered and developed by Millardet in 1885 (see p. 194). It is prepared by mixing a solution of copper sulphate with lime, forming a blue gelatinous suspension. A common recipe employs 1 kg copper sulphate crystals and 1.25 kg hydrated lime $100 \, l^{-1}$ water, but may be varied according to the requirements for disease control and sensitivity of plant tissue. The components should always be prepared freshly. It is a good, tenacious, foliar fungicide and has the valuable characteristic that the weathered remains of earlier deposits assist the adhesion of the later application (Martin and Worthing, 1976). Fungicidal activity is associated with the slow release of water-soluble copper compounds, the ultimate toxicant being the cupric ion.

In other fixed copper compounds, the copper ion is only slightly soluble and they are, therefore, less phytotoxic than Bordeaux mixture, but also less effective as fungicides.

2. *Quintozene*

Quintozene or pentachloronitrobenzene (PCNB) is active against some zoosporic soil-borne fungal pathogens like *Plasmodiophora brassicae* and *Olpidium brassicae* but it is relatively ineffective against *Pythium*, *Phytophthora* and *Aphanomyces* (Kreutzer, 1963; Rodriguez-Kabana *et al.*, 1977). It is stable in soil, non-volatile, not very soluble in water, and is more fungistatic than fungicidal.

3. *Dialkyldithiocarbamates*

Dialkyldithiocarbamates, such as thiram and ferbam, are safe fungicides in terms of phytotoxicity and are used against many pathogens as sprays or as soil and seed treatments. Unlike maneb and zineb (see below) neither thiram nor ferbam breaks down to ethylenethiourea although ferbam does leave black residual deposits that are often undesirable on edible plant parts. Thiram appears to be more active against facultative saprophytes than against obligate parasites (Kreutzer, 1963).

4. *Ethylenebisdithiocarbamates (EBDCs)*

EBDCs like maneb, zineb and nabam are the cornerstones of chemical control of downy mildews and foliar blights. Zineb, maneb and the combinations mancozeb and metiram have low mammalian toxicity, but may decompose, especially when heated, to ethylenethiourea, which has been shown to induce carcinomes in the thyroid gland (Ulland et al., 1972). Treated edible crops should not be harvested within seven days of application.

5. Captan

As a foliar fungicide against zoosporic plant pathogens, captan has never become as popular as EBDCs and against this group of pathogens it is more widely used as a seed protectant. The activity of captan as a broad-spectrum seed protectant is similar to that of thiram, although captan is more active against Pythium than against Rhizoctonia and not at all effective against Phytophthora and Aphanomyces (Kreutzer, 1963).

6. Drazoxolon

This fungicide is used as a seed treatment to control zoosporic fungi that cause damping-off and as a soil drench against zoosporic root invading fungi but because of its considerable mammalian toxicity, care should be taken to protect users.

7. Chlorothalonil

Chlorothalonil is an excellent foliar broad-spectrum fungicide used against a wide range of crop diseases including downy mildews and late blights although Jones (1978) stressed the fact that it is effective only if spray applications start before the first disease symptoms are detectable in the field.

8. Fenaminosulf

Fenaminosulf is a narrow-spectrum soil and seed fungicide, particularly effective against pathogens in the genera Aphanomyces, Phytophthora and Pythium. It is decomposed by light (Hills and Leach, 1962), but this does not impair its activity once incorporated in the soil and, as it is very water soluble, it penetrates well as a soil drench. Fenaminosulf is very stable and can persist for over one year in soil (Alconero and Hagedorn, 1968; Mitchell and Hagedorn, 1971) with very little difference in degradation rate in different soil types. It is relatively mobile in soil (Helling et al., 1974) and tends to move more in muck than in clay loam (Alconero and Hagedorn, 1968).

B. SYSTEMIC FUNGICIDES

As systemic fungicides are usually mono-site inhibitors and affect one single enzyme or structural protein in the fungal cell, and since different fungal groups may have marked quantitative and qualitative differences regarding these aspects of their biochemistry, most systemic compounds exhibit selective antifungal activity. Until recently, the Oomycetes escaped all the widely used systemic compounds: benzimidazoles had no affinity to the tubulin of their microtubuli, oxathiins did not affect their energy pathways, chitin synthetase inhibitors were not toxic because these fungi do not employ chitin in their cell walls, and compounds that interfere with sterol bio-synthesis were tolerated because Oomycetes grow well without sterols.

With the advent of a number of very selective systemic Oomycete fungicides the situation changed and control strategies could become more directed and sophisticated. Since most of these systemics are too new to have been discussed in review articles, we will describe them in more detail than the older, broad-spectrum systemic materials. The selective activity of these systemic compounds among and within the groups of zoosporic fungi is summarised in Table 1 while the general characteristics of prothiocarb, propamocarb, cymoxanil, fosetyl-Al, and the acylalanines were briefly described and compared by Schwinn (1979).

1. *Benzimidazoles*

Benzimidazole fungicides, such as benomyl, carbendazim and thiophanate-methyl are not generally active against zoosporic fungi, but their toxicity to *Plasmodiophora*, *Olpidium* and *Physoderma* warrants their mentioning here. Carbendazim, the fungitoxic degradation product, is known to bind to tubulin protein units of microtubuli, thus interfering with mitosis (Davidse, 1975) and the selective toxicity of benzimidazoles is based on differences in affinity of tubulin to these compounds among various fungal groups (Davidse and Flach, 1977).

2. *Streptomycin*

The antibiotic streptomycin is toxic to gram-negative and gram-positive bacteria. After its first isolation in 1944 from *Streptomyces griseus*, it became widely used in medicine, and on a small scale in agriculture to control certain bacterial fruit diseases. Besides it antibacterial action, it was found to be effective against *Phytophthora infestans* (Müller *et al.*, 1954; Hodgson, 1963), *Pseudoperonospora humuli* (Horner, 1963), *Peronospora tabacina* and *P. parasitica* (Anderson, 1956). Its antifungal action was first thought to be induced indirectly on the pathogens via the plant metabolism but Vörös

Table 1. Activity of systemic fungicides against various zoosporic plant pathogenic genera.

Systemic fungicide: Fungal genus	Benzimidazoles	Streptomycin	Hymexazol	Etridiazole	Chloroneb	Cymoxanil	Propamocarb	Fosetyl-Al	Acylalanines	Remarks
Myxomycota										
Plasmodiophoromycetes										
Plasmodiophora	+		±			±			−	h
Spongospora	−									−
Polymyxa	−						+		−	
Eumycota										
Chytridiomycetes										
Olpidium	+		−				+		−	t
Synchytrium	−								−	h
Physoderma	+								−	o
Oomycetes										
Saprolegniales										
Aphanomyces	−	+					+/−			h
Peronosporales										
Albugo	−							−	+	
Bremia	−					−	±	±	+	
Peronospora	−	+				−	+/−	−	+	
Peronosclerospora	−				+ s				+	
Plasmopara	−					+	−	+	+	
Pseudoperonospora	−	+				+	+ s		+	
Sclerophthora	−								+	
Sclerospora	−			+ s					+	
Pythium	−		+	+	±	±	±	+	+	
Phytophthora (on foliage)	−	+	−			+	−	−	+	
Phytophthora (on roots and stems)	−		−	+	+/−	−	+	+	+	

Abbreviations: +, high activity; + s, activity only with applications to seed or soil; −, no useful activity; ±, some activity; +/−, activity varies from species to species; blank, for no information available; h, herbicides known to affect pathogen; t, control by triadimefon reported; o, control by oxycarboxin reported.

(1965) demonstrated a direct effect of the antibiotic on the fungi. Much less streptomycin was absorbed by mycelium of insensitive than of sensitive fungi, and he suggested that the presence of chitin might make the cell walls less permeable to streptomycin. Streptomycin binds to 30S ribosomal subunits in bacteria (Kaji and Tanaka, 1968), and it was found that no binding to 30S subunits takes place in bacteria that are resistant to streptomycin. It is not clear why some Oomycetes are sensitive to this compound, since no binding occurs to their 50S subunits. Perhaps the site of action here is the mitochondrial ribosome. More detail on the various mechanisms of interference with protein synthesis is given by Siegel (1977) and the agricultural use of antibiotics was reviewed by Zaumeyer (1958).

3. *Hymexazol*

Hymexazol was introduced in 1970 as a soil fungicide and plant growth promoter. It is effective *inter alia*, against the soil-borne zoosporic pathogens *Aphanomyces* and *Pythium* (Takahi *et al.*, 1974). *In vitro*, hymexazol inhibits *Pythium* but not *Phytophthora* and can, therefore, be used in selective culture media for *Phytophthora* spp. (Tsao and Guy, 1977). The mechanism of action in sensitive organisms is possibly inhibition of nucleic acids (Kamimura *et al.*, 1976b); natural resistance in fungi is correlated with low uptake (Kamimura *et al.*, 1976a).

4. *Etridiazole*

Etridiazole, introduced in 1969, is a soil fungicide recommended for the control of *Phytophthora* and *Pythium* spp. and other pathogens of turf, vegetables, fruit, cotton, groundnuts, and ornamentals. It is also used in combination with quintozene as a soil fungicide and seed treatment for the control of pre- and post-emergence cotton seedling diseases caused by *Rhizoctonia*, *Pythium*, *Phytophthora* and *Fusarium* spp. (Worthing, 1979). Some systemic activity was observed by Al-Beldawi and Sinclair (1969) in cotton and by Muller *et al.* (1972) in tomato and, consequently, etridiazole was regarded as a slow moving systemic fungicide. Recently, Kajiwara *et al.* (1979) observed a quick uptake and a uniform distribution of etridiazole in corn.

   The biochemical mechanism of action was studied by Halos and Huisman (1976b) who observed that etridiazole inhibited respiration in *Pythium* spp. and speculated the site of action to be blockage of electron transport between cytochrome *b* and *c*. They indicated that resistance in *Pythium* spp. was due to an increased use of an alternative respiration pathway of electron transport, mediated by ubiquinone (Halos and Huisman, 1976a). Lyr *et al.* (1977) observed lysis of mitochondria in *Mucor mucedo* Bref. by etridiazole and assumed that this effect was induced by a liberation of phospholipases within the mitochondria. A significant thickening of the cell wall was also observed.

5. *Chloroneb*

Chloroneb, introduced in 1967, was the first systemic fungicide with antifungal action against Oomycetes. It is taken up by the roots and concentrated in the roots and lower stem parts (Fielding and Rhodes, 1967; Sinclair and Darrag, 1966). Zoosporic fungi vary widely in their sensitivity to chloroneb; *Phytophthora cinnamomi* was quite sensitive for instance, but species of *Pythium* were found only moderately so (Hock and Sisler, 1969) and use of the chemical has remained limited.

The mode of action is not yet elucidated, but there are indications that chloroneb interferes with DNA synthesis and mitotic cell division (Hock and Sisler, 1969; Tillman and Sisler, 1973; Kataria and Grover, 1975). Tillman and Sisler (1973) produced a mutant of *Ustilago maydis* (D.C.) Corda resistant to chloroneb. This resistance was monogenic and conferred cross-resistance to diphenyl, dicloran, *p*-dichlorobenzene, hexachlorobenzene, naphthalene and quintozene. The mode of action of chloroneb is suspected to be similar to that of the dicarboximides (vinclozolin, iprodione, procymidone) and the aromatic hydrocarbon fungicides mentioned (Georgopoulos *et al.*, 1979; Kaars Sijpesteijn, 1982). Lyr (1977) found that chloroneb caused lytic effects in the mitochondrial structure of *Mucor mucedo* similar to those caused by etridiazole, while Kataria and Grover (1975) believed inhibition of cellular and mitochondrial respiration to be the primary mode of action in *Rhizoctonia solani* Kühn.

6. *Cymoxanil*

Cymoxanil is a selective Oomycete fungicide with local systemic action. This means it may penetrate plant tissue, but is not translocated to other parts of the plant. It is exclusively used in combination with a protectant fungicide such as mancozeb, folpet or a fixed copper. Cymoxanil has curative properties when used shortly after infection, which permits the use of a more flexible spray programme than is possible solely with protectants. It also decomposes very rapidly and becomes inactive within four to six days, the principal breakdown product being glycine (Belasco, 1980). Its half-life in soil is less than two weeks (Douchet *et al.*, 1977). The formulation in combination with a protectant fungicide provides more persistent control than the short residual life of cymoxanil alone would give and the formulations and recommendations are designed to apply half rates of the protectant component together with low amounts (100 to 120 g ha^{-1}) of cymoxanil. Although excellent protection against soil-borne *Pythium* spp. was achieved in cucumber by seed and soil treatments (Diaconu, 1979), cymoxanil is mainly used in foliar applications to control *Phytophthora*. In contrast to its very safe use in foliar sprays, cymoxanil was found phytotoxic to lettuce when applied to roots at concentrations as low as 50 p.p.m. (Crute and Norwood, 1977).

7. *Prothiocarb and Propamocarb*

Prothiocarb was introduced in 1974 as a selective fungicide against diseases caused by soil-borne Oomycetes (Bastiannsen *et al.*, 1974). Kaars Sijpesteijn *et al.* (1974) found the *in vitro* antifungal spectrum confined to members of the Oomycetes, although great differences in sensitivity among these species were noted. Fungi belonging to the Zygomycetes and Chytridiomycetes tested were all insensitive. The antifungal activity of prothiocarb is greatly influenced by pH, temperature and quality of the growth medium; *in vitro*, it is less fungitoxic as the acidity of the medium increases although in soils this pH effect is apparently less pronounced. Prothiocarb was reported to be fungistatic rather than fungicidal and is more active in soil treatments than when applied as foliar sprays.

Prothiocarb exhibits two modes of action (Kerkenaar and Kaars Sijpesteijn, 1977). Its most common action is exerted by the intact molecule, which acts against members of the Peronosporales and Leptomitales and against some members of the Saprolegniales, like *Aphanomyces* species. For other members of the Saprolegniales, comprising species of *Achlya*, *Pythiopsis* and *Saprolegnia*, however, the toxic effect of prothiocarb can be antagonised by L-methionine. Apparently, prothiocarb releases the volatile substance ethylmercaptan that causes L-ethionine to be incorporated into proteins instead of L-methionine, so resulting in the production of missense-proteins. Fungitoxicity to the latter species, therefore, was ascribed to ethylmercaptan released from prothiocarb.

Propamocarb is an analogue of prothiocarb containing a propyl carbamate moiety instead of a thiocarbamic acid S-ethyl ester group. In contrast to prothiocarb it is odourless, but in many other respects they are very similar (Papavizas *et al.*, 1978). Both compounds are very water soluble ($>50\%$).

8. *Ethyl Phosphonates*

Fosetyl-Al and LS 73-1038 (sodium ethyl phosphonate) form an entire new class of fungicides exhibiting no fungitoxic activity *in vitro*, but high protective and curative activity against some downy mildews and soil-borne species of *Phytophthora* and *Pythium* when applied to plants. These compounds are translocated symplastically in plants (Bertrand *et al.*, 1977) and there are indications that antifungal action in tomato is mediated by stimulation of the defence mechanisms of the host plant by increasing the levels of phenolic substances and phytoalexins (Vo-Thi-Hai *et al.*, 1979). The systemic moiety is probably the ethyl phosphonate ion. Aluminium and sodium respectively ionise slowly and rapidly, releasing the organic moiety at the plant surface.

9. *Acylalanine-type Compounds*

This is a new, very promising group of structurally related systemic fungicides, including metalaxyl, furalaxyl, milfuram, Galben, RE 26745 and RE 26940. The antifungal spectrum of these fungicides is entirely confined to the Peronosporales (Schwinn *et al.*, 1977) and to a lesser extent to the

Hyphochytriomycetes (Bruin, 1980). Most fungi belonging to the Peronosporales are extremely sensitive to most of these compounds *in vitro* as well as *in vivo*, although vast differences exist in sensitivity between fungal species and in toxicity between different compounds (Staub *et al.*, 1979; Bruin and Edgington, 1981). The name "acylalanine compounds" as a generic name for this group of fungicides does not do justice to all members, since RE 26940 does not contain an alanine moiety. Moreover, these compounds do not appear to act biologically as amino acid analogues. However, since it does not appear possible to find a much more descriptive name that encompasses all members, and since this name is established in the literature (Staub *et al.*, 1978a, 1979), we will use the word "acylalanine-type" fungicides for a broader coverage.

Metalaxyl and furalaxyl are readily-taken up by leaves, green stems and roots, and are translocated pseudapoplastically, mainly with the transpiration stream (Staub *et al.*, 1978a; Cohen *et al.*, 1979) and foliar protection by metalaxyl within 1 hour after a soil drench of four-leaf tomato plants was reported by Cohen *et al.* (1979). Metalaxyl also gives good translaminar protection against grape downy mildew (*P. viticola*) and there are indications that some symplastic transport takes place (Staub *et al.*, 1978a). It did not inhibit motility and germination of zoospores, nor appressorium formation, penetration or initiation of the first haustorium but further fungal development was completely inhibited (Staub *et al.*, 1978b).

Acylalanine-type fungicides exhibit a remarkable, but limited curative action and data on *Albugo candida*, *Bremia lactucae* and *Phytophthora infestans* indicate that acylalanine-type fungicides are effective eradicants if applied during the first half or two-thirds of the incubation period after infection. Later applications do not inhibit lesion formation and spore production, but may have an adverse effect on spore viability (Bruck *et al.*, 1980; Hildebrand, personal communication).

After their introduction in 1977, acylalanine-type fungicides have been evaluated against a multitude of diseases caused by members of the Peronosporales, resulting in an avalanche of reports on their efficacy in controlling local and systemic infections of downy mildews and white blisters, and of many diseases caused by *Phytophthora* and *Pythium* species. They can be applied as sprays against air-borne inoculum of downy mildews and late blight, as seed dressing, in-furrow application, soil drench, soil incorporation or as granules against soil-borne parasites. Because they are taken up by roots and translocated upwards in plants, seed and soil applications of acylalanine-type fungicides can also be utilised to control foliar diseases.

Metalaxyl is not considered to be a very volatile fungicide, but it was found to have considerable activity in the vapour phase (Crute and Jagger, 1979), which probably caused a redistribution of the compound between treated and non-treated tobacco leaves during the curing process (Bruin *et al.*, 1981).

Although the mode of action of acylalanine-type fungicides is not yet elucidated, there are strong indications that they interfere with RNA biosynthesis. Davidse (personal communication) found that metalaxyl substantially inhibited incorporation of [³H]uridine into RNA in *Phytophthora megasperma* Drechsler f. sp. *medicaginis* within five minutes of application. The fact that resistance to metalaxyl usually is cross-linked to resistance to all related compounds suggests a similar mode of action for these materials (Bruin, 1980; Bruin and Edgington, 1981). There are strong indications that acylalanine-type compounds have a more complicated fungicidal action *in vivo* than *in vitro*. Several species of *Phytophthora* and *Pythium* quickly lost their sensitivity to metalaxyl *in vitro* when exposed continuously to sublethal concentrations, but *in vivo*, neither *Phytophthora infestans* nor *Peronospora parasitica* (Fr.) Tul. lost any of their sensitivity when exposed to sublethal concentrations in potato tuber tissue and cabbage seedlings, respectively. The isolate of *P. infestans* was completely insensitive to metalaxyl *in vitro*, but did not grow on potato tuber tissue containing as low a concentration as 0.015 µg metalaxyl g⁻¹ (Bruin and Edgington, 1981). Thus, in addition to its direct fungitoxicity, metalaxyl must have a plant-mediated antifungal action, possibly through stimulating host defence mechanisms. Ward *et al.* (1980) found metalaxyl to stimulate hypersensitive reactions, accompanied by accumulation of the phytoalexin glyceollin in soybean hypocotyls after inoculation of *P. megasperma* var. *sojae*, thus changing compatible into incompatible host–pathogen interactions. Metalaxyl by itself, without inoculation, did not stimulate phytoalexin accumulation. Crute and Jagger (1979) compared similar reactions of host cell death in lettuce with reactions observed when non-hosts or resistant cultivars were infected by incompatible isolates of *B. lactucae*.

The occurrence of a number of serious outbreaks of resistance to metalaxyl in cucumber downy mildew (Reuveni *et al.*, 1980) and potato late blight (Davidse *et al.*, 1981) makes the future use of acylalanine-type fungicides uncertain and stresses the need for well-balanced application schedules utilising fungicide combinations and avoiding prolonged selection pressure of one single systemic fungicide. The fact that no resistance to metalaxyl was found in 100,000 ha of potatoes where this compound was applied as Fubol (10% metalxyl plus 48% mancozeb) (Anon., 1980) is a hopeful indication that we might be able to take full advantage of these potent chemicals without running into resistance problems, as long as we apply them wisely.

C. UNIQUE CHARACTERISTICS OF OOMYCETE-SELECTIVE FUNGICIDES

1. *Water Solubility and log P-Values*

Historically, almost all fungicides used in agriculture have been sparingly soluble in water. In order for a compound to interfere with the aqueous chemistry of the fungus it must be soluble in water, but in order to penetrate the lipidic plasma membrane, a certain degree of lipophilicity is necessary (Horsfall, 1945, 1956). These conflicting characteristics triggered a great deal of research and theorising in the 1940s, which resulted in the concept of the oil/water partition coefficient as a determinant for permeation. This partition coefficient is the ratio of solubility in oil and water, usually expressed as the log P-value of a compound, indicating the logarithm of the ratio of solubility in n-octanol (simulating oil) and water. Permeation increased as P increased (Horsfall, 1945, 1956). Rich and Horsfall (1952) found that 4-nitrosopyrazole derivatives became more fungitoxic as their log P-value increased to a peak which varied with the fungus. Maximum penetration was achieved with optimum log P-values, usually in the range of 2 to 3. Often water solubility was inversely related to fungal toxicity. However, low water solubility does not always result in high lipid solubility.

These concepts are based on research with non-zoosporic fungi as test organisms, and hold for the majority of protectant broad-spectrum fungicides as well as for the systemic fungicides that do not selectively affect Oomycetes. With the advent of systemic fungicides that act selectively against Oomycetes a peculiar phenomenon is observed in that all these compounds are quite soluble in water, although at least one, metalaxyl, has a relatively high log P-value (Table 2). Unfortunately, we are unaware of the log P-values of the other compounds listed, but they are probably quite low. As Oomycetes usually live in aqueous environments, hydrophilicity could be expected to be an important asset for toxicity of selective compounds. No data are available on whether this results from a more hydrophilic plasma membrane, from different membrane transfer proteins, or from more hydrophilic sites of action in zoosporic, as compared with other fungi.

2. Side-effects

Since the Oomycete-selective systemic fungicides were introduced only recently, few data are available on their short- and long-term effects on non-target organisms and on other environmental factors. We may speculate, however, that because of the highly selective action to a group of fungi which is known to have only weak saprophytic competitive properties, and the relatively small amounts of chemical used, the general environmental impact will not be very significant. Phytotoxic effects will safeguard over-use of some of these compounds. What we can say is that through the use of these effective systemics, the use of protectants will drastically decrease, which by itself is a great step towards cleaner food production. Unconfirmed reports indicate that good control of downy mildew on crops by acylalanine-type compounds could lead to increased incidences of powdery mildew. We must keep in mind here, of course, that by controlling one pathogen effectively,

Table 2. Correlation between Oomycete selectivity and hydrophilicity of fungicides.

Fungicide	Selectivity to Oomycetes	Water solubility (μg ml^{-1})	Log P[a]
Protective			
EBDCs: maneb, metiram, zineb, mancozeb	n[b]	<25	
Dialkyldithiocarbamates: thiram, ferbam	n	30	
Fenaminosulf	o	20,000	
Systemic			
Chloroneb	n	8	
Etridiazole	n	25	
Cymoxanil	o	1,000	
Hymexazol	o	8,500	
Fosetyl-Al	o[c]	120,000	
Propamocarb	o	700,000[d]	
Acylalanines: metalaxyl	o	7,400	2.95
milfuram	o	140	2.34
RE 26745	o	20	1.61

[a] P is n-octanol/water partition coefficient.
[b] n is not selective to Oomycetes and o is selectivity confined to Oomycetes.
[c] Not directly toxic, but controls diseases caused by some Oomycetes.
[d] Formulated at 70% a.i. in water as a true solution.

another pathogen will find a greater potential for expansion. From our knowledge of the biological and chemical activity of the systemic Oomycete fungicides, a number of intriguing questions come to mind. For instance, since acylalanine-type fungicides are also toxic to Hyphochytriomycetes, several members of which are known to parasitise oospores (Sneh et al., 1977; Wynn and Epton, 1979) the effect of these compounds on oospore survival is very uncertain. Also, the effects of glycine, released from cymoxanil deposits, on the microflora of the phyllosphere are not known, and neither are the effects of aluminium ions, left behind on the leaf surface after release and penetration of the ethyl phosphonate moieties.

IV. Fungicide Resistance

Before the use of systemics, resistance to fungicides was not regarded as a major problem (Georgopoulos and Zaracovitis, 1967). The reason for this is that most non-systemic fungicides are general cell poisons, affecting many vital processes in the fungal cell. Apparently, it is almost impossible for a fungus to overcome this toxicity with a number of simultaneous mutations and such resistance as did occur was mostly due to decrease in membrane permeability or detoxification, resulting with a few exceptions, in a low degree of resistance (Dekker, 1969). We are aware of only one report of resistance in zoosporic fungi to multi-site fungicides—that to mercury in *Plasmodiophora brassicae* (Tinggal and Webster, 1981).

After the introduction of mono-site systemic fungicides the picture changed: between 1971 and 1976 at least 23 different plant pathogenic fungal species developed resistance in the field to benzimidazole fungicides (Georgopoulos, 1977), three to polyoxin, one to ethirimol, one to dimethirimol, one to oxycarboxin and one to kasugamycin (Dekker, 1976). Subsequently, many more species acquired resistance to these and other systemic fungicides (Delp, 1980), but these resistance problems had no impact on zoosporic fungi since the compounds involved are not toxic to them (with the exception of benzimidazoles to some members of the Plasmodiophoromycetes and Chytridiomycetes). The history, origin and mechanisms of resistance *in vitro* and *in vivo*, and genetical and practical aspects of the resistance are well documented in reviews by Dekker (1976), Fehrmann (1976), Georgopoulos (1977) and Uesugi (1978).

Under laboratory conditions several strains of *Phytophthora* with decreased sensitivity to toxicants were produced. Shaw and Elliott (1968) obtained spontaneous mutants of *P. cactorum* to streptomycin at high frequency (eight resistant isolates from 10^5 zoospores). About the same ratio was obtained with streptomycin resistance in *P. capsici* Leonian (Timmer *et al.*, 1970). Shaw and Khaki (1971) exposed 10^8 zoospores of *P. drechsleri* Tucker to a high concentration of *p*-fluorophenylalanine (Fpa) and obtained one viable resistant isolate. Treatment with the mutagenic agent *N*-methyl-*N'*-nitro-*N*-nitrosoguanidine (NG) resulted in four isolates resistant to chloramphenicol. Various mutagenic agents induced resistance in *P. drechsleri* to cycloheximide, Fpa, chloramphenicol and tetracycline (Khaki and Shaw, 1974). Shattock and Shaw (1975) obtained strains of *P. infestans* resistant to chloramphenicol and streptomycin using NG, and in the same way Long and Keen (1977a,b) induced resistance in *P. megasperma* var. *sojae* to cycloheximide and Fpa. This research indicates that zoosporic fungi do have a propensity to develop resistance if the conditions are conducive for them to do so and there is no reason to believe that these pathogens are less prone than fungi belonging to other groups to become resistant to mono-site toxicants under field conditions. In fact, the wide genetic variability observed between different isolates of various *Phytophthora* species, as well as their remarkable adaptability to host changes are clear signs of warning. The fact that no resistance to the compounds mentioned, nor to fenaminosulf, etridiazole, chloroneb, hymexazol, prothiocarb, propamocarb, fosetyl-Al or cymoxanil has been observed may be ascribed to the very limited use these compounds have had in agriculture until now. Cymoxanil has the advantage of being marketed only in combination with a protectant which probably decreases the threat of development of resistance.

For the acylalanine-type fungicides the picture is more complicated. Lukens (1978) subcultured *Plasmopara viticola* during nine successive transfers on grape plants containing EC_{95} concentrations of milfuram, and found no decrease in sensitivity of the pathogen to this compound. A year later the same was reported for *P. infestans* and metalaxyl on potato by Staub *et al.* (1979). However, they found a great variation in sensitivity to acylalanine-type compounds among 30 isolates of *P. infestans* with EC_{50} values ranging from 0.01 to over 100 µg ml^{-1}. It was also noted that several strains quickly lost sensitivity when cultured for four transfers on rye agar amended with 1 µg ml^{-1} furalaxyl or CGA 29212, a related acylalanine. In general though, loss of sensitivity was accompanied by decreased pathogenicity and resistant isolates were, with one exception, not able to infect tomato and potato plants that were treated with relatively high doses of CGA 29212. From these results it was concluded that resistance *in vitro* is not necessarily related to resistance *in vivo*.

Although no resistance in practice had developed by November 1979, Schwinn (1979) strongly advocated the use of acylalanines in mixture with other fungicides to decrease the chance of development of resistance. At the same time Bruin and Edgington (1980b) showed that *P. capsici* quickly lost sensitivity to metalaxyl when subcultured on agar amended with sublethal concentrations of the fungicide, resulting in a stable form of resistance without loss of pathogenicity on pepper. In fact, the resistant strains successfully infected plants that were treated with concentrations of metalaxyl high enough to be phytotoxic. Induction of resistance with UV irradiation also resulted in highly resistant strains of *P. capsici* and *Pythium ultimum* Trow without loss of pathogenicity (Bruin and Edgington, 1980b, 1981); Davidse (1980, 1981) obtained similar results with *P. megasperma* f. sp. *medicaginis*, using NG.

Although various authors strongly advised the use of systemic fungicides in combination with other materials (Schwinn, 1979; Edgington *et al.*, 1980; Delp, 1980), metalaxyl was marketed and used as a single compound, resulting in dramatic outbreaks of resistance in *Pseudoperonospora cubensis* (Reuveni *et al.*, 1980) and *P. infestans* (Davidse *et al.*, 1981). It is not hard to identify the factors and faults that led to these unnecessary disasters, and they may serve as a guide to establish strategies and recommendations to prevent new calamities.

In order to understand the mechanisms operating in the development of resistance, the following factors should be considered: the nature of the pathogen, the properties of the chemical and the characteristics of resistant mutants.

(a) *Nature of the pathogen.* Variation in morphological and physiological characteristics have been studied in detail for many species of *Phytophthora.* The tremendous variability and plasticity, especially of isolates of *P. infestans*, have puzzled researchers for many years. This pathogen reacted to the introduction of new genes for resistance in potato varieties with the rapid formation of new pathogenic races. These races themselves appear to be flexible in their pathogenicity and may change their characteristics during a few passages through host plants (Reddick and Mills, 1938; Mills, 1940a,b). The origin of the enormous asexual variation may be based on the observed high frequencies of mutation (cf. Gallegly, 1968), somatic recombination (e.g. Denward, 1970; Leach and Rich, 1969), heterokaryosis (Long and Keen, 1977a,b; Bruin and Edgington, 1981), cytoplasmic controlled variation (e.g. Caten and Jinks, 1968) and on anastomosis (Stephenson *et al.*, 1974a,b). Each of these mechanisms may operate in zoosporic fungi, making them extremely "plastic" in their parasitic behaviour. This, and their diploid constitution, which permits fungal units to carry much more genetic information than they express phenotypically, may be indicative of what to expect when we start tackling those organisms with single-site chemicals.

Staub *et al.* (1979) showed great variability in sensitivity to acylalanine fungicides among 30 isolates of *P. infestans*, and Bruck *et al.* (1981) found comparable differences among field isolates of *Peronospora tabacina*, sampled from metalaxyl-treated fields. Bruin and Edgington (1981) observed great differences in sensitivity to metalaxyl among single zoospore isolates from one hyphal tip isolate of *P. capsici*, indicating the coexistence of nuclei with different genetic expression of responses to the toxicant within a single hypha. They had reason to believe that "resistant" nuclei may pre-exist in nature and hypothesised that adaptation to this compound may simply be based on a selection for "resistant" nuclei.

(b) *Properties of the chemical.* Fungi develop high levels of resistance to mono-site much faster than to multi-site fungicides. Although the mechanisms of action of the new systemic Oomycete fungicides are not yet fully known, their systemic properties and highly selective antifungal spectra are indications of single sites of action. Nevertheless, prothiocarb may act differently on different species while fosetyl-Al probably lacks a direct fungitoxic action, which makes this compound unique among the fungicides.

(c) *Characteristics of resistant mutants.* If resistance to a fungicide is linked with loss of virulence or decreased viability, as occurred in the sterol-inhibiting fungicides triforine (Fuchs *et al.*, 1977) and fenarimol (Waard and Gieskes, 1977), resistance will probably not become a problem in the field. Therefore, growth, sporulation and pathogenic characteristics of laboratory-produced resistant strains should always be studied before extrapolations to field situations can be made. In the case of metalaxyl resistance produced in the laboratory, many resistant strains of various *Phytophthora* and *Pythium* species were still as pathogenic as the parent strains.

If we pool our knowledge on the above-mentioned factors, we arrive at the frightening conclusion that all the factors are present for the rapid development of resistance in the Peronosporales to acylalanine-type fungicides as soon as these pathogens are placed under intense selection pressure.

It is not clear, however, why resistance developed so soon in *Pseudoperonospora cubensis* and in *Phytophthora infestans*, but not in *Peronospora tabacina*, even after five years of intensive use of metalaxyl in tobacco in Australia. We may hypothesise that besides differences among the various pathogens, differences between host plant varieties may also be of significance, especially if we suppose that the antifungal action of acylalanine-type fungicides involves effects on host defence mechanisms. These mechanisms will be different in each plant species.

It seems possible for pathogens to circumvent host defence mechanisms that are governed by single genes, just as pathogens can tackle major genes for resistance. With this in mind it will be interesting to see whether resistance to fosetyl-Al will develop.

The use of combinations of fungicides with the aim of preventing or delaying the development of resistance is a matter of controversy and is only backed by circumstantial evidence and mathematical models (Delp, 1980; Kable and Jeffrey, 1980; Dekker, 1982). At least the accompanying fungicide has its own retarding effect on the development of the disease and may eliminate sporadically appearing resistant units. In case resistance develops to one component of the mixture, the second might keep the population low and hopefully below the threshold level to cause crop losses. However, this approach asks for a strong companion, which is not always available, and might have economic or registrational restraint. The powerful systemics should not be used indiscriminately in routine protective spray schedules, but their use should be restricted to those treatments where they can exercise their optimal effect, such as in seed treatments and as eradicant applications shortly after infections occur.

V. Future Considerations

Beyond doubt, the introduction of systemic Oomycete fungicides was a significant step towards better directed, more subtle, cleaner and more efficient control of plant diseases caused by zoosporic fungi. This does not mean, however, that the protectant blunderbuss compounds are entirely outdated; they are still badly needed, for instance, to help safeguard systemics against resistance.

Very few data are available on ways to avoid or delay the onset of resistance; this area of research is of vital importance to secure a prolonged use of these systemics. It would be a shame to lose some or all of these potent compounds due to resistance resulting from ignorant usages. New compounds should be subjected to tests before commercial release to estimate the chance of development of resistance in target organisms. Understandably, agrochemical companies have traditionally been very individualistic and secluded in developing and marketing of fungicides. Fortunately, this attitude is slowly eroding under the pressure of practical resistance problems for not only it is vital to pool knowledge on the subject, but it is slowly realised that pooling of resources (e.g. recommending combined use of two or more highly specific fungicides), may prolong the useful life of each compound involved. In this respect governmental registration agencies should become more receptive to changing needs.

The development of modern weather and disease prediction systems provides an additional valuable aid to effective disease control. They permit the farmer to wait until the need for fungicide applications arises; a need that is determined by economic disease threshold levels, inoculum pressure and environmental conditions. The systemic compounds even allow him to wait until infection has taken place and still effectively protect the crop. It might be clear that, in order to take optimal advantage of these new chemicals, monitoring disease and weather, as well as utilisation of disease prediction systems either via regional communication media or on an individual basis, is indispensable. Schwinn (1979) rightly stressed the need for improved application techniques to deliver the fungicidal molecules precisely to the sites where they are needed and old-fashioned methods such as we use with protectants-seed infusion techniques, soil drenches, granules, slow release granules or pills, should be carefully evaluated and compared with other application methods.

Undoubtedly, breeding for resistance, and especially breeding for horizontal resistance, is a valuable tool to combat diseases, and although this road is long and difficult, it should lead to the development of varieties with long-lasting disease resistance. In this area, systemic fungicides can play a key role. Hitherto, completely resistant varieties containing major genes for resistance were considered non-target plants for the diseases concerned. Whilst protecting these resistant varieties with fungicides may appear wasteful, it could lengthen their life significantly by minimising the likelihood of new pathogen races arising.

Breeding for horizontal resistance will almost always result in varieties with less than complete levels of disease resistance. Here, low rates of fungicides can complement this resistance to acceptable levels. Partial resistance can, therefore, be considered as a substitute for the bulk of fungicide use. This substitution effect, already indicated by Van der Plank (1963), was clearly demonstrated by Crute and Gordon (1980a) who obtained better control of downy mildew of lettuce (*Bremia lactucae*) with eight times lower concentrations of metalaxyl on a variety with horizontal resistance than with full rates on a susceptible variety.

The development of the systemic Oomycete fungicides showed that zoosporic fungi are not untouchable and can be controlled effectively with selective chemicals. A vast area of research lies in front of us to refine application techniques and schedules, and to find compounds that specifically act on sites that are unique to zoosporic fungi: inhibition of zoosporangium formation or release, motility, encystment or germination of zoospores for instance.

A very promising area of plant disease control, especially of biotrophic parasites, lies in the specific enhancement of natural resistance to pathogens such as the stimulation of host defence mechanisms; an area that is timidly set foot on by the ethyl phosphonates, compounds that also usher in the era of so badly needed symplastically translocated fungicides.

7

Zoosporic Fungi and Viruses

Double Trouble

DAVID S. TEAKLE

*Department of Microbiology, University of Queensland,
St. Lucia, Australia*

I.	Introduction	233
II.	The Vectors	234
	A. Chytridiomycetes	235
	B. Plasmodiophoromycetes	235
III.	The Viruses	236
	A. The Isometric Viruses	236
	B. The Straight Tubular Viruses	236
	C. The Filamentous Viruses	238
	D. The Viruses of Unknown Morphology	238
IV.	Virus-Vector Relationships	239
	A. The Non-Persistent Relationship	239
	B. The Persistent Relationship	240
V.	Epidemiology and Control	241
	A. Geographical Coincidence of Vector and Virus	241
	B. Host Range of Vector and Virus	242
	C. Effect of Moisture	244
	D. Effect of Temperature	245
	E. Effect of Soil pH	246
	F. Effect of Biocides	246
VI.	Conclusion	247
	Acknowledgement	248

I. Introduction

Two groups of zoosporic fungi, the chytrids and the plasmodiophorids, contain members which are both plant parasites and plant virus vectors. Although it is valid to study these fungi solely as plant parasites, it is often in their vector roles that they are economically more important. Truly these zoosporic vector fungi represent "double trouble" for the farmer faced with

ZOOSPORIC PLANT PATHOGENS
ISBN 0 12 139180 9

disease losses and for the plant pathologist who must integrate knowledge concerning the virus, vector and host in order to design control measures.

The five proven vectors of plant viruses are two closely related members of the Class Chytridiomycetes, *Olpidium brassicae* (Woronin) Dang. and *O. radicale* Schwartz and Cook (syn. *O. cucurbitacearum* Barr), and three plasmodiophorids, *Polymyxa graminis* Led., *P. betae* Keskin and *Spongospora subterranea* (Wallr.) Lagerh. All are obligate plant parasites. Of these five fungi, only *Spongospora subterranea* is of major economic importance in its own right, causing the powdery scab disease of potatoes and the crookroot disease of watercress, although *Polymyxa betae* gives rise to concern as the cause of root necrosis and stunting in heavily infected spinach plants (Barr, 1979). The three other fungi may sometimes be abundant in the roots of plants and are of wide geographical distribution, but are of little significance in disease unless associated with the viruses which they transmit.

Four groups of viruses can be transmitted by zoosporic fungi. These virus groups have particles which morphologically are either isometric (roughly spherical), straight tubular, filamentous, or of unknown shape. Viruses in the isometric group have a relationship with their vectors which may be termed *non-persistent*, whereas viruses in the other three groups have a *persistent* relationship with their vectors. The non-persistent viruses are borne on the surface of zoospores and do not persist in the resting sporangia, whereas the persistent viruses are borne internally in zoospores and survive, often for many years, in dormant resting sporangia.

In this review only the well-established zoosporic vectors and their viruses mentioned above will be discussed although it should be mentioned that other claims for fungal transmission of viruses have been made. For instance, *Synchytrium endobioticum* (Schilb.) Perc. was associated with the transmission of potato virus X by Nienhaus and Stille (1965), but this was not confirmed by Lange (1978). Moreover, air-borne spores of rusts and mildews have been associated with transmission of tobacco mosaic virus (Yarwood, 1971; Yarwood and Hecht-Poinar, 1973), and wheat stem rust spores with transmission of brome mosaic virus (Wechmar, 1980), but these, which do not involve zoosporic fungi, are regarded as outside the scope of this review.

II. The Vectors

Solving the problems concerning the life histories and the taxonomic position of the fungal vectors has been difficult for several reasons. Firstly, these fungi are obligate plant parasites, largely confined to the underground parts of their hosts. Secondly, all stages, except the dormant resting spores or resting sporangia, are relatively delicate, and thirdly, the work has often been done with cultures which are not genetically pure.

A. CHYTRIDIOMYCETES

Members of the Class Chytridiomycetes are inconspicuous zoosporic fungi, most of which attack freshwater algae, fungi, pollen grains, microscopic animals, or living or decaying plant tissues (Sparrow, 1960). The only known virus vectors in the group are two plant infecting species, *Olpidium brassicae* and *O. radicale*. A possible life cycle of *O. brassicae* was suggested by Teakle (1969), and the life cycle of *O. radicale* is probably similar.

A single thin-walled sporangium of a vector strain of *O. brassicae* can produce zoospores which, after infecting roots, develop into either thin-walled or resting sporangia (Sahtiyanci, 1962; Campbell and Lin, 1976). Therefore, vector strains of *O. brassicae* appear to be homothallic whereas at least some non-vector crucifer infecting strains require paired single sporangial isolates for resting sporangium production and hence are probably heterothallic. Presumably resting sporangia are produced following the fusion of either zoospores or thalli, whereas thin-walled sporangia result from infection by a single zoospore. Whether *O. radicale* is homothallic or heterothallic has not been reported.

Both *O. brassicae* and *O. radicale* comprise a large number of strains which vary in morphology and preferred hosts. *O. brassicae* is reported to differ from *O. radicale* in having smaller, more spherical zoospores (2.8–5.6 μm diameter compared with 4–5 × 8 μm), a rapid, jerky zoospore motion instead of a slow, steady motion, and a stellate instead of a smooth wall of the resting sporangium (Hiruki and Alderson, 1976; Lange and Insunza, 1977). Especially with *O. brassicae*, the occurrence of isolates with morphological differences, such as in the length and number of exit tubes on zoosporangia and in the shape of resting sporangia, has led some workers to distinguish more than one species (e.g. Sahtiyanci, 1962). However, since the comparative study of different isolates from one host species has shown that morphological variation is considerable (Garrett and Tomlinson, 1967), most workers have continued to regard *O. brassicae* as a broad species containing diverse strains.

B. PLASMODIOPHOROMYCETES

Members of the Class Plasmodiophoromycetes are obligate parasites of algae, aquatic fungi and higher plants (Karling, 1968). The only known plasmodiophorid virus vectors are three plant-infecting species, *Polymyxa graminis*, *P. betae* and *Spongospora subterranea*. Possible life cycles of *P. graminis* and *S. subterranea* were suggested by Teakle (1969); the morphology and life cycle of *P. betae* are similar to those of *P. graminis*. Despite their similarity it has been suggested that these two *Polymyxa* spp. be retained as separate entities because of their distinctive host ranges (Barr, 1979).

The *Polymyxa* spp. have zoosporangia which are elongate, lobed and irregular in shape, and which produce long exit tubes. They are either septate or non-septate, but at maturity any internal walls partially or completely dissolve and the entire thallus becomes a single zoosporangium (Barr, 1979). The resting spores occur as irregularly-shaped aggregates or cystosori. *Spongospora subterranea* differs from the *Polymyxa* spp. in having zoosporangia which occur either singly or in loose groups, and in having cystosori in the form of hollow spore balls.

III. The Viruses

There are 15 viruses for which reasonably convincing evidence of transmission by zoosporic fungi has been published. Of these viruses, 12 can be placed in three groups on the basis of gross morphology, while the other three have not been characterised morphologically by electron microscopy and, pending this, may be placed in a fourth group (Table 1). All of the viruses have been sap transmitted by the leaf-rubbing technique except two members of the fourth group, lettuce big-vein and freesia leaf necrosis viruses.

A. THE ISOMETRIC VIRUSES

The three isometric (approximately spherical) viruses have stable particles, i.e. they usually remain infectious for some days in sap at room temperature. Tobacco necrosis virus and satellite virus are transmitted by *O. brassicae*, while cucumber necrosis virus is transmitted by *O. radicale*. Usually the viruses remain localised in the roots of their hosts, and, except when they cause excessive root necrosis, are responsible for plant damage. However, if tobacco necrosis or cucumber necrosis viruses spread to the tops of plants, death may result. Lethal diseases of tobacco, cucumber, bean or tulip have resulted from partially systemic infection by tobacco necrosis virus, and of cucumber by cucumber necrosis virus (Kassanis, 1970a; Dias and McKeen, 1972).

Satellite virus is of special scientific interest because it is non-infectious by itself, and depends for its replication on simultaneously infecting plant roots with tobacco necrosis virus. Therefore, it occurs naturally only with tobacco necrosis virus. Replication and spread of tobacco necrosis virus may be retarded by the presence of satellite virus (Kassanis, 1970b).

B. THE STRAIGHT TUBULAR VIRUSES

The particles of these viruses are composed of a helically-wound RNA molecule with an associated protective protein coat. Usually the particles of

Table 1. The viruses transmitted by zoosporic fungi

Group and particle shape	Virus	Particle size (nm)	Vector	Virus–vector relations	References[c]
Isometric	Satellite	17	Olpidium brassicae	Non-persistent	15
	Tobacco necrosis	26	O. brassicae	Non-persistent	14
	Cucumber necrosis	31	O. radicale	Non-persistent	82
Straight tubular	Beet necrotic yellow vein	65-105 + 270 + 390 × 20	Polymyxa betae	Persistent	144
	Potato mop-top[a]	150 + 300 × 20	Spongospora subterranea	Persistent	138
	Wheat mosaic[a]	– 110-160 + 300 × 20	P. graminis	Persistent	77
	Peanut clump	190 + 245 × 20	P. graminis	Persistent	235
	Oat golden stripe	150 + 300 × 20	P. graminis	? Persistent	Plumb and Macfarlane (1977)
Filamentous	Barley yellow mosaic[b]	275 + 550 × 13	P. graminis	Persistent	143
	Oat mosaic	600-750 × 13	P. graminis	Persistent	145
	Rice necrosis mosaic[b]	275 + 550 × 13	P. graminis	Persistent	172
	Wheat spindle streak mosaic[b]	190-1975 × 13	P. graminis	Persistent	167
Unknown	Lettuce big-vein		O. brassicae	Persistent	Campbell (1962)
	Tobacco stunt		O. brassicae	Persistent	Hiruki et al. (1975)
	Freesia leaf necrosis		O. brassicae	? Persistent	Dorst (1975)

a and b Those viruses having the same letter are reported to be serologically related.
c Numbers refer to CMI/AAB Descriptions of Plant Viruses.

each virus can be shown to fall into two or three classes according to their length (Table 1). With wheat mosaic virus, infectivity depends on the presence of both short (100–160 nm) and long (300 nm) particles (Tsuchizaki et al., 1975). This complementation of the genome by RNA species in different sized particles possibly occurs with other members of the group (Table 1), but convincing evidence is so far lacking.

Morphologically the straight tubular viruses transmitted by fungi resemble tobacco mosaic virus, which is a highly stable, mechanically-transmitted virus sometimes transmitted without a vector in soil (Broadbent, 1965). Further, a distant serological relationship exists between two fungus-borne members, potato mop-top and wheat mosaic viruses, and tobacco mosaic virus (Gibbs, 1977; Table 1). Therefore despite the relative instability of their particles, which may reflect their protected internal carriage in their fungal vectors, these fungus-transmitted viruses are regarded as possible members of the Tobamovirus (tobacco mosaic virus) Group (Matthews, 1979).

C. THE FILAMENTOUS VIRUSES

The four viruses in this group have flexuous particles which structurally resemble the viruses with straight tubular particles in comprising a helically-wound RNA protected by a similarly-wound protein coat (Table 1). Considerable variation in particle length has been observed, but so far it has not been established whether infectivity is associated with a particular length class or whether complementation between particles of different length classes is required for infectivity. Studying such questions has been difficult because these viruses have particles which are somewhat labile and subject to fragmentation. Three members of the group have been reported to show a distant serological relationship (Usugi and Saito, 1976; Table 1). Also the group may have affinities with the aphid-transmitted Potyviruses and the mite-transmitted wheat streak mosaic virus, since similar proteinaceous "pinwheel" inclusion bodies are induced in infected plant cells by all these viruses.

D. THE VIRUSES OF UNKNOWN MORPHOLOGY

The three disease agents placed in this group are called viruses in this review although their viral nature is still not fully established. The tobacco stunt virus is unique amongst members of this group in that it is mechanically transmissible by sap inoculation, although special conditions (i.e. presence of 1mM 1-phenylthiosemicarbazide in 0.01 M phosphate buffer, pH 6.8, during tissue grinding and inoculation) are important in achieving this (Hiruki, 1975). Since adding ribonuclease to preparations of tobacco stunt virus rapidly abolishes the ability of the virus to be sap transmitted whereas adding deoxyribonuclease has little effect, this virus may contain RNA (Hiruki *et al.*, 1975).

The interrelationships of the three viruses in this group have not been determined. However, a possible relationship between freesia leaf necrosis and lettuce big-vein viruses was indicated by the development of big-vein disease when lettuce was planted in soil containing freesia leaf necrosis virus (Dorst, 1975). These two viruses, as well as tobacco stunt virus, share the same fungal vector, *O. brassicae*.

IV. Virus–Vector Relationships

Two distinctly different relationships are recognised between the vector fungi and the viruses they transmit. These relationships may be termed *non-persistent* and *persistent*. A summary of the characteristics distinguishing the two relationships is given in Table 2.

Table 2. Characteristics indicating the differences between non-persistent and persistent virus–vector relationships of zoosporic fungi

Characteristic	Non-persistent relationship	Persistent relationship
Zoospores acquire virus-transmissibility *in vitro*	Yes	No
Virus seen on zoospore surface using negative-stain electron microscopy	Yes	No
Virus transmissibility of zoospores reduced or eliminated by treatment with specific virus antiserum	Yes	No
Virus transmissibility greatly reduced or eliminated by air-drying or acid treatment of resting spores	Yes	No
Virus present in thin sections of the protoplast of zoospores or resting spores	No	Yes[a]

[a] So far observed only for beet necrotic yellow vein virus (Tamada, 1975).

A. THE NON-PERSISTENT RELATIONSHIP

This applies only to the isometric viruses (tobacco necrosis and cucumber necrosis viruses and satellite virus) and their specific *Olpidium* vectors. Each virus is acquired only by the zoospore stage of its vector, and this *zoospore acquisition* can occur *in vitro* when an appropriate zoospore–virus mixture is prepared. Under natural conditions in soil, zoospore acquisition no doubt occurs when zoospores, released from either thin-walled sporangia in roots or resting sporangia in root debris or soil, contact virus released from infected roots or root debris. At least initially the zoospore-acquired virus is surface-borne, and can be seen on the plasmalemma which surrounds both the head and flagellum of the vector (Temmink *et al.*, 1970). No such adsorbed virus is seen when non-vector *Olpidium* strains and the transmitted viruses are mixed, or when vector *Olpidium* strains and morphologically similar but non-transmitted viruses are mixed. Further, little or no transmission of tobacco necrosis virus occurs when the virus is added to zoospores which have already encysted on roots (Teakle and Gold, 1963). Apparently, adsorption of virus by zoospores depends on the surface properties of both members of the combination; the nature of the interaction is unknown.

The exact manner by which zoospore-acquired virus enters the plant root is not clear. Entrance of the virus occurs at about the same time as the protoplast of the encysted zoospore enters (Teakle, 1962). Possibly the virus on the surface of the zoospore contaminates the protoplast at the time of zoospore encystment, when the flagellum is wrapped around the zoospore head and the limiting membranes of the flagellum and head merge (Temmink, 1971; Alderson and Hiruki, 1977). Thus, this virus could be suitably placed to enter the root when the zoospore cyst protoplast flows into the root. How the virus is released from the protoplast to infect the root cell is not clear, but this must occur quickly; multiplication of tobacco necrosis virus can be detected 10 hours after *Olpidium* and the virus are added to roots, and lesions of tobacco necrosis virus occur within 20 hours (Teakle, 1962). *Olpidium* thalli developing within tobacco necrosis virus-infected roots apparently do not acquire the virus internally, since zoospores released from these roots lose the ability to transmit when exposed to virus antiserum (Campbell and Fry, 1966).

Although the association between *Olpidium* and tobacco necrosis virus is intimate, it also appears to be transient and the term *non-persistent* is appropriate. Presumably the same relationship applies to the other isometric viruses, cucumber necrosis and satellite viruses, and their *Olpidium* vectors.

B. THE PERSISTENT RELATIONSHIP

This applies to the three other groups of viruses, i.e. those with straight tubular, filamentous or unknown particles, and their fungal vectors. In contrast to the situation with the isometric viruses, the *in vitro* mixing of virus and zoospores does not result in virus acquisition and transmission. These viruses apparently are acquired only by thalli developing in virus-infected roots or, in the case of potato mop-top virus, tubers (Campbell and Fry, 1966; Tamada, 1975). This *thallus acquisition* of virus later results in the release of virus-transmitting zoospores which carry the virus internally instead of externally as occurs with the isometric viruses.

Almost nothing is known about the process by which the fungal thalli acquire the viruses. Presumably the viruses multiply in cells containing immature thalli, and the viruses pass through the limiting membrane of the thallus, accumulate and later are released in zoospores from the fruiting thallus (i.e. either a thin-walled sporangium, a resting sporangium or a group of resting spores).

Although the process of thallus acquisition of virus is apparently efficient, it is not completely so. Lettuce big-vein virus was not transmitted by some washings of big-vein lettuce roots which contained few zoospores (Campbell and Grogan, 1964), and this virus was carried by only about 50% of cultures of *Olpidium* obtained from single sporangia which developed in big-vein lettuce (Campbell and Lin, 1976).

A remarkable feature of the interaction of persistent viruses and their vectors is the high stability of the virus when in the resting stage of the fungus. Apparently the fungus protoplasm affords a highly protective environment, and the virus can persist for many years in the resting sporangium or spore without losing infectivity. Soil or debris, presumably containing the viruliferous fungus, can be air-dried and stored for years, and then placed near the roots of plants to recover the virus (McKinney, 1937; Pryor, 1946; Hidaka *et al.*, 1956). Evidence that lettuce big-vein virus is not surface-borne by resting sporangia is the retention of virus infectivity despite hydrochloric acid treatment, a procedure which would be expected to destroy exposed virus (Campbell, 1962). Evidence for the internal carriage of beet necrotic yellow vein virus by *Polymyxa betae* is provided by the detection of particles resembling the virus in ultrathin sections of zoospores of the vector (Tamada, 1975).

V. Epidemiology and Control

A number of factors affect the occurrence of the diseases caused by the fungus-transmitted viruses. These factors include the geographical coincidence of the virus and its vector, the presence of favourable host plants, suitable environmental conditions such as soil moisture, atmospheric humidity and temperature, and the use of biocides. Control measures aimed at lowering the incidence or severity of the diseases can operate against the vector, against the virus, or against both.

A. GEOGRAPHICAL COINCIDENCE OF VECTOR AND VIRUS

A notable feature of the geographical distribution of the viruses and their vectors is that, although a vector fungus may occur without the viruses it transmits, the viruses are apparently dependent on their vector for long-term survival. This conclusion is based on the fact that significant natural occurrence of the viruses has always been associated with occurrence of the specific zoosporic vector, where this has been adequately searched for. Apparently these viruses have not acquired efficient alternative methods of natural transmission. In this regard the fungus-transmitted viruses can be contrasted with certain other soil-borne viruses; for instance, the Nepoviruses are often transmitted by seed or pollen as well as nematodes (Lister and Murant, 1967).

An exception to the generalisation that the fungus-transmitted viruses lack alternative methods of natural transmission is the tuber transmission of potato mop-top virus (Jones and Harrison, 1969). However, the vector, *S. subterranea*, is also tuber-transmitted, and so far there have been no reports of the potato mop-top virus occurring without its vector.

When the geographical distribution of the fungus vectors and the viruses that they transmit is looked at broadly (Table 3), it is seen that the five fungus vectors have moderate to wide geographical distributions. However, because these fungi, except for *S. subterranea*, cause inconspicuous symptoms and are found only in plant roots, records of their occurrence are relatively rare in less developed countries. Probably their geographical distribution is much wider than is so far reported.

Each of the viruses has a reported geographical distribution which is usually less extensive than that of its vector (Table 3). For instance, tobacco necrosis virus is reported to occur in four continents, whereas its vector, *O. brassicae*, occurs in six continents. Also, peanut clump virus is reported only from Africa, although its vector *P. graminis* is reported from five continents.

B. HOST RANGE OF VECTOR AND VIRUS

All of the five fungus vector species have a moderate to wide host range. However, the vector situation is complicated by the existence of biotypes and varieties, some of which have narrow host ranges and others wide. For instance, morphologically similar, but physiologically different races of *O. brassicae* were isolated, along with *Pythium* and tobacco necrosis virus, from carrot with the rusty root disease in Canada (Barr and Kemp, 1975). In England some big-vein-transmitting lettuce isolates of *O. brassicae* infected *Chrysanthemum* spp., whereas others did not; none multiplied in cabbage and, conversely, non-vector cabbage isolates failed to multiply in lettuce (Garrett and Tomlinson, 1967). Lettuce isolates of *O. brassicae* are usually vectors of lettuce big-vein and tobacco necrosis viruses, whereas crucifer isolates are usually not (Teakle and Hiruki, 1964). Physiological races have also been described in *O. radicale*. In Canada, cucumber necrosis virus is transmitted by isolates of *O. radicale* ($\equiv O.$ *cucurbitacearum*) which are found only in curcurbitaceous hosts (Barr, 1968), whereas isolates in other countries may have broad host ranges (Lange and Insunza, 1977).

Physiological races also occur with plasmodiophorid vector species. For instance, *Polymyxa betae* f. sp. *amaranthi* develops in *Amaranthus reflexus*, but not *Chenopodium album*, spinach or beets which are known hosts of *P. betae* (Barr, 1979). Similarly *Spongospora subterranea* var. *nasturtii* infects watercress but not tomato, whereas *S. subterranea* from potato infects tomato but not watercress (Tomlinson, 1958a,b). The vector ability of the new physiological races of these species has not been described.

Table 3. References to papers indicating the geographical distribution of the fungus vectors and the fungus-transmitted viruses

Vectors and viruses	References to their occurrence in:					
	Africa	Asia	Australasia	Europe	North America	South America
Olpidium brassicae	1	2	3	4	5	6
Tobacco necrosis		7	3	8	9	
Tobacco necrosis satellite				10	11	
Lettuce big-vein		12	13	14	15	16
Tobacco stunt		2				
Olpidium radicale		17		18	19	
Cucumber necrosis		17			19	
Polymyxa graminis	20	21		22	23	24
Wheat mosaic		25		22	26	27
Oat golden stripe				28		
Peanut clump	20					
Barley yellow mosaic		29		30		
Oat mosaic				28	31	
Rice necrosis mosaic		32				
Wheat spindle streak mosaic		33		34	35	
Polymyxa betae		36		36	36	
Beet necrotic yellow vein		37		38	39	
Spongospora subterranea	40	40	40	40	40	40
Potato mop-top				41		41

References: 1. Commonwealth Mycological Institute, 1979; 2. Hiruki *et al.*, 1975; 3. Shukla *et al.*, 1979; 4. Lange and Olson, 1976a,b; 5. Barr and Kemp, 1975; 6. Lin, 1979; 7. Ramachandraiah *et al.*, 1979; 8. Lange, 1976; 9. Uyemoto and Gilmer, 1972; 10. Kassanis, 1970b; 11. Uyemoto *et al.*, 1968; 12. Iwaki *et al.*, 1978; 13. Fry, 1958; 14. White, 1980; 15. Campbell *et al.*, 1980; 16. López and López, 1977; 17. Komuro, 1971; 18. Lange and Insunza, 1977; 19. Dias and McKeen, 1972; 20. Thouvenel and Fauquet, 1981; 21. Inouye and Saito, 1975; 22. Canova, 1966; 23. Barr, 1979; 24. Tocchetto, 1974; 25. Tsuchizaki *et al.*, 1975; 26. Brakke, 1971; 27. Caetano *et al.*, 1978; 28. Catherall and Boulton, 1979; 29. Inouye and Saito, 1975; 30. Huth, 1979; 31. Hebert and Panizo, 1975; 32. Inouye and Fujii, 1977; 33. Ahlawat *et al.*, 1976; 34. Signoret *et al.*, 1977; 35. Slykhuis and Barr, 1978; 36. Commonwealth Mycological Institute, 1978; 37. D'Ambra and Mutto, 1977; 38, Putz, 1977; 39. Al Musa and Mink, 1981; 40. Commonwealth Mycological Institute, 1974; 41. Harrison, 1974.

Ability to transmit a virus to a plant is not dependent on the vector being able to multiply in that plant. For instance, although *O. brassicae* can transmit tobacco necrosis virus to tulip causing a necrotic disease, the fungus usually is unable to propagate in tulip roots (Mowat, 1970; Lange, 1976). Similarly, *P. graminis* can transmit the peanut clump virus to peanut but does not multiply in peanut (Thouvenel and Fauquet, 1981). In such cases the fungi and often the viruses are maintained by parasitising the roots of weeds or other crops.

Susceptibility of host cultivars varies and control is sometimes possible by altering the cultivar. For instance, in New Zealand glasshouses tobacco necrosis virus may cause 20–50 % loss in cucumbers planted after tomatoes, which are highly susceptible to both the virus and its vector, *Olpidium brassicae*. The cucumber cultivar Marketer is highly susceptible, while Triumph is least susceptible (Thomas and Fry, 1972). Similarly, if French beans are transplanted at the 3-leaf-stage between fruiting tomato plants, they may become infected with tobacco necrosis virus and *O. brassicae* (Thomas, 1973). Altering the rotation tomatoes–beans–tomatoes–cucumbers–tomatoes may give some control of the virus in the cucumber and bean crops.

Host resistance has also been observed with other fungus-transmitted virus diseases. For instance, the lettuce cultivars Calmar and Salinas have field resistance to big-vein disease in California in that they seldom develop vein-banding symptoms before the plants are 35 days old and ready for thinning (Campbell *et al.*, 1980). Resistance to soil-borne wheat mosaic in wheat cultivars in the United States has long been recognised (McKinney, 1948).

C. EFFECT OF MOISTURE

All of the five vectors reproduce by zoospores which require free moisture for their release. Where soil moisture is low, poor fungus infection will occur; conversely, frequent irrigation or rain will result in numerous crops of zoospores. In Scotland, the incidence of potato mop-top virus was not related to soil type, but was shown to increase with increase in total annual rainfall (Cooper and Harrison, 1973). In California, soil drainage was shown to be important in the incidence of lettuce big-vein disease; "big-vein-prone" soils were poorly drained, whereas "big-vein-suppressive" soils were rapidly drained (Westerlund *et al.*, 1978). An explanation of this was that resting sporangia of the vector *O. brassicae* required wetting in nearly saturated soil for approximately six days before they were capable of germination.

A drop in atmospheric relative humidity from 80–90 % to 40 % without an appreciable drop in temperature has been correlated with improved growth and development of *P. graminis* in roots of *Sorghum arundinaceum* in the Ivory Coast of Africa (Thouvenel and Fauquet, 1980). The reason for this improved fungal growth during a time of low humidity is not known.

D. EFFECT OF TEMPERATURE

Usually, cool conditions favour build-up of the fungi, both because the optimum temperatures for reproduction of the fungi are low and because cool conditions favour high soil moisture. However, the fungi may still operate during warm to hot conditions. For instance, although *P. graminis* has an optimum temperature for growth of approximately 18°C (Ledingham, 1939), it develops in Africa at field temperatures of 27°C (Thouvenel, personal communication) and at glasshouse temperatures of 23–30°C (Thouvenel and Fauquet, 1980). Since the zoospores of *P. graminis* rapidly become immobile at 28°C (Ledingham, 1939), the fungus must be operating at temperatures close to its maximum. Possibly a faster generation time of the resting spore stage (of 10–12 days at 27°C (Thouvenel and Fauquet, 1980) compared with 21–28 days at 17°C (Rao, 1968)), partly compensates for lower zoospore motility at the higher temperature. A similar speeding up of the life cycle of the similar vector *P. betae* with higher temperature has also been observed; infection occurred in seedlings inoculated with homogenised root preparations containing the fungus in 13 days at 23°C but in 23 days at 10°C (Abe, 1974).

A low temperature is favourable for the build-up of *O. brassicae*. This fungus multiplied better in lettuce at 10, 13 or 16°C than at 22 or 27°C (Fry and Campbell, 1966). Zoospore motility is retained far longer at temperatures below 20°C than above 20°C (Teakle, 1962) and this would allow longer for zoospores to find and infect a root.

Temperature is also important in the development of virus diseases transmitted by *O. brassicae*. Big-vein is more severe and incidence is greater in lettuce grown and harvested during the cooler part of the year in California (Campbell *et al.*, 1980). Further, glasshouse tests showed that big-vein was severe if the tops of lettuce were kept at 14°C, whereas little disease developed at 24°C. Translocation of the agent from roots to the tops is slow at 10°C but is rapid at 18°C or 22°C. Translocation also occurred at 24°C, although no symptoms developed (Westerlund *et al.*, 1978).

The effect of temperature on diseases caused by tobacco necrosis virus is variable. On the one hand, glasshouse-grown cucumber in New Zealand is more severely affected by the virus in winter and early spring than in late spring or summer when mildly-affected plants may recover (Thomas and Fry, 1972). On the other hand, systemic infection of field-grown French bean in Europe is most prevalent in July and August, the hottest months (Bawden and Want, 1949). Studies on mechanically-inoculated plants in England have shown that the virus multiples 20 times more abundantly in tobacco in winter than in summer (Bawden, 1964), and that it multiplies to a greater extent in French bean at 22°C than at 26°C (Harrison, 1956). However, when French bean was mechanically inoculated on the leaves in Australia it developed a high incidence of systemic infection at 27–31° but not at 17–21°C (Behncken,

1968). It seems likely that lower temperatures of about 22°C will favour transmission by the vector *O. brassicae* and local multiplication of the virus. However, systemic infection by the virus may be favoured by higher temperatures which increase the rate of virus multiplication.

Viruses transmitted by plasmodiophorid vectors vary considerably in their optimum temperatures. For instance, wheat mosaic virus develops favourably at soil temperatures of 16–18°C (Linford and McKinney, 1954), at which temperatures infection by the vector is also favoured (Ledingham, 1939). Similarly, with mechanical infection, symptoms occur when wheat seedlings are incubated at 20°C or less, with an optimum temperature of approximately 16°C (Brakke, 1971). Potato mop-top virus also requires a low temperature to produce symptoms. In naturally infected potato plants, the shoot symptoms develop best at 5–15°C, whereas symptoms in mechanically inoculated tobacco leaves occur regularly only below 20°C (Harrison, 1974).

In contrast, the peanut clump virus has the high optimum temperature for multiplication of 35°C (Thouvenel *et al.*, 1976). It appears to be suited to the hot tropical, African environment where it occurs, whereas wheat mosaic and potato mop-top viruses are suited to cool temperate regions.

E. EFFECT OF SOIL pH

With the plasmodiophorid vectors a neutral pH favours fungus infection. With *P. graminis* a pH value near 7 gave the best growth of the fungus in roots of plants grown in a glasshouse soil mix (Thouvenel and Fauquet, 1980). Similarly *P. betae* developed abundantly at pH 7–8, but poorly below pH 6; inoculum appeared to lose viability below pH 5.5 (Abe, 1974).

With *S. subterranea*, plant infection by the fungus and transmission of the potato mop-top virus are inhibited if the pH value of the soil is lowered to 5.0 by applying sulphur. However, this does not kill the virus-carrying resting spores, and virus transmission resumes when the pH value of soil is later raised (Harrison, 1977).

B. EFFECT OF BIOCIDES

Chemical control of the fungus-transmitted virus diseases is possible, although at present it is economically feasible only with high value crops. Lettuce big-vein incidence in California and England was reduced to near zero by methyl bromide soil fumigation, a treatment which also improved the uniformity and rate of plant growth (Campbell *et al.*, 1980; White, 1980). Benomyl applied as a transplant drench to lettuce was systemic in the roots and prevented *Olpidium* infection or reproduction, whereas chloropicrin plus D–D, Vorlex or metalaxyl were not effective (Campbell *et al.*, 1980) even though they sometimes improved plant growth. Surfactant compounds at low concentrations ($1\text{--}10\ \mu g\ ml^{-1}$) were toxic to zoospores in an aqueous solution (Tomlinson and Faithfull, 1979b), and controlled big-vein in lettuce grown by the nutrient film technique (Tomlinson and Faithfull, 1980).

For biocidal control of *S. subterranea*, zinc oxide or zinc sulphate may be added to the soil to kill the zoospores and prevent virus transmission. However, the resting spores survive and when the zinc content of the soil drops because of leaching by rain, the virus is again transmitted. Formaldehyde or organic mercurial fungicidal dips are used to treat infected tubers in order to prevent *S. subterranea* establishment at new sites (Harrison, 1977).

Considerable scope exists for improving the biocidal control of the vectors of fungus-transmitted viruses under commercial conditions.

VI. Conclusion

The decade starting in 1960 can be regarded as the "Golden Age" in the study of virus-transmission by zoosporic fungi. Before 1960, associations between *P. graminis* and wheat mosaic (Linford and McKinney, 1954) and *O. brassicae* and lettuce big-vein (Fry, 1958; Grogan *et al.*, 1958) had been reported, but no claims for virus transmission had been made. However, in the 1960s *O. brassicae*, *O. radicale*, *P. graminis* and *S. subterranea* were recognised as vectors of a wide range of viruses, and our concepts of virus–vector relationships were formed. In the 1970s, the number of viruses known to be transmitted by zoosporic fungi was expanded and *P. betae* (morphologically similar to *P. graminis* but with a different host range) was shown to be an additional vector. Presumably additional viruses will continue to be shown to be fungus-transmitted. One possible candidate of potential importance is the maize white line mosaic virus, with which 35 nm diameter non-infectious particles and both *P. graminis* and an *Olpidium* sp. have been associated (Boothroyd and Israel, 1980; Zoeten *et al.*, 1980; T. A. Zitter, personal communication).

Will the viruses of undetermined morphology causing tobacco stunt, lettuce big-vein and freesia leaf necrosis soon be characterised? In this connection it is interesting that straight tubular particles resembling wheat mosaic and other plasmodiophorid-transmitted viruses (Table 1, Group 2) are associated with both tobacco stunt (Kuwata and Kubo, 1981) and lettuce big-vein (Haeske, 1958; Ragozzino and Furia, 1974; Chod *et al.*, 1976). Will further work show that these are the particles of the viruses which cause these diseases? This would mean that viruses with straight tubular particles are transmitted in a persistent manner by either the chytridiomycetous fungus *O. brassicae*, or one of the three plasmodiophorid fungi.

What are the physiological interactions which occur when fungal vector and virus doubly infect a host plant and how do these differ from those which occur during single infection? In this connection it has been shown that *P. betae* and beet necrotic yellow vein virus are associated with the rhizomania disease of beet, which is characterised by a yellowing, crinkling and wilting of the leaves, a stunting of the whole plant and an abnormal proliferation of the secondary roots which later become necrotic (D'Ambra and Mutto, 1977). *Polymyxa betae*, when free of virus, fails to induce rhizomania, but causes slight root necrosis (Tamada *et al.*, 1974), whereas the virus, when inoculated mechanically to leaves of seedling beet, causes foliage and root symptoms similar to those produced by mixed natural infection with both *P. betae* and the virus (Tamada, 1975; Fujisawa and Sugimoto, 1976). Does this mean that the fungus plays little part in the rhizomania disease except as a vector? What advances will be made in our knowledge of epidemiological factors involved in disease outbreaks and what new control measures will be devised? These and other questions constitute the challenge and interest for the future.

Acknowledgement

I wish to thank the people who provided advice and information during the writing of this review, including V. D'Ambra, H. J. M. van Dorst, R. T. Plumb, J. C. Thouvenel, T. A. Zitter and G. A. de Zoeten.

8

Zoosporic Fungal Pathogens of Lower Plants

What Can be Learned From the Likes of *Woronina*?

CHARLES E. MILLER and DANIEL P. DYLEWSKI

Department of Botany, Ohio University, Athens, Ohio, USA

I.	Introduction	249
II.	Experimental Procedures	251
	A. The Zoosporic Fungal Host	251
	B. Isolation and Endoparasite/Host Culture	252
	C. Maintenance of Host/Parasite Culture	253
	D. Method of Preparing and Examining Synchronised Infections of Host Hyphae	254
	E. Production of Zoospores	255
III.	Studies of *Woronina pythii*	257
	A. Introduction	257
	B. Development of Sporangiogenous Plasmodia	258
	C. Ultrastructure of Secondary Zoospores	268
	D. Development of Cystogenous Plasmodia	276
	E. Host Range of *Woronina pythii*	279
	F. Cyst Viability of *Woronina pythii*	281
IV.	Conclusion	282

I. Introduction

Zoosporic fungi that are parasitic on lower plants, especially algae and other fungi, are not uncommon; even a cursory survey of the literature reveals more than 225 parasitising algae and more than 66 parasitising other fungi (Sparrow, 1960). Even discounting the often laissez-faire attitudes regarding the erection of new taxa and the fact that most of the described parasitic associations have not been studied from a pathological standpoint or even proved to be of a parasitic nature, it is still a large number. Most parasitic zoosporic fungi growing on or in algae and other fungi belong to the order Chytridiales although parasitic taxa are also found in the orders

Lagenidiales, Saprolegniales, Plasmodiophorales and Hyphochytriales. Such zoosporic pathogens of lower plants have usually been studied by mycologists who were primarily interested in the organisms *per se* and not in the diseases they caused.

These zoosporic parasites of lower plants may be holocarpic, with all the protoplasm comprising the thallus being converted at maturity into reproductive propagules, or eucarpic, with some differentiation of the thallus into reproductive and vegetative structures. Zoosporic fungal parasites are also described as either monocentric, in developing a single reproductive zoosporangium or resting body, or polycentric, in developing more than one (usually several) reproductive rudiments. The parasite may be epibiotic (growing on the host) or endobiotic (growing within the host) although epibiotic parasites actually have epibiotic reproductive rudiments (the zoosporangia or resting bodies) and endobiotic absorbing and anchoring structures (the rhizoids). In some taxa the zoosporangium may be epibiotic and the resting body, as well as the rhizoids, endobiotic. All holocarpic fungi are endobiotic and are usually monocentric. Epibiotic forms are all eucarpic and can be either monocentric or polycentric.

The advantages of studying obligate intracellular parasitism in lower plants, especially fungi, in contrast to higher plants, include host-tissue simplicity, rapidity of host development and growth, rapid development and completion of the parasite life cycle, complete control of the environmental conditions in which the host and parasite are grown (exemplified by host–parasite axenic culture), ease of application of experimental procedures and repeatability of experiments. The usually simple thalloid body structure of lower plants permits ease of microscopic observation of the host–parasite association, frequently in an active state, while the lack of complex, highly differentiated tissue permits easy isolation of host or parasite organelles or extraction of substances for analyses. Most algal hosts are unicells or simple colonial forms and most fungal hosts are usually simple coenocytic or septate hyphal filaments with structural differentiation only into single-celled gametangia, sporangia or conidia.

The thalli of lower plants, especially zoosporic fungi, grow very rapidly and, in many associations, the host and/or parasite, sometimes complete their development and life cycles within 24 hours; certainly within three to four days. *Woronina pythii* Goldie-Smith (the parasite to be discussed in detail in this account) completes its vegetative life cycle in two-membered axenic culture with its *Pythium* host within 60 hours and produces viable cysts (resting spores) within three to four weeks. Such rapid growth and development of bifungal host–parasite pairs provides the opportunity for repeated experimentation and observation of the disease or host–parasite relationships under study. Zoosporic lower plants can easily be grown in two-membered axenic liquid or solid culture, thus removing the possibility of other organisms influencing the development of the disease under study. In some cases a chemically defined medium may be used for growth of the host–parasite pair and with these relatively simple and controlled culture conditions, the removal for analysis of substances produced by the organisms or the addition of experimental substances to the medium may be performed easily.

II. Experimental Procedures

A. THE ZOOSPORIC FUNGAL HOST

1. *Collection*

Endobiotic, zoosporic fungal parasites frequently parasitise other zoosporic water-mould fungi, especially members of the Saprolegniaceae, Lepto-mitaceae and Pythiaceae (Sparrow, 1960). Water moulds are found both in water and soil, growing on dead vegetable or animal matter. In aquatic habitats, they have been found on dead twigs, fruits, seeds, leaves, algae and dead insects of various kinds as well as on dead and dying fish. The mycelium of the water mould forms delicate, whitish threads around the substratum just beneath the surface of the water. It is unnecessary to rely upon collecting such water moulds in the vegetative condition, however, as the fungal propagules (zoospores and resting bodies) are produced in such great abundance, and are so thoroughly disseminated throughout the substratum, that a few cubic centimetres of soil or water brought into the laboratory will almost invariably furnish one or more species if a bait is added and allowed to remain for two or three days. Water and soil samples may be collected in sterilisable containers of any convenient size—pre-sterilised sealable plastic bags are excellent for collecting soil whereas wide-mouth glass or plastic bottles with screw or snap-caps are best for water collection. Water moulds may also be collected directly from natural habitats by submerging baits in retrievable traps for two or three days.

2. *Gross Culture*

Water and soil samples may be gross cultured in standard petri plates. Glass-distilled water should be added to soil samples and sterile hemp-seed halves (achenes of *Cannabis sativa* L.) or whole sesame seeds (which are easier to obtain) should be added to the water and soil collection. The seeds should be sterilized by autoclaving (15 p.s.i. for five to eight minutes) or by exposing them to propylene oxide vapour for 12 hours. The baits may be placed on debris in the sample or merely floated on the water. As soon as growth becomes visible to the naked eye the entire mould and substratum should be transferred to a sterile Petri dish, and washed thoroughly by squirting a stream of distilled water from a wash bottle.

3. *Isolation and Axenic Culture*

To obtain bacteria-free cultures, single hyphal threads should be cut from the cultures on sesame seed and placed on the relatively dry surface either of a very low nutrient agar medium or 2% water agar. Antibiotics may also be added to the medium, but these are not normally necessary as the water-mould hyphae will usually grow rapidly away from the slowly enlarging colonies of bacteria at the point of inoculation. Small blocks of agar containing hyphal tips from the periphery of the advancing mycelium can be cut from the agar surface. The blocks of agar can then be transferred to new agar-containing petri plates or culture tubes and, in this way, an axenic, stock culture of the host can be established.

Water moulds will usually only grow vegetatively on agar medium and do not produce zoosporangia. To prepare aquatic, axenic water cultures of water moulds for study or for infection with a parasite, sterilised sesame seeds or hemp-seed halves should be placed on the surface of the agar culture in which the water mould is growing. Within two or three days the hyphae of the fungus will have grown into the cut surfaces of the seeds which should then be removed aseptically from the culture and placed in sterile water in petri dishes. Useful publications with methods for collection and isolation of water moulds and other aquatic fungi include those by Johnson (1956), Emerson (1958) and Miller and Ristanović (1969).

B. ISOLATION AND ENDOPARASITE/HOST CULTURE

Critical studies on the biology of these obligately endobiotic zoosporic parasites require the establishment of two-membered host/parasite axenic cultures and the assumed difficulty in establishing such cultures is probably responsible for the lack of developmental and experimental investigations of these organisms. Emerson (1950) was the first to report the establishment of two-member host/parasite axenic culture of aquatic fungi: *Rozella achlyae* Shanor and *R. allomycis* Foust (Chytridiales) were grown in their water-mould hosts *Allomyces* spp. (Blastocladiales) in two-membered bacterium-free culture. Emerson (1950) did not, however, report the techniques used to establish the cultures and Slifkin (1961, 1962) was the first to publish methods for establishing host/parasite axenic cultures of two zoosporic fungi. She established host/parasite cultures of *Olpidiopsis incrassata* Cornu (Lagenidiales) in *Saprolegnia ferax* (Gruith) Thuret (Saprolegniaceae). About the same time, Mullins (1961) also established two-membered axenic cultures of *Dictyomorpha dioica* Mullins (Chytridiales) in *Achlya flagellata* Coker (Saprolegniaceae).

The methods used by Slifkin and by Mullins were to inoculate pieces of parasitised host hyphae onto petri plates containing 2% water agar, or agar with very low nutrients. It was important to keep the agar surface free of water in order to inhibit motile bacteria. The infected mycelium was previously washed with a stream of sterile water to remove as many bacteria as possible. The infected host hyphae were permitted to grow several millimetres beyond the points of inoculation and, hopefully, away from the bacteria. Small agar blocks with infected hyphal tips in or on them, were cut out and placed in axenic hemp-seed cultures of the host in water or agar cultures. New infection in host hyphae occurred upon liberation of zoospores from the transferred infected hyphal tips. This method does not differ from that described above for isolation of the host except that the hyphal tips are infected with the endoparasite. *Woronina pythii* (Plasmodiophorales) and its host *Pythium* sp., the endoparasite/host study presented here as an example, were isolated into bifungal culture using the methods of Slifkin (1961).

C. MAINTENANCE OF HOST/PARASITE CULTURE

Because the zoospores of the parasite in aquatic culture are restricted to host hyphae within the confined area of the culture tube or dish, most hosts are rapidly killed by the heavy infections. It is commonly necessary, therefore, to provide new host hyphae periodically. The frequency of adding host hyphae to maintain the bifungal culture depends on how rapidly the parasite develops and whether the host itself produces zoospores. If zoospores are produced regularly by the host or if the host is vigorous enough to continue hyphal growth under these culture conditions, it may be necessary to add only sterilised sesame or hemp seeds for it to maintain itself after the culture is established. Eventually, under these culture conditions, the parasite and host cease developing vegetatively and produce resting bodies. Frequently, both host and parasite can be retrieved from such cultures by replacing the staled medium and adding both fresh host and seeds to the culture.

In our study, axenic stock cultures of *Woronina pythii* and its hosts were maintained in screw-cap culture tubes (Kimax or Pyrex, 20 × 125 mm) each containing 10 ml of sterilised Emerson's water (two parts glass-distilled water and one part lake water). The host inoculum was prepared by placing sterilised sesame seed or hemp-seed halves on the surface of the host growing in petri dishes containing agar; we used M_3 agar (17 g potato-dextrose agar, 5 g dextrose, 5 g soluble starch, 1 g peptone, 1 g yeast extract and 1000 ml glass-distilled water) but any agar medium on which the host will grow is adequate. The *Pythium* culture was three days old when the seeds were placed at the growing edge of the hyphae. After 48 hours at room temperature,

Pythium hyphae had grown into the seeds, which were then removed aseptically and placed into the culture tubes containing Emerson's water. After 24 hours, during which time the host became well established on the seeds, zoospores of the parasite were pipetted into the culture tube, or, more frequently, a small mass of infected host hyphae with mature and discharging sporangia of the parasite was placed in the test tube. The parasite inoculum was obtained from stock cultures ranging in age from 54 hours to 21 days and incubated at 18°C. One sesame seed or one hemp-seed half was added each week to each culture tube.

Held (1972) improved a method developed by Emerson (1958) for maintaining two-membered stock cultures of aquatic fungal hosts and parasites on agar media. The bifungal culture growing on 10–12 ml of dilute medium in slants (20 × 150 mm test tubes) or 50 ml in standard petri dishes and incubated at 23°C, only needed transferring every three to four months for *Rozella/Allomyces* and every 2 months for *Olpidiopsis varians/Achlya flagellata*, as opposed to an interval of two to three weeks if the fungi were maintained in aquatic culture (Held, 1972). The slants were stored horizontally so that the free liquid was spread over the entire agar surface, thus enhancing zoospore escape and host dispersal. We have found Held's method to work excellently in bifungal cultures of *Olpidiopsis varians* Shanor parasitising the water mould *Aplanopsis terrestris* Hohnk.

D. METHOD OF PREPARING AND EXAMINING SYNCHRONISED INFECTIONS OF HOST HYPHAE

For developmental studies of both sporangiogenesis and cystogenesis, we used the following methods to establish synchronous host/parasite cultures.

1. *Sporangiogenous Phase*

Host sesame-seed cultures prepared as described above were grown in 12 ml of Emerson's water in standard petri dishes for 24 hours at room temperature. These cultures were then inoculated with masses of infected host hyphae containing mature and discharging sporangia of *W. pythii*. This inoculum was removed from the petri-dish cultures after one hour and this was considered the datum in timing the development of the resulting sporangial infections. The petri-dish cultures were incubated at 18°C and host/parasite material was removed at six-hour intervals for up to 60 hours. In preparation for observations with the transmission electron microscope, the material was fixed in 2% glutaraldehyde in 0.1 M phosphate buffer at pH 6.8 for one hour at room temperature, rinsed in phosphate buffer for 20 minutes, and embedded in 2% water agar. Small cubes containing the infected material were cut from the agar plate and further processed in 1 % osmium tetroxide in 0.1 M phosphate buffer at pH 6.8 for one hour at 4°C, dehydrated in an ethanol-propylene oxide series, and embedded in an Araldite–Epon mixture.

2. *Cystogenous Phase*

At first, only sporangia were formed in the timed petri-dish cultures as prepared above. As the cultures aged, however, and staling substances accumulated, the environmental conditions became less favourable and the parasite switched from the sporangiogenous to the cystogenous phase. To time development of the cystogenous plasmodia more accurately and reduce the number of sporangiogenous plasmodia in the culture, stale rather than fresh Emerson's water was used for the medium in petri-dish cultures. Emerson's water from stock cultures at least five weeks old was membrane filtered (Miller and Dylewski, 1981) and sesame seed cultures of *Pythium* grown for 24 hours at room temperature in 12 ml of this filtered water. Each *Pythium* culture was then inoculated with a mass of infected host hyphae containing mature and discharging parasite sporangia. The infected host hyphae were removed after one hour and the *Pythium* cultures incubated at 18°C. After 50–60 hours, very large numbers of discharging zoosporangia and zoospores were apparent in the cultures. During maximum sporangial discharge (approximately 57 hours post-inoculation) the infected *Pythium* seed cultures were removed and replaced with new 24-hour-old uninfected *Pythium* cultured in stale Emerson's water on sesame seeds. (This change was considered the datum in timing the resulting developments.) Infection by the secondary zoospores then took place, and approximately 95 % of the infections formed cystosoral plasmodia. Developing cystosoral thalli were removed at six-hour intervals for a period of 72 hours and prepared for transmission electron microscopy as described above for sporangial thalli.

E. PRODUCTION OF ZOOSPORES

Large quantities of zoospores of approximately the same age were required as inoculum for further experiments or for structural study with the electron microscope. Highly concentrated suspensions of secondary zoospores of *W. pythii* were relatively easy to obtain. Large masses of heavily infected *Pythium* hyphae containing mature and discharging sporangia were placed in petri dishes containing 12 ml of Emerson's water for one hour. Aliquots of zoospores were then pipetted from the petri dishes to be used as inocula for various experiments. Zoospores were prepared for microtomy and electron microscopy by adding 2 % glutaraldehyde in 0.1 M phosphate buffer at pH

6.8 to the suspension which was then centrifuged into a pellet and embedded in 2% water agar to facilitate further processing.

The age of the zoospores in a suspension obtained as described above is reasonably synchronised because of the nature of *W. pythii* and other aquatic zoosporic fungi. Under conditions of liquid culture, zoospores mature in zoosporangia but usually do not emerge unless disturbed. Within a mass of heavily infected hyphae therefore, there will be hundreds of mature sporangia containing thousands of zoospores ready to emerge. Removing them from an aging culture medium and placing them into fresh Emerson's water or glass-distilled water, usually results in massive zoospore emergence. Most mature zoospores will usually emerge from sporangia within 30 minutes; thus, all the swimming zoospores are of similar age. With host/parasite two-membered cultures on agar medium in tubes or petri dishes, the method for obtaining masses of parasite zoospores is essentially the same. After it is determined that mature sporangia are present, 1–5 ml of sterile glass distilled water is pipetted into the culture tube or dish. After at least 30 minutes, the liquid and suspended zoospores are poured off or pipetted out. The density of zoospores in suspension or in a measured aliquot can be determined by using a Howard Mould Counting chamber or a Petroff–Houser bacterial counter. In our studies the collected zoospores of *W. pythii* were killed and fixed in very dilute glutaraldehyde before determining the density of the suspension. For additional methods of obtaining zoospores for study and experimentation, and for information relating to cultures of other host/parasite pairs (*Rozella/Allomyces* and *Olpidiopsis/Achlya*) see Held (1972, 1973).

In host/parasite culture of two zoosporic fungi, the zoospores of host and parasite emerge under similar environmental conditions. Obtaining a suspension of zoospores of only one partner for further experimentation requires special separation of host and parasite propagules. This was not a problem with the original host of *W. pythii*, for the *Pythium* formed no sporangia, even after being exposed to various treatments known to be conducive to their formation. Mullins and Barksdale (1965) separated the zoospores of *Dictyomorpha* (3.5 μm in diameter) from the host zoospores of *Achlya flagellata* (8.5–10.5 μm in diameter) by filtering the spore suspension through a sterile sintered glass filter. The pore size was such that only the smaller zoospores of the parasite passed through. Held (1972) used a Millipore filter (5 MWP; pore size 5 μm) to separate zoospores of the parasite *Olpidiopsis* from those of its *Achlya* host. It should be noted moreover that on agar medium a pure suspension of parasite zoospores may be obtained relatively easily because many water moulds, especially members of the Saprolegniaceae, do not usually produce zoospores unless grown in water on suitable substrates.

III. Studies of Woronina pythii

A. INTRODUCTION

Woronina pythii and other members of the order Plasmodiophorales are aquatic, biflagellated fungal endoparasites of vascular plants, algae and other fungi. Included in this class are nine genera and about 35 species. At least five of the taxa are of economic importance, parasitising food plants of man: *Plasmodiophora brassicae* Woronin, cause of the clubroot disease of cabbage and other Cruciferae; *Spongospora subterranea* (Wallr.) Lagerh. var. *subterranea* Tomlinson cause of the powdery scab disease of potatoes; *S. subterranea* var. *nasturtii* Tomlinson, cause of crook root disease of watercress; *Polymyxa graminis* Ledingham, a parasite in the roots of wheat and oats; and *Polymyxa betae* Keskin, a root parasite and possible virus vector in sugar beets.

Not for any of these parasites has the life cycle been completely documented, but generally all taxa have two vegetative phases apparent in their life cycles: a sporangiogenous phase which culminates in the formation of thin-walled zoosporangia and zoospores, and the cystogenous phase which terminates in the formation of thick-walled cysts (resting spores) arranged, except in *P. brassicae*, in characteristic cystosori.

Because of the interest in this group of fungal parasites by many phytopathologists, we have decided to discuss, as an example of a zoosporic pathogen of a lower plant, some of our work with *Woronina pythii*, a parasite of the aquatic fungus *Pythium*. Six species of *Woronina* have been described although three of them are very doubtful. *Woronina polycystis* Cornu, the most frequently studied taxon in this genus, is parasitic on members of the Saprolegniaceae (Oomycetes), and forms irregularly-shaped cystosori (Cornu, 1872; Fischer, 1882; Petersen, 1910; Cook and Nicholson, 1933; Goldie-Smith, 1954). *Woronina glomerata* (Cornu) A. Fischer is parasitic on the green alga *Vaucheria* (Chrysophyta) and forms loosely united, irregularly-shaped cystosori (Fischer, 1892; Zopf, 1894; Scherffel, 1925; Tokunaga, 1933). *Woronina pythii* (Goldie-Smith, 1956) is parasitic on species of *Pythium* and produces cystosori and sporangia which are spherically or irregularly-shaped.

The developmental cycle in *W. pythii* follows a pattern common to many plasmodiophoralean fungi. Each cyst during germination usually produces one primary biflagellate zoospore. These primary zoospores cease swimming, retract their flagella, encyst on a suitable host cell, penetrate the host cell as a uninucleate unwalled, amoeboid protoplast via an infection tube and develop therein into a multinucleate sporangiogenous plasmodium. Sporangiogenous plasmodia cleave into thin-walled, multilobed sporangia, within each lobe of which secondary zoospores are cleaved. After sporangial discharge, second-ary zoospores functioning similarly to primary zoospores on susceptible hosts, give rise to additional generations of sporangia, when environmental conditions are favourable for the continued growth of the host.

Under conditions unfavourable for growth of the host, cystogenesis begins. Young cystogenous plasmodia are multinucleate and morphologi-cally indistinguishable from sporangiogenous plasmodia of similar age. At the conclusion of plasmodial growth, the cystogenous plasmodia are cleaved into uninucleate, incipient cysts which become thick-walled at maturity.

B. DEVELOPMENT OF SPORANGIOGENOUS PLASMODIA

1. *Young Plasmodia*

Encysted zoospores are found within two hours of inoculating a susceptible host culture with secondary zoospores of *W. pythii* (Figs. 1a and 1b). The protoplasmic contents of the encysted zoospore penetrate the cell wall and plasma membrane of the host and the empty, hyaline cyst remains attached to the hyphal cell wall. Within six hours of inoculation, infections by the parasite appear as small swellings on the hyphal filaments (Fig. 1c) containing young uninucleate plasmodia (Fig. 2a). By 24 hours after inoculation, plasmodia have increased in size and are multinucleate (see Figs. 1d and 1e and Fig. 3a).

The interface between the *Pythium* sp. host and *W. pythii* is a single unit membrane, the plasmodial membrane of the parasite, 19–24 nm thick (see Figs. 2 and 3). During the first 24–30 hours of growth of the plasmodium, pseudopodium-like structures grow into and partially surround host or-ganelles and cytoplasm (Figs. 2a and 2b). Serial section analysis of what appear to be food vacuoles show that masses of host protoplasm are never completely surrounded by the plasmodium of the parasite. During this vegetative phase, the plasmodium increases in size, and the single nucleus undergoes five synchronous cruciform mitotic divisions yielding a total of 32 nuclei. After 24–30 hours of growth, only a narrow, peripheral band of host protoplasm remains between the plasmodium and the host cell wall (Fig. 3a), as the sporangiogenous plasmodia almost completely fill host-cell swel-lings. Pseudopodium-like structures are no longer formed, but small finger-like extensions of the plasmodial membrane still project into host cytoplasm (Fig. 3b). The irregularly-shaped interphase nuclei and darkly staining lipoid globules are the most prominent structures in 24-hour and older, pre-cleavage plasmodia (Fig. 3a).

2. Cleavage of Plasmodia

The transformation of vegetative plasmodia into multi-lobed sporangia begins approximately 30 hours post-inoculation. By the time cleavage furrows appear, the thickness of the plasmodial membrane is 8–9 nm, and the nucleoli begin to disperse within the nucleoplasm of each nucleus. The nuclei undergo one synchronous "transitional" division (division number 6). The nuclei during this division lack the persistent nucleoli characteristic of cruciform divisions, but are not devoid of nucleolar material, unlike the nuclei during the sporangial mitosis. Eventually, the nucleoli disperse completely, and the nuclei undergo at least one or two sporangial mitoses within the individual lobes. Nuclei in small lobes undergo one sporangial mitotic division (division number 7); those in large lobes undergo two (divisions 7 and 8).

Cleavage furrows are composed of double-unit membranes separated by a clear space and generally free of host debris (Fig. 3c). By 42 hours after inoculation, each plasmodium is cleaved into 25–35 multi-nucleate, irregularly-shaped sporangial masses (Fig. 3c). Nuclei measure 1.6 μm in diameter, lack nucleoli and are surrounded by rough endoplasmic reticulum (Fig. 3c). Numerous membrane-bound vesicles, almost completely filled with a light staining material are now distributed throughout the protoplasm of each lobe (Fig. 3c). Between 42 and 48 hours post-inoculation, the irregularly-shaped sporangial lobes become spherical [3.5–10 ($\bar{x} = 6.5$) μm in diameter] as they contract slightly, and interlobular spaces are formed (Fig. 4a).

3. Zoosporogenesis

The protoplasmic contents of each sporangial lobe cleave into 4 to 12 incipient biflagellate zoospores approximately 48 hours after inoculation (Figs. 4a and 4b). The two-unit membranes of the zoospore cleavage furrows are separated by long, slender, tubular-shaped vesicles (Figs. 4b and 4c). Large lipoid globules positioned between the sporangium and the host cell wall are the only apparent remains of disintegrated host protoplasm (Fig. 4a). Mature, discharging, and empty sporangia are present in cultures 48–60 hours old (see Figs. 1f–i and Figs. 4c and 5a–c). Sporangia are spherical to subspherical in shape, $12 - 36$ ($\bar{x} = 27$) μm in diameter, and each consists of 25–35 interconnected lobes. Each lobe is surrounded by an electron-dense cell wall (30 nm thick). Continuous lobes are separated by thin electron-translucent walls (18 nm thick) that break down prior to the release of zoospores (see Figs. 4 and 5).

Sporangial discharge soon follows the formation of exit tubes from small, peripherally situated lobes (see Figs. 1f and 1g and Fig. 5). These exit lobes

Scale bars represent 1.0 μm unless otherwise indicated in caption.
Abbreviations used in Figs. 1–13.

parasite

ar	= apical region	mb	= electron-dense, membrane-bound body
b	= basal body	n	= nucleus
c	= centriole	nu	= nucleolus
cf	= cleavage furrow	p	= parasite
d	= dictyosome	pv	= pleomorphic vesicle
ds	= dark speck	r	= ribosome
er	= endoplasmic reticulum	s	= intercellular space
f	= flagellum	t	= microtubule
l	= lipoid globule	v	= vacuole
m	= mitochondrion	v_1	= membrane-bound vesicle (Type I)
		v_2	= membrane-bound vesicle (Type II)
		w	= sporangial cell wall

host

ER	= endoplasmic reticulum	N	= nucleus
H	= host	NU	= nucleolus
L	= lipoid globule	PM	= plasma membrane
M	= mitochondrion	V	= vacuole
MB	= microbody	W	= cell wall

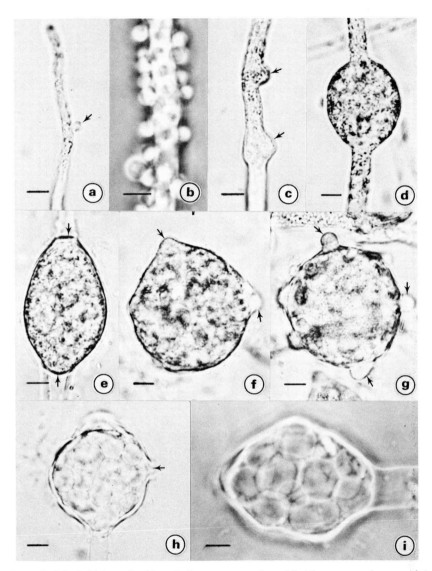

Fig. 1. Bright field (a and c–h) and phase contrast (b and i) microscopy of sporangial development. Scale bar = 5.0 μm. (a) Encysted secondary zoospore (arrow); (b) numerous encysted secondary zoospores; (c) two young infections, six hours old (arrows); (d) 18 hours old infection; (e) intercalary infection (24 hours old) with both host septa visible (arrows); (f) partially expanded exit papillae (arrows) projecting from mature sporangium (48 hours old); (g) mature sporangium (48 hours old) with fully expanded exit papillae (arrows); (h) empty multilobed sporangium (54 hours old) with single exit lobe (arrow) in focal view; (i) empty sporangium (54 hours old) illustrating the "bubbly" aspect.

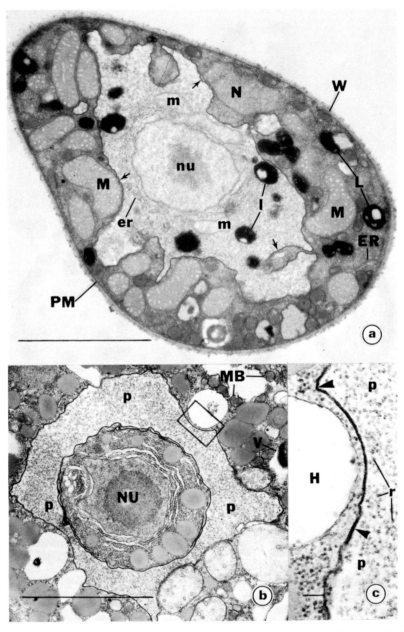

Fig. 2. Early plasmodial growth (a and b) and host/parasite interface (c). (a) Six hours old uninucleate sporangiogenous plasmodium; (b) pseudopodium-like structure of parasite (p) surrounding a uninucleate mass of host protoplasm; (c) interface of *W. pythii* and *Pythium* sp. Enlargement of enclosed area in (b) showing parasite membrane morphology (arrows). Scale bar = 0.1 μm.

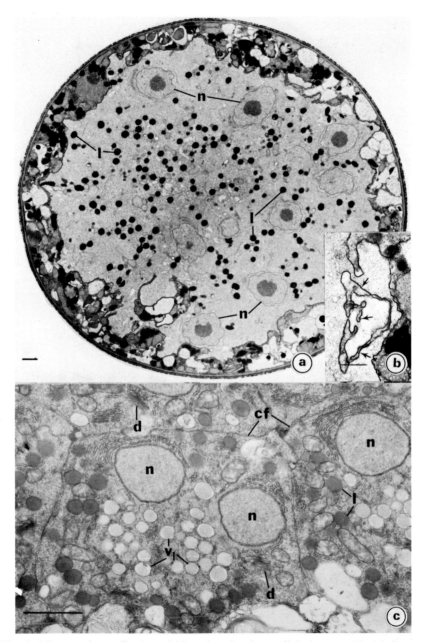

Fig. 3. Mature plasmodia (a and b) and early plasmodial cleavage (c). (a) Survey micrograph of 24 hours old infection containing a large multinucleate plasmodium surrounded by a narrow band of host protoplasm; (b) host/parasite interface of 24 hours old infection showing small finger-like projections (arrows) of the outer plasmodial membrane (scale bar = 0.5 μm); (c) portion of 42 hours old plasmodium in early cleavage.

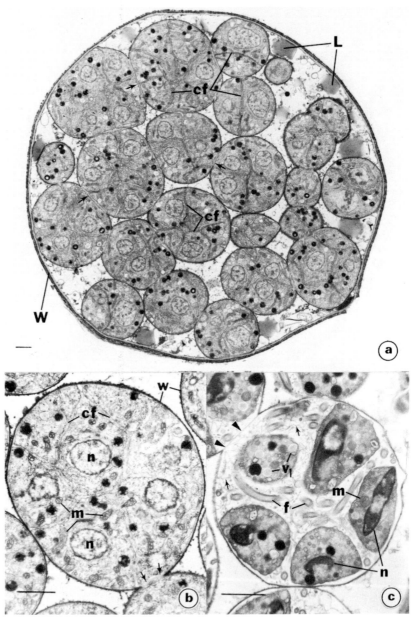

Fig. 4. Secondary zoospore cleavage. (a) Survey micrograph of multilobed sporangium (48 hours old) undergoing zoospore cleavage: cleavage furrows, and thin walls (arrows) between contiguous lobes; (b) single lobe of 48 hours old sporangium: note the thin wall (arrows) between contiguous lobes, zoospore cleavage furrows with tubular-shaped vesicles in between, and sporangial cell wall; (c) mature zoospores in 54 hours old sporangial lobe with tubular-shaped vesicles situated between zoospores (small arrows), thin wall (large arrows) between contiguous lobes, and membrane-bound vesicles.

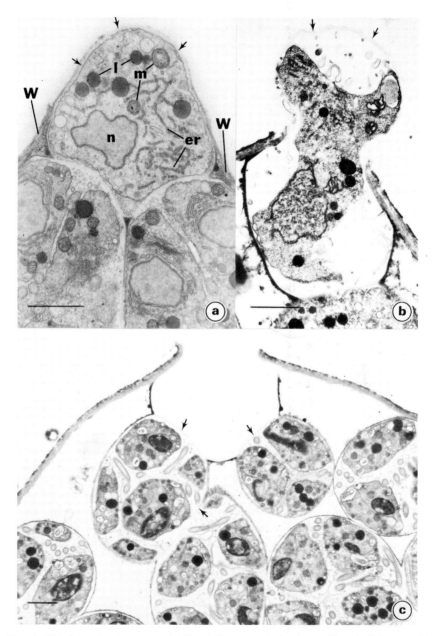

Fig. 5. Exit-lobe development. (a) Partially expanded exit papilla of 48 hours old sporangium (note the ruptured host cell wall); (b) fully expanded exit papilla surrounded by expanded thin wall (arrows); (c) empty exit lobe of mature sporangium. The thin walls (arrows) that separate contiguous lobes become more electron-translucent as the sporangium matures.

are 2.4–3.5 ($\bar{x} = 3.1$) μm in diameter, uninucleate, and their contents do not differentiate into zoospores (see Figs. 5a and 5b). Each sporangium has an average of eight exit lobes evenly distributed on its periphery. Exit-lobe surfaces in contact with the host cell wall are bounded by a thin wall that appears similar in profile to those found between contiguous lobes (see Fig. 5). As the sporangium matures, the exit lobes gradually increase in size and form short exit tubes. Exit tube enlargement is accomplished through the apparent stretching and expansion of their thin outer walls (Fig. 5) and eventually perforation of the contiguous host cell wall occurs.

Exit tubes usually attain a maximum length of 2.2–3.1 ($\bar{x} = 2.8$) μm. The thin wall surrounding the exit tube ruptures and its protoplasmic contents are discharged into the external environment (Figs. 5b and 5c). The first swimming zoospores escape from lobes contiguous with the exit lobes. Sporangial discharge proceeds centripetally from lobe to lobe until the entire sporangium is empty, and zoospores from lobes other than those peripherally situated must swim through adjacent lobes to escape the sporangium (Fig. 5c). Each sporangium takes approximately one hour to empty (Figs. 1h and 1i).

4. *Discussion*

The sporangiogenous plasmodia of *Woronina pythii* are remarkably similar in ultrastructural morphology to the sporangiogenous plasmodia of *P. betae* (D'Ambra and Mutto, 1975) and to the cystogenous plasmodia of *Sorosphaera veronicae* (Braselton *et al.*, 1975) and *Plasmodiophora brassicae* (Williams and McNabola, 1967). Many of the organelles, especially mitochondria and endoplasmic reticulum, are similar in size, shape, and ultrastructural morphology in the three parasites.

The thallus–host interface varies in the above three plasmodiophoralean taxa. The interface between the cystogenous plasmodia of *P. brassicae* and cabbage is a seven-layered structure approximately 23 nm thick and is interpreted as the closely appressed tripartite plasma membranes of the host and parasite (Williams and McNabola, 1970; Williams and Yukawa, 1967). The interface between the cystogenous plasmodia of *S. veronicae* and *Veronica persica* Poir (Braselton and Miller, 1975) is a single unit membrane, approximately 9–12 nm across, and interpreted to be the plasma membrane of the parasite. This membrane also exhibits a typical tripartite lamellar structure. An electron micrograph of the interface between the sporangiogenous plasmodia of *P. betae* and *Beta vulgaris* L. cv. "Alba P" (D'Ambra and Mutto, 1975) depicts a parasite plasma membrane which is similar in thickness to that of *S. veronicae*. The sporangiogenous plasmodial envelope of *W. pythii* is a single electron-dense band and does not exhibit the tripartite lamellar structure found in cystogenous plasmodia of *P. brassicae* and *S. veronicae*.

The host–parasite interface between *W. pythii* and *Pythium* sp. is unique among plasmodiophoralean taxa because of the exceptional thickness of the plasmodial envelope during early vegetative growth and its subsequent reduction in thickness near the cessation of plasmodial growth. The vegetative plasmodia of *W. pythii* have a single enveloping membrane, approximately 19–24 nm thick but reduced in thickness to 8–9 nm prior to cleavage. This dimorphism of the plasmodial envelope may be associated with the digestive role in parasite metabolism. Williams and McNabola (1967) reported a reduction in thickness of the cystogenous plasmodial envelope of *P. brassicae* at the cessation of vegetative plasmodial growth, but they used the term plasmodial envelope, to describe the appressed plasma membranes of both host and parasite. The reduction in thickness of the "plasmodial envelope" in *P. brassicae* resulted, therefore, from the disintegration of the outer, host plasma membrane and not from a change in thickness of the plasmodial envelope of the parasite.

The pseudo-food vacuoles of *W. pythii* are formed by pseudopodium-like protoplasmic projections which grow into and partially surround portions of host protoplasm. These protoplasmic projections significantly increase the digestive surface area to volume ratio of the plasmodium and may function as a feeding mechanism and facilitate vegetative growth of the parasite. The engulfment of the host protoplasm and the formation of true-food vacuoles by plasmodia has been reported in *P. brassicae* by Williams and McNabola (1967) (and see Chapter 5 of this volume).

D'Ambra and Mutto (1975) described the fine structure of cleavage-furrow formation in the sporangiogenous phase of *P. betae*. They found that the cleavage furrows in *P. betae* were free of contents and suggested that they were formed by the "lining up of and fusion of vesicles." In *P. betae* the cleavage furrows initially delineate the individual lobes of the multi-chambered sporangia, and through their continued formation, later define the boundaries of each individual zoospore. The cleavage furrows in *W. pythii* are composed of two parallel unit membranes. Cytoplasmic cleavage is accomplished through an increase in length of these membranes until sporangial lobes, and eventually individual zoospores, are differentiated. During zoospore cleavage in *W. pythii*, the two unit membranes of each furrow are separated by long, slender, tubular-shaped vesicles.

Sporangial wall morphology of *W. pythii* is similar to that described for *S. veronicae* (Talley *et al.*, 1978). In both taxa, the electron-dense regions of the sporangial wall are monolayered and approximately 30 nm across. The walls of the multi-chambered sporangia of *P. betae* (D'Ambra and Mutto, 1977) are bilayered, having outer electron-dense, and inner less electron dense layers. The thin cross walls that separate the contiguous lobes of *W. pythii* are, however, similar in morphology to those found in *S. veronicae* and *P. betae*.

In both *W. pythii* and *P. betae* (D'Ambra and Mutto, 1977), exit tubes arise from uninucleate exit lobes whose contents do not differentiate to form zoospores. These uninucleate lobes are situated on the periphery of the parasite sporangium and are usually in physical contact with the wall of the host cell. The exit lobes form exit tubes which perforate the host cell wall. D'Ambra and Mutto (1977) speculate that the host cell wall of *Beta vulgaris* is structurally degraded by digestive enzymes contained within "membrane bound osmiophilic vesicles" present within the cytoplasm of the exit lobes, and together with physical pressure created by the enlargement of fungal cell vacuoles, perforation of the host cell wall is accomplished (D'Ambra and Mutto, 1977).

Although membrane bound osmiophilic vesicles are not apparent in the exit lobes of *W. pythii*, this does not eliminate the possibility that digestive enzymes are present in the cytoplasm of these lobes. In *W. pythii*, pressure on the host cell wall is accomplished through the expansion in size of the entire exit lobe.

C. ULTRASTRUCTURE OF SECONDARY ZOOSPORES

Mature secondary zoospores within sporangia of *W. pythii* are 2.2–2.8 ($\bar{x} = 2.4$) μm in diameter (Fig. 7a). Each zoospore has a single baciliform nucleus 0.5×1.9 μm in size and the nucleoplasm contains much electron-dense material, but electron-translucent regions with membrane profiles are apparent also (see Figs. 6b,c and Fig. 7a). The cytoplasm contains endoplasmic reticulum, free ribosomes, lipoid globules, spherically-shaped mitochondria, pleomorphic vesicles, membrane-bound vesicles, and small electron-dense membrane-bound bodies (see Figs. 6 and 7).

Secondary zoospores of *W. pythii* are biflagellate, and both flagella are of the whip-lash type. Basal bodies of the flagella (see Figs. 6a–c and Fig. 7a) are 0.45–0.55 ($\bar{x} = 0.5$) μm in length, 0.15–0.23 ($\bar{x} = 0.21$) μm in width, and each exhibits a 3×9 pinwheel configuration of microtubules in cross sectional view. Basal body length is thus less than the 1.7 μm in secondary zoospores of *S. veronicae* (Talley *et al.*, 1978) and the 0.6 μm in *Polymyxa betae* (Keskin, 1964). Basal bodies of the secondary zoospores of *S. veronicae* contain nine triplet microtubules and appear morphologically similar in transverse section to the centrioles in sporangiogenous plasmodia of *W. pythii*. The length of the basal body in primary zoospores of *P. brassicae* is 0.8 μm and the basal bodies are composed of nine doublet microtubules (Aist and Williams, 1971).

A basal plate is positioned at the proximal end of each basal body (Fig. 6b), and a terminal plate at the distal end (see Figs. 6c and 7a). Basal bodies of mature zoospores (see Figs. 6b,c and 7a) contain two distinct rows of dark specks which are not present in basal bodies of immature zoospores (Fig. 6a). These rows of dark specks are also present in basal bodies of secondary zoospores of *P. betae*, but not in *S. veronicae*. The angle of insertion between

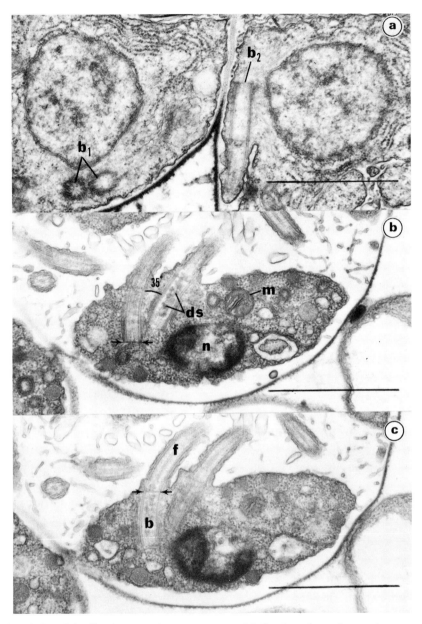

Fig. 6. Basal bodies in secondary zoospores. (a) Section through two immature zoospores showing two grazed basal bodies (b_1) in one zoospore and a longitudinally sectional basal body (b_2) in the other. Note the absence of dark specks within the basal bodies of immature zoospores; (b) portion of mature zoospore showing paired basal bodies describing a 35° angle; basal plate (arrows), dark specks, spherically-shaped mitochondrion, and grazed electron-dense nucleus; (c) adjacent section to (b); terminal plate (arrows) of basal body.

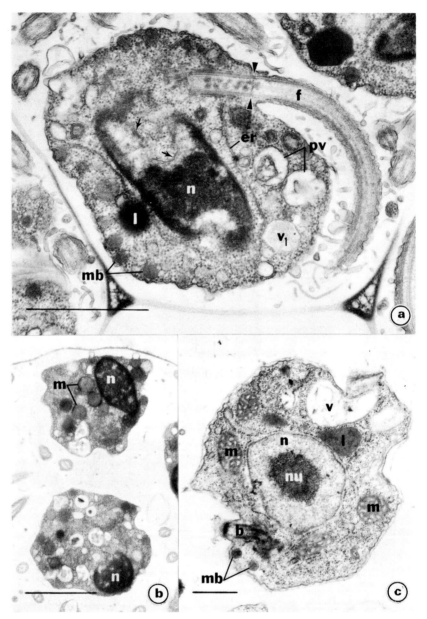

Fig. 7. Mature zoospores within sporangia (a and b) and swimming zoospore (c). (a) Terminal plate (large arrows) of basal body, with intranuclear membrane (small arrows); (b) portion of sporangial lobe just prior to discharge showing two zoospores in maximum state of contraction; (c) swimming zoospore. Scale bar = 0.5 μm.

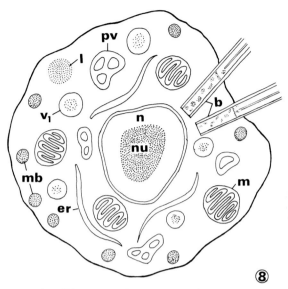

Fig. 8. *Woronina pythii.* Diagrammatic representation of a swimming secondary zoospore.

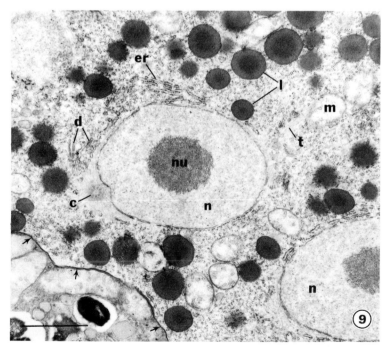

Fig. 9. Portion of 30-hours-old cystogenous plasmodium. Note the thick outer plasmodial membrane (arrows).

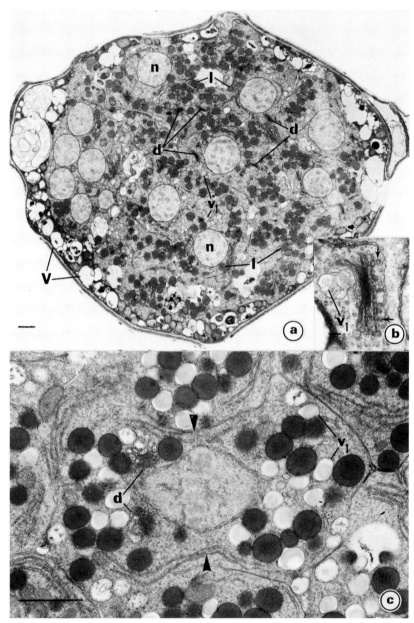

Fig. 10. Cystogenous plasmodial cleavage. (a) Survey micrograph of early plasmodial cleavage (36 hours post-inoculation): cleavage furrows (arrows) and akaryotic nuclei; (b) dictyosome morphology during the akaryote nuclear stage with type I vesicles. Note the spherically-shaped vesicles (arrows) between endoplasmic reticulum and dictyosome. Scale bar = 0.1 μm; (c) dumb-bell-shaped protoplasmic mass (42–48 hours post-inoculation) containing metaphase nucleus, and numerous type I dictyosomal vesicles. Cytoplasmic pinching occurs in the same plane as that of the chromatin plate (arrows). Paired dictyosomes are apparent at the left nuclear pole.

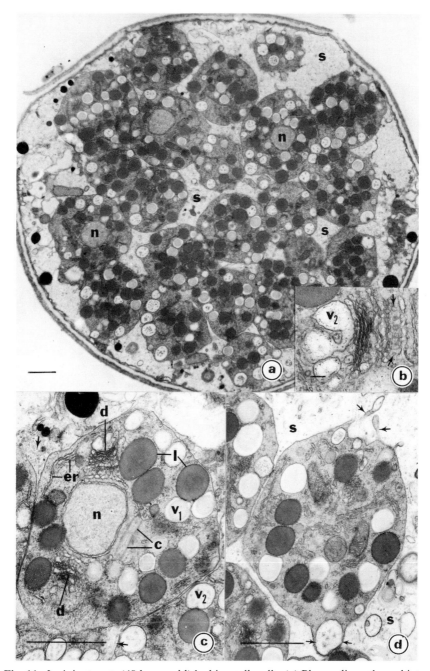

Fig. 11. Incipient cysts (48 hours old) lacking cell walls. (a) Plasmodium cleaved into uninucleate cells; (b) dictyosome morphology between late prophase (42 hours post-inoculation) and cyst wall formation (54 hours post-inoculation) with type II vesicles, and spherically-shaped vesicles (arrows) situated between the endoplasmic reticulum and dictyosome. Scale bar = 0.1 μm; (c) incipient cyst with tubular-shaped vesicles (arrows) and other organelles situated between the two membranes of the cleavage furrows; (d) incipient cyst with two peripherally situated type II dictyosomal vesicles in sectional view. The plasma membrane of the cell appears to be locally distended by one vesicle (small arrows) and ruptured by the other (large arrows). The vesicular contents are present in the intercellular spaces.

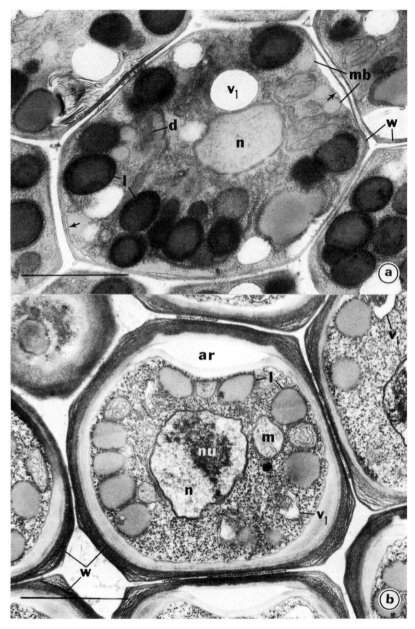

Fig. 12. Incipient and mature cysts. (a) Incipient cyst (54 hours old): nucleus, dictyosome, lipoid globules, type I dictyosomal vesicles, plasma membrane (arrows), membrane-bound bodies, and thin, electron-dense cell wall; (b) mature cyst (72 hours old): nucleus, nucleolus, type I dictyosomal vesicles, lipoid globules, mitochondria, thick bilayered cell wall, and electron-translucent apical region.

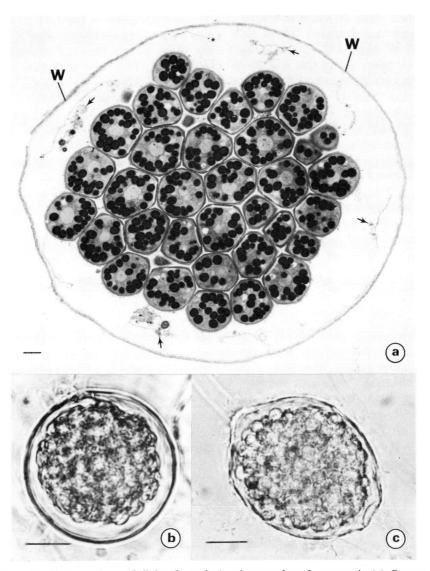

Fig. 13. Electron (a) and light (b and c) micrographs of cystosori. (a) Survey micrograph of 60 hours old infection showing hyphal cell wall and degenerating host organelles (arrows). Note the absence of the extra-cellular contents of the type II dictyosomal vesicles in the intercellular spaces; (b) immature cystosorus (60 hours old) illustrating the "lumpy" aspect of incipient cysts during this developmental stage under the light microscope. Scale bar = 10.0 μm; (c) mature cystosorus (72 hours old). Scale bar = 10.0 μm.

paired basal bodies in secondary and primary zoospores of plasmodiophor-alean taxa is similar. Those of *W. pythii* (secondary zoospores) describe a 30–35° angle with respect to each other (Figs. 6b and 6c); in *S. veronicae* (secondary zoospores) it is 40–45° (Talley *et al.*, 1978) and in *P. brassicae* (primary zoospores) 45° (Aist and Williams, 1971).

Just prior to sporangial discharge zoospores undergo cytoplasmic and nucleoplasmic contraction from 2.2–2.8 ($\bar{x} = 2.4$) μm to 1.5–1.7 ($\bar{x} = 1.65$) μm in diameter (Fig. 7b). The tubular-shaped cristae of the mitochondria are almost indistinguishable, and the electron-translucent regions in the nucleoplasms are no longer apparent (Fig. 7b). Free-swimming zoospores are spherical to slightly elongate in shape and expand to 2.4–3.0 ($\bar{x} = 2.6$) μm in diameter (Fig. 7b and Fig. 8). Nuclei are spherical in shape, averaging 0.91–1.2 ($\bar{x} = 1.0$) μm in diameter, and each contains a single, centrally situated nucleolus approximately 0.37–0.62 ($\bar{x} = 0.51$) μm in diameter and vacuoles are apparent in the cytoplasm (Fig. 7c and Fig. 8).

D. DEVELOPMENT OF CYSTOGENOUS PLASMODIA

1. Young Plasmodia

Cystogenous plasmodia of *W. pythii* are developmentally and morphologi-cally indistinguishable from sporangiogenous plasmodia during the first 24 hours of growth. Both types of plasmodia are bounded by a thick plasmodial membrane (19–24 nm) and both contain the same complement of cellular organelles. Cystogenous plasmodia (Fig. 9) more than 24 hours old contain more lipoid globules and appear slightly denser under the light microscope than do sporangiogenous plasmodia of the same age. The plasmodial membrane is less thick (8–9 nm) and nucleolar dispersal is completed by 36 hours after inoculation of host cultures (Fig. 10a).

2. Cleavage of Plasmodia

Cystogenous plasmodia begin to cleave 30–36 hours after inoculation (Fig. 10a). Cleavage is closely correlated with nuclear division. Between 36 and 48 hours post-inoculation, cysts begin to form as nuclei undergo one synchron-ous, meiosis 1-like division (Figs. 10a and 10c). Pachytene nuclei containing synaptonemal complexes are apparent (Fig. 10a). Cleavage furrows, mor-phologically identical to those apparent during secondary zoospore cleavage, first appear in plasmodia after nucleolar dispersal is complete (the so-called akaryote stage, Fig. 10a). During this stage of development, dictyosomes are 0.22 × 0.71 μm in cross-sectional view and are composed of five to eight very closely stacked cisternae (Figs. 10a and 10b). Cisternae on the maturing surface of the dictyosome are dilated, and appear to blebb-off membrane-bound vesicles (type I). These vesicles are spherical in shape, average 0.21–0.27 ($\bar{x} = 0.25$) μm in diameter and are almost completely filled with lightly staining material (Fig. 10). During late prophase and early metaphase, plasmodia cleave into protoplasmic bodies that appear rectangular in section, each containing a dividing nucleus. In late metaphase, the protoplasmic contents of each section begins to pinch in half in the same plane as that of the chromatin plate (Fig. 10c). After telophase, when cleavage is complete, each interphase nucleus is 0.91–1.2 ($\bar{x} = 1.0$) μm in diameter, and is surrounded by an irregularly-shaped mass of cytoplasm.

3. *Formation of Vesicles*

Between late prophase (42 hours post-inoculation) and cyst wall formation (54 hours post-inoculation), dictyosomes have six to eight loosely stacked cisternae and average 0.52 × 0.53 μm in cross section (Figs. 11b and 11c). A second type of membrane-bound vesicle (type II) blebbs off from dilated cisternae located on the maturing surface of the dictyosome, as well as from the dilated peripheral surfaces of other cisternae (Figs. 11b and 11c). These vesicles are spherical, 0.19–0.24 ($\bar{x} = 0.22$) μm in diameter, and are partially filled with amorphous masses of electron-dense, fibrilloid material (Figs. 10c, 11a–d).

By 48 hours post-inoculation, the uninucleate cells (incipient cysts) of the parasite have rounded-up and are 2.8–3.5 ($\bar{x} = 3.0$) μm in diameter (Figs. 11a, c and d). Nuclei lacking nucleoli are centrally situated within the cells (Figs. 11a and 11c). A single pair of centrioles, aligned end-to-end, is positioned parallel to the nuclear envelope of each nucleus: each centriole is 0.32–0.37 ($\bar{x} = 0.35$) μm in length (Fig. 11c). Both types of dictyosomal vesicle increase in size; type I vesicles now average 0.35 μm in diameter and are evenly distributed throughout the cytoplasm of each cell (Figs. 11a, b and c). Type II vesicles now averaging 0.37 μm in diameter, migrate toward the peripheral surfaces of each cell, and come in contact with the plasma membrane (Figs. 11a and 11c). The vesicles continue their outward migration until they locally distend the plasma membrane into the intercellular spaces (Figs. 11a and 11d). Very little cytoplasm remains between the vesicle and the distended plasma membrane (Fig. 11b). Eventually, the plasma membrane and the membrane which surrounds each type II vesicle ruptures, thereby releasing the vesicular contents into the intercellular spaces between the cysts (Figs. 11a and 11d). Membrane located on the inner facing surface of each vesicle now becomes plasma membrane (Fig. 11d).

4. *Cyst Wall*

The cyst wall is bilayered, 0.12–0.25 ($\bar{x} = 0.22$) μm in thickness, and formed between 54 and 72 hours post-inoculation (Fig. 12 and Fig. 13a). The electron-dense outer surface of each cyst wall is formed first (Fig. 12a) and the contents of the type II dictyosomal vesicles probably contribute to its formation. The vesicular material has a similar fibrillar composition to the outer cyst wall, and during formation of this wall, there is a gradual reduction in the amount of vesicular material in the intercellular spaces. A space of at least 25 nm usually separates the outer cell walls of contiguous cysts (Figs. 12a and 12b). Prior to cyst inner wall formation, small membrane-bound bodies filled with electron-dense material, are peripherally situated within each cell, and many of them appear to be fused with the plasma membrane (Fig. 12a). The composition and staining properties of these bodies appear similar to those of the inner wall of the cyst, and the contents of these bodies might possibly be precursors or components of the inner cyst wall. Dictyosomes are no longer positioned at the nuclear poles and appear to be degenerating (Fig. 12a). The contents of the type I dictyosomal vesicles now appear more electron-translucent (Fig. 12a). During this developmental stage, lipoid globules stain darkly and are the most noticeable cytoplasmic structures (Fig. 12a).

The hyphal cell wall and a small number of degenerating organelles are all that remain in host cells in cultures 60 hours after inoculation (Fig. 13a). A peripheral space is now apparent between the cystosorus and host cell wall (Fig. 13a) and immature cystosori appear lumpy under the light microscope (Fig. 13b). Cystosori (Figs. 12b and 13c) present in cultures 72 hours post-inoculation are golden-brown in colour under the light microscope. Cystosori of this age are mature and liberate primary zoospores after drying and rehydration. Most cystosori of *W. pythii* in *Pythium* sp., are spherical to subspherical in shape, 12–42 ($\bar{x} = 27$) μm in diameter, and are composed of 75–600 ($\bar{x} = 215$) cysts (Fig. 13c) The uninucleate cysts measure 2.7–3.2 ($\bar{x} = 3.0$) μm in diameter and are spherical, subspherical, ovoid or polyhedral in shape. Nuclei are centrally situated within the cysts, and 0.8–1.1 ($\bar{x} = 0.9$) μm in diameter; each nucleus contains a darkly staining irregularly-shaped nucleolus (Fig. 12b). Protoplasm of mature cysts lacks dictyosomes and centrioles. The contents of the type I dictyosomal vesicles are now electron-dense and seemingly granular in composition (Fig. 12b). In the apical region of each cyst, an electron-translucent area occurs between the two layers of the cell wall (Fig. 12b).

5. *Discussion*

During cytoplasmic cleavage in cystogenous plasmodia of *W. pythii*, the multinucleate plasmodium is cleaved first into many bodies that appear rectangular in section, each containing a single dividing nucleus. The cleavage furrows are composed of double-unit membranes separated by long, slender, tubular-shaped vesicles and are generally free of cytoplasmic debris. The rectangular-sectioned protoplasmic bodies then pinch in half at the conclusion of the meiosis I-like division resulting in the formation of 210–220 uninucleate protoplasmic masses or incipient cysts. Cytoplasmic cleavage and meiosis I also occur concurrently during the cystogenous development of *S. veronicae* (Harris *et al.*, 1980; Braselton and Miller, 1978), *P. betae* (D'Ambra and Mutto, 1975) and *P. brassicae* (Garber and Aist, 1979b). The cleavage furrows in the latter two species are formed, however, by the alignment and fusion of intracellular vacuoles around uninucleate masses of protoplasm.

During cytoplasmic cleavage of cystogenous plasmodia of *W. pythii*, there is a coordinated intracellular association among dictyosomes, membrane-bound vesicles, and cyst wall formation. Plasmodial cleavage begins concurrently with the start of an increase in dictyosomal activity (vesicle formation for instance). Type I vesicles are produced during the initial stages of cleavage and are morphologically similar to those synthesised during sporangiogenesis (see Fig. 3c). Although type I vesicles persist throughout cleavage and are present in mature cysts (as well as in secondary zoospores), their ultimate fate is unknown. During later stages of cleavage, type II dictyosomal vesicles are apparent and are partially filled with the extracellular precursors of the cyst wall. By fusion of the vesicles with plasma membrane, the vesicular material is deposited in the spaces between incipient cysts and contributes to the formation of the outer cyst wall. Increase in vesicle formation is known to occur during cleavage of cystogenous plasmodia in *P. brassicae* (Williams and McNabola, 1967), *S. veronicae* (Braselton and Miller, 1978) and *P. betae* (D'Ambra and Mutto, 1975). The origin of these vesicles was not reported, but the vesicles in cystogenous plasmodia in these three taxa may also have dictyosomal origins and play a role in cell wall synthesis, as in *W. pythii*.

In the mature cystosori of *W. pythii*, the cell walls of contiguous cysts are separated by a space of at least 25 nm. This cyst arrangement suggests why the cystosorus fragments so easily into individual cysts. In *S. veronicae* (Braselton and Miller, 1978) and *P. betae* (D'Ambra and Mutto, 1975), the cyst walls are fused and are difficult to separate. In *P. brassicae*, cysts are not aggregated into cystosori.

The cysts (resting spores) of many species of the Plasmodiophorales are reportedly difficult to germinate and many species require special conditions for germination (Karling, 1968). The cysts of *W. pythii*, however, are easy to germinate. After a one-week period of dehydration, cysts liberate primary zoospores within 48 hours of being rehydrated.

E. HOST RANGE OF *WORONINA PYTHII*

The host range of *Woronina pythii* in axenic host/parasite culture was investigated. Sesame-seed cultures of host species to be tested were grown axenically for 24 hours at room temperature in petri dishes containing 12 ml of Emerson's water. Five cultures of each host to be tested were inoculated with parasitised mycelium of the original host (*Pythium* sp.) containing discharging sporangia and swimming secondary zoospores of *W. pythii*. The inoculum was removed from the petri dish after 1.5 hours, and the cultures were incubated at 18°C under intermittent light. This method of inoculation was possible because the original *Pythium* host of *W. pythii* from Michigan never produced sporangia or zoospores and could not therefore become established in test cultures and produce confusion in results. Test host species unsuccessfully infected by the first inoculation were reinoculated at 24 hour intervals for 72 hours; six weeks after the original inoculation, all test host species were checked to determine whether or not they were parasitised. Parasitised host species were maintained by adding a sterile sesame seed to each culture every seven days for four weeks.

Woronina pythii parasitised all 15 species of *Pythium* tested: *Pythium acanthicum* Drechsler, *P. aphanidermatum* (Edson) Fitzp., *P. aristosporum* Vanterpool, *P. arrhenomones* Drechsler, *P. debaryanum* Hesse, *P. dissotocum* Drechsler, *P. irregulare* Buisman, *P. iwayamai* Ito, *P. okanoganense* Longale, *P. proliferum* de Bary, *P. pulchrum* von Minden, *P. spinosum* Saw., *P. torulosum* Coker and Patt., *P. ultimum* Trow and the *Pythium* sp. that was the original host of this isolate. Four species of the closely related genus *Phytophthora* were not parasitised.

Other taxa of the Plasmodiophorales parasitise other water moulds (*Woronina polycystis* in *Saprolegnia; Sorodiscus cokeri* Goldie-Smith in species of *Pythium; Octomyxa achlyae* Couch, Leitner and Wiffen in *Achlya*; and *O. brevilegniae* Pendergrass in *Brevilegnia*) and because of the morphological similarities of *W. pythii* to some of these fungi, several representatives of the Saprolegniaceae were tested as possible hosts including 28 species of *Achlya*, four species of *Saprolegnia*, three species of *Brevilegnia* and one each of *Aphanomyces* and *Dictyuchus*. None of them were parasitised by *W. pythii*.

Our results in these host range studies therefore confirm the work of Goldie-Smith (1956) who reported that *W. pythii* will only parasitise species of *Pythium* although her isolate of *W. pythii* was unable to parasitise all species of *Pythium* which she tested. Goldie-Smith (1956), however, did not use bacteria-free two-membered cultures as we did, and this may have influenced her results. On the other hand, our isolate parasitised all species of *Pythium* tested, but to different degrees of severity. These differences between our isolate and that of Goldie-Smith suggests the existence of physiological specialisation in *W. pythii*. The demonstration of such specialisation would require the use of differential hosts and the standardisation of experimental procedures. Physiological races could then be classified according to their reactions with different hosts. Such a system is now being successfully employed in the characterisation of physiological races of *Plasmodiophora brassicae* (Buczacki *et al.*, 1975).

F. CYST VIABILITY OF *WORONINA PYTHII*

Cyst age at the time of dehydration and its influence on viability were investigated. Cysts ranging in age from two weeks to 14 months were dried at room temperature for 10 days on individual sterilised paper tabs which were later individually rehydrated at room temperature in 12 ml of Emerson's water in petri dishes containing a 24 hour-old *Pythium* host culture on sesame seeds. Cultures were checked for infection every 24 hours for a period of two weeks and 25 stock cultures were tested for each age interval. The effect of storage temperature on the viability of dehydrated cysts was also investigated. All stock cultures tested were two months old prior to dehydration. They were dehydrated on paper tabs for 10 days at room temperature, then stored in air-tight containers at 0°C, 6°C, 20°C, or at ambient room temperature: 90 stock cultures were tested for each temperature (i.e. 10 for each storage length). Paper tabs with dried cysts were rehydrated as described above.

The age of the stock culture prior to dehydration influenced culture survival and cyst germination of *W. pythii* (Table 1). Stock cultures 0–3 weeks old contained only sporangial stages of the parasite, and were incapable of surviving dehydration but mature cysts were formed in all stock cultures older than three weeks. The thick-walled cysts were capable of surviving dehydration in axenic laboratory culture when the stock cultures were between three weeks and four months old before drying. Cyst germination did not occur when laboratory stock cultures were five months or older before drying. The largest number of host infections was produced from rehydrated stock cultures approximately two-months old prior to drying. In these cultures, swimming zoospores were liberated from germinating cysts within 36–48 hours after rehydration.

Table 1. The influence of storage temperature on survival of *W. pythii*

Storage temperature (°C)	Storage time (months) of dehydrated stock cultures prior to rehydration								
	1	2	3	4	6	8	10	12	14
0	+	+	+	+	+	+	+	+	+
6	+	+	+	+	+	+	+	+	+
20	+	+	+	+	+	+	−	−	−
ambient	+	+	+	+	+	−	−	−	−

+ denotes evidence of cyst germination; − denotes no evidence of cyst germination.

Differential effects of storage temperature on the viability of dehydrated cysts were not apparent during the first six months of storage as the parasite was recovered from all rehydrated paper tabs tested (Table 1). After six months of storage, dehydrated cysts stored at 20°C or at ambient room temperature lost their ability to germinate and no germination occurred after 10 months (Table 1). Dehydrated cystosori stored at 0°C and 6°C remained viable for the duration of the experiment (14 months). Cystosori used in this experiment were taken exclusively from two-month-old stock cultures.

IV. Conclusion

The life cycles of zoosporic plant pathogens are not unlike those exhibited by saprophytic fungi although their substrates are certainly different. All zoosporic fungi have functionally similar phases in their life cycles: a vegetative or non-sexual reproductive phase and a phase during which resting bodies are produced. During the former, abundant propagules which may germinate immediately are produced and rapidly dispersed; during the latter, resistant spores or cysts are formed, usually through a sexual reproductive process. The production of resistant spores usually commences as substrate or other factors necessary for continued vegetative growth become limiting or because metabolic waste substances detrimental to continued vegetative growth accumulate. It appears clear from recently published cytological and ultrastructural studies on five taxa of the Plasmodiophorales (*Sorosphaera veronicae*, *Polymyxa betae*, *Plasmodiophora brassicae*, *Woronina pythii* and *Spongospora subterranea*) that patterns of development and life cycles are strikingly similar. A knowledge of the host–parasite relationships (cytology, ultrastructure, physiology, biochemistry and genetics) in one of the taxa might therefore be applicable to an understanding of the relationships in related taxa. Plant pathologists have tended in the past, however, to exclude from their consideration those organisms pathogenic on economically unimportant hosts and, most pertinently, those pathogenic on other, lower organisms. *Woronina pythii* in *Pythium* is an excellent example of the latter type and the relative ease in handling the bifungal pair in two-membered axenic culture, the rapidity of growth of host and parasite and the structural simplicity of the fungal host all facilitate experimental study.

Woronina pythii in two-membered axenic culture in *Pythium* responds to culture manipulation as do some zoosporic, saprophytic fungi, such as *Chytriomyces hyalinus* (Chytridiales) or *Olpidiopsis* (*Lagenidiales*), the latter a parasite of several species of water-mould fungi. We have been able to initiate resting body formation in *C. hyalinus* (Miller and Dylewski, 1981) and cyst formation in *W. pythii* by growing the fungi in spent culture medium

(medium with accumulated staling substances and greatly reduced nutrients). This response in change from the sporangial phase (zoospore production) to the cystogenic phase (resistant spores or cysts) suggests that these phases may be controlled by the medium or host cell physiology, and not necessarily by an obligatory nuclear state (either n, $n + n$ or $2n$) of the parasite. A change in the nuclear condition occurs, but in response, perhaps, to host or culture chemistry rather than at a predetermined stage in the life cycle. The vegetative and/or asexual, zoosporic phase may be continued indefinitely if supportive conditions of culture are present and maintained. It is unlikely that this situation occurs under "natural conditions", whether the fungus is a parasite or a saprophyte. Host–plant physiology changes naturally with age and, perhaps, also in response to parasite intrusion; *in vitro* culture conditions alter as food sources are reduced and metabolic waste substances accumulate.

We have just begun to study host–parasite relations between *W. pythii* and *Pythium* but even with the relative ease of handling the two-membered axenic cultures, results are not easily obtained. Nonetheless, we have presented here cytological and ultrastructural information regarding secondary zoospores and their penetration into the host, sporangiogenous plasmodial development, sporangiogenesis, zoosporogenesis and emergence of secondary zoospores. We have also presented cytological and ultrastructural details of development and cleavage of cystogenous plasmodia and cysts, and cyst wall formation in the cystosorus. Not reviewed here but presently in manuscript are details of the ultrastructure of cruciform division, synaptonemal complexes, what appears to be a meiosis I-like division, and an interesting procedure for nuclear elimination prior to cleavage in cystogenous plasmodia.

The more we study *Woronina*, the more are we persuaded that the behaviour of this previously neglected organism can teach us much about economically important members of related taxa. If this is true of the Plasmodiophorales moreover, it may be true of other groups too and plant pathologists concerned with particularly "difficult" zoosporic fungi may find it profitable to interest themselves in such seemingly unimportant lower plant parasites.

Appendix I

Validation of the Class Name Hyphochytriomycetes

M. W. DICK

Department of Botany, the University of Reading,
Whiteknights, Reading, Berks, UK

Hyphochytriomycetes orthgr. emend.
synonym: Hyphochytridiomycetes Sparrow (1958)

It has been pointed out that no formal statement has been written to support the above orthographic change, although it has been used (Dick, 1976). The name is derived from the genus *Hyphochytrium* Zopf. The family name Hyphochytriaceae was given by Karling (1943) and the family was placed by him in the order Anisochytridiales Karling. Sparrow (1960) used the ordinal name Hyphochytriales ("It seems more appropriate here to make use of the long-existing family name"). Sparrow (1958) was presumably employing the same philosophy when constructing the class name, but allowed an unnecessary "id" to creep in (perhaps a slip because of familiarity with Chytridiomycetes based on the genus *Chytridium* and family Chytridiaceae).

ZOOSPORIC PLANT PATHOGENS
ISBN 0 12 139180 9

Appendix II

Some Methods for Studying Zoosporic Plant Pathogenic Fungi

LENE LANGE and LAURITZ W. OLSON

*Institute of Seed Pathology for Developing Countries, Hellerup, Denmark
and Institute of Genetics, University of Copenhagen, Copenhagen, Denmark*

I. Detection and Culture	285
A. Factors of Importance	285
B. Detection of Plant and Seed-Borne Inoculum	286
C. Examples of Various Culturing Techniques	286
II. Axenic Cultures of Zoosporic Plant Parasitic Phycomycetes	288
III. Preparation for Light Microscopy	288
IV. Preparation for Electron Microscopy	289
A. Preparation of Uniflagellate Zoospores (After Lange and Olson, 1979)	289
B. Preparation of Biflagellate Zoospores	290

No attempt is made in this brief account to survey fully all the procedures used at various times in the collection, culture and examination of zoosporic plant pathogens. We shall, however, describe a few procedures based on our experience of handling these organisms that we have found to be useful.

I. Detection and Culture

A. FACTORS OF IMPORTANCE

No uniformly standard procedure can be given for the detection and cultivation of obligate zoosporic fungal parasites. However, several general approaches can be useful in obtaining the different stages of the life cycle in sufficient quantities and on a suitable substrate.

For infection experiments carried out in a growth chamber, laboratory, glasshouse or in the field, select the optimum host and the optimum conditions of light, temperature and humidity and be aware of the possible use of diurnal changes in these factors. If possible, try to infect leaves as these are easiest to handle, especially for electron microscope preparations. Determine the optimum time for harvesting the material for the various life

ZOOSPORIC PLANT PATHOGENS
ISBN 0 12 139180 9

cycle stages. Obtaining the synchronous release of a large number of zoo-spores is often the most serious problem and to optimise zoospore release the effects of a change in temperature, shaking versus still cultures, mechanical rupture of the host tissue, enzymic treatment, and mechanically induced cracking of resting sporangia should be investigated.

B. DETECTION OF PLANT AND SEED-BORNE INOCULUM

The detection and determination to the generic or specific level of plant and seed-borne fungi can often be done simply by examining the organism *in situ* in its host (see the section below on preparation for light microscopy) although incubation of the host tissue or transfer of the fungus to another host or substrate may be necessary. A standard procedure (Safeeulla, personal communication) for obtaining growth of zoosporic plant-parasites is mentioned below in the section dealing with the downy mildews. The detection of seed-borne Phycomycetes is difficult as the standard seed health testing procedures *a priori* exclude the zoosporic fungi. Attempts to use selective media for *Pythium* and *Phytophthora* have failed due to the fact that nutrients present in the seed apparently allow other fungi to overcome the selectiveness of the substrate. Other alternative methods to detect and isolate Phycomycetes from seeds are currently being developed.

C. EXAMPLES OF VARIOUS CULTURING TECHNIQUES

1. *Olpidium*

For the most extensively studied species, *O. brassicae* (Woronin) Dang. the following technique has been used. *Solanum villosum* seeds are sown among small, cleaned stones separated from normal gravel in a glass terrarium with a close fitting lid to minimise the possibility of secondary infection spread by the air. The stones are watered with sterile, distilled water, and the zoospore inoculum is then added. Sealing the lid of the terrarium with tape prevents evaporation.

2. *Synchytrium*

The only species of *Synchytrium* of major importance is the causal agent of potato wart disease, *S. endobioticum* (Schilberszky) Percival. This species normally infects the germinating buds of the potato tuber and classical methods for studying the disease were described by Lemmerzahl (1930) and by Spieckermann and Kotthoff (1924). However, potato tubers are not well-suited for microscopic studies and an alternative, simple method has been developed (Lange, 1978). Small droplets of a zoospore suspension are placed on the upper surface of the cotyledons of tomato seedlings. The cotyledons should be young but expanded, a stage recognised by the true leaves just beginning to appear. The tomato seedlings are then kept for about 48 hours in darkness under still and extremely humid conditions, thus ensuring that droplets of the inoculum remain on the cotyledons and do not evaporate. After this initial period of incubation the seedlings can be kept under normal glasshouse conditions. The inoculated leaves will later show local wart-like protrusions and all stages of the life cycle of *S. endobioticum* can be found in the tomato cotyledon tissue.

3. *Physoderma*

In spite of what is said in the literature (see Tisdale, 1919) *Physoderma maydis* Shaw, the only economically important species of this genus, can infect non-meristematic tissue if kept under constant humid conditions (Olson *et al.*, 1980). This can be achieved by the simple method of growing the maize host in a wooden frame covered with polyethylene film and placed on the laboratory bench. The plants are watered daily by using an atomizer-sprayer. The inoculum of pre-soaked resting sporangia of *P. maydis*, is sprayed onto the plants or transferred by means of a fine camel hair brush and constant high humidity is maintained. All stages of the life cycle develop on the plant and resting sporangia can be detected as brown spots or stripes, normally starting in the mid-vein of the leaf.

4. *Downy Mildews*

The most important of the zoosporic downy mildews are found in the genera *Plasmopara*, *Pseudoperonospora* and *Sclerospora*. The vegetative part of the life cycle, zoosporangium development and zoospore discharge, can best be studied after incubation of infected leaves in a moist chamber. The major steps in the incubation procedure developed by Professor Safeeulla (Mysore, India) for studying the cereal downy mildews, are the following: leaves which show downy mildew growth are harvested from the field (note that the time of day for harvest of some species may be critical). The leaves are immediately brought to the laboratory, rinsed gently but thoroughly with running tap water, and then placed on filter paper to dry with the lower surface upwards. The underside is then carefully dried with blotting paper although care must be taken not to injure the leaf surface. The leaves are then cut into pieces, of *c.* 2–5 cm², excluding those leaf portions which have started to decay or contain the larger veins. The leaf pieces are then placed (lower side up) on 3–4 layers of saturated blotting paper in Petri dishes. The lid of the Petri dish is then sealed with a piece of moist blotting paper and the dishes carefully transferred to a cool place (for most species below 20°C) and covered to exclude light. The optimum time for incubation is found by first inspecting the leaves with the unaided eye to detect the start of sporangiophore growth, and, later, with a stereo or compound microscope to identify the stage of zoosporogenesis of interest. For zoospore release, the leaves on which mature zoosporangia have been observed are carefully scraped with a dull razor blade and the zoosporangia suspended in water. Zoospore release usually takes place in less than 1 hour.

The culture can be maintained by placing pre-germinated seeds (with roots c. 0.5 cm in length) on the leaves prior to removal of the zoosporangia, or by floating the pre-germinated seeds in a zoospore suspension. After one to two days the seedlings are potted and kept under normal growth conditions (Safeeulla, 1976). Zoospore encystment and the early stages of infection may also be studied by examining such pre-germinated seeds which have been placed in a zoospore suspension.

5. *Plasmodiophora*

Considerable study has been made of the optimum conditions for the maintenance of *Plasmodiophora brassicae* Woronin in host plants. For a discussion of these conditions and the problems involved, see Buczacki *et al.* (1975), Dixon (1976), and Chapter 5 of this volume.

II. Axenic Cultures of Zoosporic Plant Parasitic Phycomycetes

Most of the non-obligate zoosporic plant pathogenic species are in the genera *Pythium* and *Phytophthora*. The maintenance of these genera in culture is described in detail in standard texts (for example, Stevens, 1974) and will not be dealt with here. Isolates of *Aphanomyces euteiches* Drechsler, an important pathogen of garden peas (*Pisum sativum*) can be easily cultured on artificial media by the techniques described by Sundheim (1972) and Llanos and Lockwood (1960). Axenic cultures of some obligately parasitic zoosporic organisms such as *Plasmodiophora* may be achieved in host callus (see Buczacki, 1980).

III. Preparation for Light Microscopy

The zoospore is best studied using phase contrast or interference contrast optics on unstained living material. The main problem with such studies is that zoospores are very active swimmers and they are extremely small.

Because of their small size the use of the high power oil immersion objective may be necessary. For photographing zoospores unlimited patience is required in waiting for the rare occasions when they do not move. Alternatively, a flash attachment fitted to the light source is required. If neither of these two criteria can be fulfilled it may prove useful to try to coat the glass slide with a thin layer of agar to slow down the zoospores or to try to paralyse them with a mixture of lysine–leucine (Olson and Fuller, 1968).

Light microscopic studies of zoosporic plant pathogens *in situ* in their hosts are often most successful with leaf strips. The preparation of such material is easy but the exact performance of the technique will vary with the host: a diseased leaf is rolled tightly around a forefinger of one hand while a pair of very fine forceps is held with the other hand. A small piece of the leaf is taken with the forceps and the strip detached from the leaf by gently pulling in the direction of the leaf's veination. A leaf strip prepared in this way will have a thick portion at one end and will consist of only the epidermal layer at the other. The leaf strip is mounted so that the inner cell layer is uppermost and the optimum thickness for studying the fungus can be selected under the microscope after the strip is mounted in water or in a weak solution of cotton blue. The preparation can be converted into a semi-permanent slide by placing a drop of lactophenol at the edge of the cover slip, and then allowing the lactophenol to substitute for the water as the water evaporates. The cover slip is then ringed with nail varnish.

IV. Preparation for Electron Microscopy

A. PREPARATION OF UNIFLAGELLATE ZOOSPORES (AFTER LANGE AND OLSON, 1979)

Zoospores require a mild and gentle fixation to be well preserved. Uniflagellate zoospores should be fixed in purified, unbuffered glutaraldehyde (Anderson, 1967) by adding a 2% solution of glutaraldehyde, drop by drop over a three to five minute period to an equal volume of spore suspension and mixing gently. The spores are fixed for 20 minutes at 23°C and then pelleted by centrifugation. The spore pellet is washed three times in distilled water (23°C) over a 15 minute period and then post-fixed in 1% OsO_4 in 0.05 M potassium phosphate buffer for 20 minutes (23°C, pH 7.0). After post-fixation the spores are washed as described above. To enhance contrast, the spores are stained in a nearly saturated aqueous solution of uranyl acetate for 24 hours (23°C). The zoospores are then washed as described above and the spore pellet mixed with a small volume of 1% agar which has been melted and cooled to 45°C, and then centrifuged in a conical centrifuge tube. After centrifugation the pellet is cooled to 2°C for 10 minutes and 30% ethanol is overlaid over the agar pellet. The agar pellet is released from the bottom of the centrifuge glass by gently forcing the ethanol down between the agar pellet and the side of the centrifuge glass by means of a pasteur pipette. The zoospore-agar pellet is then dehydrated in a graded ethanol series (30, 50, 70, 96%; 15 minutes each), cut into 1 mm^3 pieces in 100% ethanol, washed four times in 100% ethanol, four times in propylene oxide, and then infiltrated with Spurr's resin (Spurr, 1969).

B. PREPARATION OF BIFLAGELLATE ZOOSPORES

The preparation of biflagellate zoospores for electron microscopy is the same as described above for uniflagellate zoospores with the exception that the primary fixation in gluteraldehyde must be made in buffered gluteraldehyde at pH 7.0 using 0.05 M Sorensen's phosphate or similar buffer.

Appendix III

The Genera of Phytopathogenic Zoosporic Fungi

DONALD J. S. BARR

Biosystematics Research Institute, Agriculture Canada,
Central Experimental Farm, Ottawa, Ontario, Canada

Achlya (Saprolegniaceae)

In China and Japan some species may be opportunistic parasites, attacking rice plants that have been physiologically weakened by low temperature and poor aeration (Hashioka, 1941; Salmon, 1951; Wei *et al.*, 1955). In California, *A. klebsiana* Pieters appears to infect the rice seed coat and endosperm, thus there is a failure of seeds to germinate, or once germinated, failure of seedlings to emerge. Dogma (1975) reported that laboratory tests established several species as pathogens on rice in the Philippines, and discussion on the species reported to occur on rice, together with classification of the genus was given by Johnson (1983).

Albugo (Albuginaceae)

Basipetal succession of sporangia (Khan, 1977; Thakur, 1977) distinguishes this genus and family from other taxa of the Peronosporales. Biga (1955) included 30 species and a number of varieties in his monograph on these obligate parasites and, since then, seven additional species have been described. Several occasionally reach epidemic proportions on crops, but *A. candida* (Pers.) Kuntze, the cause of white rust on crucifers, is probably the most important plant pathogenic species. It consists of morphological and physiological races (Mukerji, 1975) and may be heterothallic (Sansome and Sansome, 1974). *A. tragopogonis* (DC) Gray, which infects many species of the Compositae, also occurs as physiological races (Whipps and Cook, 1978).

Aphanomyces (Saprolegniaceae)

The pathology of this genus was reviewed by Johnson (1983). There are seven phytopathogens, *A. euteiches* Drechsler, *A. cochlioides* Drechsler, *A. raphani* Kendrick, *A. cladogamus* Drechsler, *A. camptostylus* Drechsler, *A. brassicae*

Table 1. Classification of the Mastigomycotina

Class	Order	Family	Genera of phytopathogens
Oomycetes	Thraustochytriales	Thraustochytriaceae	
	Saprolegniales	**Saprolegniaceae** Leptolegniellaceae Ectrogellaceae Haliphthoraceae	*Achlya* *Aphanomyces*
	Leptomitales	Leptomitaceae Rhipidiaceae	*Lagena*
	Lagenidiales	**Lagenidiaceae** Olpidiopsidaceae Sirolpidiaceae	*Peronophythora* *Phytophthora* *Pythium* *Trachysphaera*[b]
	Peronosporales	**Pythiaceae** **Albuginaceae** **Peronosporaceae**	*Albugo*

Basidiophora
Bremia[a]
Bremiella
Peronosclerospora[b]
Peronospora[b]
Plasmopara
Pseudoperonospora
Sclerophthora
Sclerospora

Class	Order	Family	Genus
Hyphochytriomycetes	Hyphochytriales	Anisolpidiaceae Rhizidiomycetaceae Hyphochytriaceae	
Chytridiomycetes	Chytridiales	**Chytridiaceae**	*Rhizophydium*
		Endochytriaceae	
		Synchytriaceae	*Synchytrium*
		Cladochytriaceae	
	Spizellomycetales	Spizellomycetaceae	*Olpidium*
		Olpidiaceae	
	Monoblepharidales	Gonapodyaceae Monoblepharidaceae	
	Blastocladiales	Coelomomycetaceae	
		Physodermataceae	*Physoderma*
		Catenariaceae Blastocladiaceae	
Plasmodiophoromycetes	Plasmodiophorales	**Plasmodiophoraceae**	*Ligniera* *Membranosorus* *Plasmodiophora* *Polymyxa* *Sorodiscus* *Sorosphaera* *Spongospora* *Tetramyxa*

[a] Genus that does not normally produce zoospores.
[b] Strictly non-zoosporic genera.

Singh & Pavgiand and *A. pisci* Srivastava; the first three being the most serious, causing diseases of peas, sugar beet and radish, respectively although many other species are parasitic or saprophytic on a wide range of hosts and substrata. The genus is predominantly aquatic and those occurring as crop pathogens are favoured by high soil moisture. *Aphanomyces euteiches* is parasitic in both soil and aquatic conditions; apart from its occurrence as a pathogen in agriculture, it grows on the amazon sword plant (*Echinodorus brevipedicellatus*) in aquaria with devastating loss (Ridings and Zettler, 1973). The genus was classified by Scott (1961) and by Johnson (1983). There are few good characteristics for distinguishing species; oogonium size, antheridial morphology and host range are the principal ones for the phytopathogenic species. Collections often appear to be named primarily on the basis of the host on which they were found but this practice is unsound because most species have broad host ranges. This genus is ripe for thorough taxonomic study using such aids as electrophoresis or serology.

Basidiophora (Peronosporaceae)

Although *B. entospora* Roze & Cornu occasionally reaches epidemic proportions on ornamental asters, the genus is relatively unimportant. There are three species, two occur on members of the Compositae, and a third, *B. butleri* (Weston) Thirum. & Whitehead on species of Gramineae, was transferred from *Sclerospora* only after the discovery of its sporangial state (Thirumalachar and Whitehead, 1952).

Bremia (Peronosporaceae)

"Sporangia" normally germinate by germ tubes and a report of zoospores (Milbrath, 1923) requires verification. *Bremia* causes downy mildew on members of the Compositae and there is some difference of opinion as to whether *B. lactucae* Regel should be subdivided into morphological varieties or split into distinctive species. On lettuce there are numerous physiological races (Crute and Johnson, 1976). Species reported on members of the Gramineae need critical re-examination according to Waterhouse (1973) to ensure that they are, in fact, *Bremia*.

Bremiella (Peronosporaceae)

An economically unimportant genus with two recognised species (Constantinescu, 1979). The resemblance of the sporangia and sporangiophores to those of *Plasmopara*, *Peronophythora* and *Phytophthora* indicate it may be a primitive genus of the Peronosporaceae (Shaw, 1978).

Dicksonomyces

A *nomen confusum* comprising the perfect state of *Peronosclerospora sorghi* Weston & Uppal and a Hyphomycete (Waterhouse, 1968a).

Lagena (Lagenidiaceae)

There is one species, *L. radicicola* Vanterpool & Ledingham which occurs inside the root cells of many plants including both monocotyledons and dicotyledons (Barr, unpublished work). It is an obligate parasite but conclusive evidence that it causes root rot remains in doubt because *Pythium* spp. are usually detectable in *Lagena*-infected roots. The original description (Vanterpool and Ledingham, 1930) included the oogonial state of a *Pythium* (Vanterpool, personal communication 1977). The fungus has been re-described by Macfarlane (1970). There are thick-walled resting spores but no evidence of a sexual state; however, the presence of bean-shaped zoospores suggests that this fungus belongs to the Oomycetes.

Lagenocystis

A superfluous synonym of *Lagena* (q.v.) proposed by Copeland (1956). Although *Lagena* has been used earlier for a genus of Rhizopoda, a name change is not required by the Botanical Code (Art. 65) when another organism with the same name is an animal.

Ligniera (Plasmodiophoraceae)

Karling (1968) described this genus as scarcely more than a convenient dumping ground for species which cause little or no hypertrophy and develop cystosori of indefinite shape, size and structure. Several of the seven species are poorly described and some reports of *Ligniera* may refer to *Polymyxa*. A critical examination of the type *L. junci* (Schwartz) Maire & Tison on *Juncus* may provide much needed information on the relationship of this genus to *Polymyxa* and *Sporosphaera*.

Membranosorus (Plasmodiophoraceae)

A monotypic genus of doubtful validity until recently (Braselton, 1983).

Olpidium (Olpidiaceae)

Seven species among the total of 44 are parasitic on roots or leaves of vascular plants (Litvinov, 1959). *O. brassicae* (Woronin) Dang. and *O. radicale* Schwartz & Cook are important because they are vectors of several plant viruses (Chapter 7) but alone they appear to do little or no harm even when infection is intense.

Peronophythora (Pythiaceae)

There is one species, *P. litchii* Chen ex Ko *et al*. (Ko *et al*., 1978) which causes fruit rot of litchi (*Litchi chinensis*) in Taiwan. It can be cultured and its sexual morphology, including an inconspicuous or absent oogonial periplasm, are clearly characteristic of the Pythiaceae whereas the determinate growth of its branched sporangiophores equally clearly resemble the Peronosporaceae. Consequently, Ko *et al*. (1978b) proposed a new family, the Peronophythoraceae for this monotypic genus.

Peronoplasmopara

A synonym of *Pseudoperonospora* (Waterhouse, 1973).

Peronosclerospora (Peronosporaceae)

This was a subgenus in *Sclerospora* until raised to generic status to accommodate species that produce conidia but never sporangia or zoospores (Shaw, 1978). It includes eight species that occur largely on tropical and subtropical members of the Gramineae, including *P. sacchari* (Miyake) Shaw on sugar cane (Mukerji and Holliday, 1975), *P. maydis* (Racib.) Shaw on maize and *P. sorghi* (Weston & Uppal) Shaw on sorghum (Kenneth, 1975a).

Peronospora (Peronosporaceae)

This is a non-zoosporic genus with conidia that lack discharge pores. The principles of species concepts being developed for this genus may serve as a basis for classifying other downy mildew genera. At one time, each record of a new host species was justification for a new species of *Peronospora* because it was generally assumed that the pathogen was highly specialised. On that basis over 250 species have been recognised. Yerkes and Shaw (1959) proposed that angiosperm families serve as arbitrary taxonomic divisions for pathogen species and this view was supported by the host range studies of Dickinson and Greenhalgh (1977) who concluded that a morphological difference was required before more than one species of *Peronospora* was recognised within any given host family. The genus includes a number of destructive phytopathogens such as *P. tabacina* Adam the cause of blue mould of tobacco, *P. destructor* (Berk.) Casp. and *P. parasitica* (Fr.) Fr. the causes of downy mildews on onion and crucifers, respectively.

Physoderma (Physodermataceae)

This genus has been transferred from the Chytridiales to the Blastocladiales because ultrastructural studies have demonstrated that they have blastocladiaceous zoospores (Lange and Olson, 1980b). The genus was monographed by Karling (1950) who reduced *Urophlyctis* to synonymy. Since this work, the late Professor F. K. Sparrow has published over 30 papers on *Physoderma* and this genus now needs revision again. Several occur on crops but only *P. maydis* Miyabe reaches minor importance as a pathogen.

Phytophthora (Pythiaceae)

Those working on this genus are fortunate in having available a substantive source book of information and references recently compiled by Ribeiro (1978), as well as keys to species (Newhook *et al.*, 1978), synoptic keys (Ho, 1981) and a compilation of original diagnoses and descriptions (Waterhouse, 1956). Serological and electrophoretic studies have been useful in the development of species concepts.

Plasmodiophora (Plasmodiophoraceae)

Karling (1968) recognised five species and two more have been recently described. However, *P. brassicae* Woronin the cause of club root of crucifers, is the only important species among these and is the subject of intensive research. Isolates of *P. brassicae* have been classified for their physiological specialisation based on infectivity to differential hosts (Buczacki *et al.*, 1975; and Chapter 5).

Plasmopara (Peronosporaceae)

The most important species are *P. viticola* (Berk. & Curt.) Berl. & De Toni and *P. halstedi* (Farlow) Berl. & De Toni which cause downy mildew on grape and sunflower, respectively. The genus is a large one with over 80 species, many of which occur on members of the Compositae, Umbelliferae and Ranunculaceae and the need for a monographic treatment of this now complex group is pressing, although the mycologist who undertakes this job will require equal expertise as a vascular plant taxonomist. Wilson (1907) proposed splitting the genus; he included with *Plasmopara sensu stricto*, those species with sporangiophore branches more nearly dichotomous than monopodial, and with zoospores produced externally to the sporangium. Wilson used *Rhysotheca* for those species with monopodially branched sporangiophores that never appear nearly dichotomous, and with zoospores produced inside the sporangium. *Rhysotheca* has never been widely accepted but the differences in the method of zoospore discharge are significant phylogenetically (Shaw, 1970). It is worth noting that if the genus does receive recognition, the species *Plasmopara viticola*, which was designated the type of *Rhysotheca* by Wilson (1970), and *P. halstedi* will belong to it, thus leaving *Plasmopara* as relatively insignificant.

Polymyxa (**Plasmodiophoraceae**)

Barr (1979) considered this genus distinct from *Ligniera* and recognized two morphologically similar species, *P. graminis* Ledingham and *P. betae* Keskin and two *formae speciales* for the latter.

Pseudoperonospora (**Peronosporaceae**)

This is the zoosporic counterpart of *Peronospora*; however, the sporangia are more delicate and branches are straight and never dichotomous. Waterhouse (1973) recognised five species and, of these, *P. cubensis* (Berk. & Curt.) Rostow. and *P. humuli* (Miy. & Tak.) Wils., the causes of downy mildew on cucurbits (Palti, 1975) and hops, respectively, are destructive parasites of economic concern.

Pythium (**Pythiaceae**)

This genus is discussed in Chapter 4. Middleton's (1943) monograph has long been the most authoritive taxonomic reference on this genus; however, a new monograph by van der Plaats-Niterink (1981) replaces it. Recent accounts on the systematics include those by Waterhouse (1967), Hendrix and Papa (1974), and Robertson (1980) while original diagnoses or descriptions have been compiled by Waterhouse (1968b). Some progress has been made on defining species limits with electrophoretic and serological techniques, and much more should be done using such methods because of the economic importance of these fungi as phytopathogens.

Rhizophydium (**Chytridiaceae**)

R. graminis Ledingham is the only phytopathogenic species in this large and complex genus. It occurs as a parasite in root cells of monocotyledons and dicotyledons (Macfarlane, 1970; Barr, 1973) but does not appear to do any harm to infected cells. It is probably much more widely distributed in temperate zones than is realised but escapes detection because of its superficial likeness to protozoa and its evanescent nature.

Rhysotheca (**Peronosporaceae**)

See *Plasmopara*.

Sclerophthora (Peronosporaceae)

This genus has been placed in the Pythiaceae by many workers (Waterhouse, 1973; Ribeiro, 1978), but Shaw (1978) pointed out that the type of sporangial development clearly relates this taxon to the Peronosporaceae. Moreover, oogenesis and symptomology are typical of the Peronosporaceae (Thirumalachar, 1969). Four species, all parasites on species on Gramineae, were recognised by Shaw (1970). *S. macrospora* (Sacc.) Thirum. (synonym *Phytophthora macrospora* (Sacc.) Ito & Tanaka) occurs widely on cereals in temperate and warm climates (Payak and Renfro, 1970).

Sclerospora (Peronosporaceae)

Species once classified in this genus that produce conidia but never produce zoospores, have been included in *Peronosclerospora* by Shaw (1978) and *S. graminicola* (Sacc.) Schroet. is the only remaining species described with an asexual reproductive stage. It is widely distributed in temperate and tropical parts of the world where it is a pathogen on cereals, and cultivated and wild grasses (Kenneth, 1975b).

Sorodiscus (Plasmodiophoraceae)

This is a poorly understood genus of doubtful validity. The four species, which are all aquatic, could be accommodated in *Sorosphaera*.

Sorosphaera (Plasmodiophoraceae)

If it were not for recent studies by Miller (Chapter 8) and his associates this would be considered a poorly known genus. There are two species, *S. veronicae* Schroet. and *S. radicalis* Cook but the latter may be identical to *Ligniera pilorum* Fron & Gaillat.

Spongospora (Plasmodiophoraceae)

There are three species, of which one, *S. subterranea* (Wallr.) Lagerh. is of economic importance. It comprises two morphologically indistinguishable varieties, one the cause of powdery scab of potato, and the other crook root of commercially grown watercress (*Nasturtium officinale*). The genus has not received taxonomic consideration since the treatment by Karling (1968).

Synchytrium (Synchytriaceae)

This is the largest genus of phytopathogenic Chytridiomycetes and by far the most important because of one species, *S. endobioticum* (Schilberszky) Percival, the cause of wart disease of potato. Several other species do, however, occur on crops and one of these *S. psophocarpi* (Raciborski) Gäumann, the cause of false rust disease of the winged bean, may become troublesome if interest in the growing of this crop continues (Vera-Chaston, 1977; Drinkall and Price, 1979). This genus was monographed by Karling (1964).

Tetramyxa (Plasmodiophoraceae)

This is a poorly studied genus of no economic importance that has not received any recent attention. Karling (1968) listed four species.

Trachysphaera (Pythiaceae)

This genus is not known to produce zoospores. There is one species, *T. fructigena* Tabor & Bunting which is a destructive pathogen on banana, coffee and cacao (Holliday, 1970). The low hexosamine content and absence of glucuronic acid and fucose in its cells place this organism in the Oomycetes (Plaats-Niterink *et al.*, 1976), and its echinulate sexual spores suggest a relationship to the Pythiaceae.

Urophlyctis

A synonym of *Physoderma* (q.v.).

References

Abe, H. (1974). *Bull. Hokkaido Prefect. Agric. Exp. Stn.* **30**, 95–102.
Adegbola, M. O. K. and Hagedorn, D. J. (1969). *Phytopathology* **59**, 1484–1487.
Adegbola, M. O. and Hagedorn, D. J. (1970). *Phytopathology* **60**, 1477–1479.
Aerts, J. (1975). *Meded. Fac. Landbouww. Rijksuniv. Gent.* **40**, 631–641.
Agnihotri, V. P. and Vaartaja, O. (1968). *Can. J. Bot.* **46**, 1135–1141.
Agrios, G. N. (1978). "Plant Pathology". Academic Press, New York.
Ahlawat, Y. S., Majumdar, A. and Chenulu, V. V. (1976). *Plant Dis. Reptr.* **60**, 782–783.
Ahmad, M., Mozmadar, A. and Hëndler, S. (1968). *Genet. Res.* **12**, 103–107.
Ainsworth, G. C. (1973). *In* "The Fungi, An Advanced Treatise", Vol. IVB (G. C. Ainsworth, F. K. Sparrow and A. S. Sussman, eds), pp. 635–648. Academic Press, New York and London.
Ainsworth, G. C. (1976). "Introduction to the History of Mycology". Cambridge University Press, Cambridge.
Aist, J. R. (1978). *In* "Woronin + 100 International Conference on Clubroot (S. T. Buczacki and P. H. Williams, eds), pp. 11–15. Department of Plant Pathology, University of Wisconsin, Madison.
Aist, J. R. and Williams, P. H. (1971). *Can. J. Bot.* **49**, 2023–2034.
Al-Beldawi, A. S. and Sinclair, J. B. (1969). *Phytopathology* **59**, 68–70.
Alconero, R. and Hagedorn, D. J. (1968). *Phytopathology* **58**, 34–40.
Alderson, P. G. and Hiruki, C. (1977). *Phytopath. Z.* **90**, 123–131.
Alexopoulos, C. J. (1962). "Introductory Mycology". John Wiley and Sons, New York.
Alexopoulos, C. J. and Mins, C. W. (1979). "Introductory Mycology" 3rd ed. John Wiley, New York, Chichester, Brisbane, Toronto.
Allen, M. W. (1948). *Phytopathology* **38**, 612–627.
Al Musa, A. M. and Mink, G. I. (1981). *Phytopathology* **71**, 773–776.
Amerson, H. V. and Bland, C. E. (1973). *Mycologia* **65**, 966–970.
Anderson, P. J. (1956). *Phytopathology* **46**, 240.
Anderson, P. J. (1967). *J. Histochem. Cytochem.* **15**, 652–661.
Angell, H. R., Hill, A. V. and Allan, J. M. (1935). *J. Coun. scient. ind. Res. Aust.* **8**, 203–213.
Anonymous (1977). *Fm. Chem.* **140**, 38–43.
Anonymous (1980). *Grower* (November 6), p. 7.
Arens, K. (1929a). *Phytopath. Z.* **1**, 169–193.
Arens, K. (1929b). *Jahrb. Wiss. Botan.* **70**, 57–92.
Aronson, J. (1965). *In* "The Fungi" Vol. 1 (G. C. Ainsworth and A. S. Sussman, eds), pp. 49–76. Academic Press, New York and London.
Aronson, J. M. and Lin, C. C. (1978). *Mycologia* **70**, 363–369.
Ashizawa, M. (1977). *JARQ* **11**, 163–168.
Atkinson, G. F. (1909). *Annls Mycol.* **7**, 441–472.
Ayers, W. A. and Lumsden, R. D. (1977). *Can. J. Microbiol.* **23**, 38–44.
Baker, J. R. (1974). *In* "Evolution in the Microbial World" (M. J. Carlile and J. J. Skehel, eds), pp. 343–366, 24th. Sym. Soc. Gen. Microbiol., Cambridge University Press, London and New York.

Barksdale, A. W. (1968). *J. Elisha Mitchell scient. Soc.* **84**, 187–194.
Barr, D. J. S. (1968). *Can. J. Bot.* **46**, 1087–1091.
Barr, D. J. S. (1970a). *Mycologia* **62**, 492–503.
Barr, D. J. S. (1970b). *Can. J. Bot.* **48**, 2279–2283.
Barr, D. J. S. (1973). *Can. Plant Dis. Surv.* **53**, 191–193.
Barr, D. J. S. (1975). *Can. J. Bot.* **53**, 164–178.
Barr, D. J. S. (1978). *BioSystems* **10**, 153–165.
Barr, D. J. S. (1979). *Can. J. Plant Pathol.* **1**, 85–94.
Barr, D. J. S. (1980a). *Can. J. Bot.* **58**, 2380–2394.
Barr, D. J. S. (1980b). *Can. J. Plant Pathol.* **2**, 116–118.
Barr, D. J. S. (1981). *BioSystems* **14**, 359–370.
Barr, D. J. S. and Allan, P. M. E. (1981). *Can. J. Bot.* **59**, 649–661.
Barr, D. J. S. and Allan, P. M. E. (1982). *Can. J. Bot.* (in press).
Barr, D. J. S. and Hadland-Hartmann, V. E. (1978). *Can. J. Bot.* **56**, 887–900.
Barr, D. J. S. and Hartmann, V. E. (1977). *Can. J. Bot.* **55**, 1221–1235.
Barr, D. J. S. and Kemp, W. G. (1975). *Can. Plant Dis. Surv.* **55**, 77–82.
Barrett, J. T. (1948). *Phytopathology* **38**, 2.
Barron, G. L. (1980). *Mycologia* **72**, 1186–1194.
Bartnicki-Garcia, S. (1968). *Ann. Rev. Microbiol.* **22**, 87–108.
Bartnicki-Garcia, S. (1970). *In* "Phytochemical Phylogeny" (J. B. Harborne, ed.), pp. 81–103. Academic Press, New York and London.
Bastiaansen, M. G., Pieroh, E. A. and Aelbers, E. (1974). *Meded. Fac. Landbouww. Rijksuniv. Gent.* **39**, 1019–1025.
Bawden, F. C. (1964). "Plant Viruses and Virus Diseases". Ronald Press, New York.
Bawden, F. C. and Want, J. P. H. van der (1949). *Tijschr. PlZiekt.* **55**, 142–150.
Beach, W. S. (1946). *Soil Sci.* **61**, 37–46.
Beach, B. C. W., Chalandon, A., Galinelli, G. and Horriére, D. (1979). *Proc. Br. Crop Prot. Conf.* pp. 319–329.
Bean, G. A., Patterson, G. W. and Motta, J. J. (1972). *Comp. Biochem. Physiol.* **43B**, 935–939.
Behncken, G. M. (1968). *Aust. J. Agric. Res.* **19**, 731–738.
Belasco, I. J. (1980). *J. Agr. Food Chem.* **28**, 1106–1108.
Benson, D. M. (1979). *Phytopathology* **69**, 174–178.
Benson, D. M. and Jones, R. K. (1980). *Plant Disease* **64**, 687–691.
Berg, A. A. (1926). *West Virginia Agr. Expt. Sta. Tech. Bull.* **205**, 1–31.
Berkenkamp, B. (1980). *Can. J. Pl. Sci.* **60**, 1039–1040.
Berlese, A. N. (1898). *Jb. wiss. Bot.* **31**, 159–196.
Berner, K. E. and Chapman, E. S. (1977). *Mycologia* **69**, 1232–1236.
Bernstein, L. B. (1968). *J. Elisha Mitchell scient. Soc.* **84**, 84–93.
Bertrand, A., Ducret, J., Debourge, J. C. and Horriére, D. (1977). *Phytiat. Phytopharm.* **26**, 3–17.
Bessey, E. A. (1942). *Mycologia* **34**, 355–379.
Bessey, E. A. (1950). "Morphology and Taxonomy of Fungi". The Blakiston Company, Philadelphia, Toronto.
Bielenin, A., Borecki, Z. and Millikan, D. F. (1976). *Phytopathology* **66**, 127–129.
Biesbrock, J. A. and Hendrix, F. F. (1967). *Mycologia* **59**, 943–952.
Biesbrock, J. A. and Hendrix, F. F. (1970). *Phytopathology* **60**, 880–882.
Biga, M. L. B. (1955). *Sydowia* **9**, 339–358.
Bimpong, C. E. and Hickman, C. J. (1975). *Can. J. Bot.* **53**, 1310–1327.
Bird, L. S. and Presley, J. T. (1965). *In* "Proceedings of the Twenty-Fifth Meeting Cotton Disease Council", Atlanta, Georgia, pp. 88–89.
Black, W. (1952). *Proc. R. Soc. Edinb.* B **65**, 36–51.

Black, W. (1954). *Am. Potato J.* **31**, 93–100.

Black, W., Mastenbroek, C., Mills, W. R. and Peterson, L. C. (1953). *Euphytica* **2**, 173–179.

Blackwell, E. (1943). *Trans. Br. mycol. Soc.* **26**, 71–89.

Blackwell, E. M. and Waterhouse, G. M. (1931). *Trans. Brit. mycol. Soc.* **15**, 294–310.

Bland, C. E. and Amerson, H. V. (1973). *Arch. Mikrobiol.* **94**, 47–64.

Blok, I. (1973). *Neth. J. Pl. Path.* **79**, 266–276.

Boccas, B. (1972). *C.r. hebd. Séanc. Acad. Sci., Paris* **275** D, 663–666.

Boccas, B. R. (1981). *Phytopathology* **71**, 60–65.

Boccas, B. and Zentmyer, G. A. (1976). *Phytopathology* **66**, 477–484.

Bollen, G. J. and Fuchs, A. (1970). *Neth. J. Pl. Pathol.* **76**, 299–312.

Bolton, A. T. (1978). *Can. J. Pl. Sci.* **58**, 569–570.

Boothroyd, C. W. and Israel, H. W. (1980). *Plant Disease* **64**, 218–219.

Bosc, M. (1946). *C.r. hebd. Séanc. Acad. Sci., Paris* **223**, 584–586.

Bouck, G. B. (1972). *In* "Advances in Cell and Molecular Biology", Vol. 2. (E. J. DuPraw, ed.), pp. 237–271, Academic Press, New York and London.

Bouck, G. B. and Brown, D. L. (1973). *J. Cell Biol.* **56**, 340–359.

Bourrelly, P. (1957). *Revue algol. Mém. Hors Sér.* **1**, 1–142.

Boyd, A. E. W. (1972). *Rev. Plant Pathol.* **51**, 297–321.

Brach, E. J. and Crête, R. (1981). *Can. J. Plant Pathol.* **3**, 106–109.

Brakke, M. K. (1971). Soil-borne wheat mosaic virus. *C.M.I./A.A.B. Descriptions of Plant Viruses* No. 77.

Brandenburg, E. (1948). *Z. PflKrankn. PflSchutz* **55**, 129–138.

Braselton, J. P. (1982). *Can. J. Bot.* **60**, 403–408.

Braselton, J. P. (1983). *Can. J. Bot.* (in press).

Braselton, J. P. and Miller, C. E. (1973). *Mycologia* **65**, 220–226.

Braselton, J. P. and Miller, C. E. (1975). *Arch. Microbiol.* **104**, 97–99.

Braselton, J. P. and Miller, C. E. (1978). *Micron* **9**, 37–38.

Braselton, J. P., Miller, C. E. and Pechak, D. G. (1975). *Amer. J. Bot.* **62**, 349–358.

Brasier, C. M. (1971). *Nature, New Biology* **231**, 283.

Brasier, C. M. (1972). *Trans. Br. mycol. Soc.* **58**, 237–251.

Brasier, C. M. (1983). *In* "Phytophthora: its Biology, Taxonomy, Ecology and Pathology" (D. C. Erwin, S. Bartnicki-Garcia and P. H. Tsao, eds). American Phytopathological Society, St. Paul, Minnesota. In press.

Brasier, C. M. and Griffin, M. J. (1979). *Trans. Br. mycol. Soc.* **72**, 111–143.

Brasier, C. M. and Sansome, E. (1975). *Trans. Br. mycol. Soc.* **65**, 49–65.

Braun, A. (1847). *Flora* **33**, 17–29.

Braun, A. (1856). *Abh. Berlin Akad.* **1855**, 21–83.

Braun, H. (1924). *J. agric. Res.* **29**, 399–419.

Broadbent, L. (1965). *Ann. appl. Biol.* **55**, 57–66.

Brooks, J. L., Given, J. B., Baniecki, J. F. and Young, R. J. (1974). *Pl. Dis. Reptr.* **58**, 291–292.

Broyles, J. W. (1956). *Phytopathology* **46**, 8.

Bruck, R. I., Fry, W. E. and Apple, A. E. (1980). *Phytopathology* **70**, 597–601.

Bruck, R. I., Gooding, Jr., G. V. and Main, C. E. (1981). *Phytopathology* **71**, 558.

Bruehl, G. W. (1953). *Technical Bulletin* **1084**, 1–23.

Bruin, G. C. A. (1980). Ph.D. thesis, University of Guelph. 110 pp.

Bruin, G. C. and Edgington, L. V. (1980a). *Phytopathology* **70**, 459.

Bruin, G. C. and Edgington, L. V. (1980b). *Phytopathology* **70**, 459–460.

Bruin, G. C. A. and Edgington, L. V. (1982). *Phytopathology* **72**, 476–800.

Bruin, G. C. A., Ripley, B. D. and Edgington, L. V. (1981). *Tob. Sci.* **25**, 128–130.

Bryant, T. R. and Howard, K. L. (1969). *Am. J. Bot.* **56**, 1075–1083.

Buczacki, S. T. (1973a). *Ann. appl. Biol.* **74**, 85–90.

Buczacki, S. T. (1973b). *Ann. appl. Biol.* **75**, 25–30.

Buczacki, S. T. (1978). *In* "Woronin + 100 International Conference on Clubroot" (S. T. Buczacki and P. H. Williams, eds), pp. 33–43. Department of Plant Pathology, University of Wisconsin, Madison.

Buczacki, S. T. (1980). *In* "Tissue Culture Methods for Plant Pathologists" (D. S. Ingram and J. P. Helgeson, eds), pp. 145–149. Blackwell, Oxford.

Buczacki, S. T. and Cadd, S. E. (1976). *Ann. appl. Biol.* **84**, 43–50.

Buczacki, S. T. and Moxham, S. E. (1979). *Rep. natn. Veg. Res. Stn.* for 1978, pp. 74–75.

Buczacki, S. T. and Moxham, S. E. (1980). *Trans. Br. mycol. Soc.* **75**, 439–444.

Buczacki, S. T., Moxham, S. E. and Clay, C. M. (1980). *Rep. natn. Veg. Res. Stn.* for 1979, p. 69.

Buczacki, S. T., Moxham, S. E. and Turner, R. H. (1979). *Trans. Br. mycol. Soc.* **73**, 343–347.

Buczacki, S. T. and Ockendon, J. G. (1978). *Ann. appl. Biol.* **88**, 363–367.

Buczacki, S. T., Ockendon, J. G. and Freeman, G. H. (1978). *Ann. appl. Biol.* **88**, 229–238.

Buczacki, S. T. and Stevenson, K. (1981). *Rep. natn. Veg. Res. Stn.* for 1980, p. 75.

Buczacki, S. T., Toxopeus, H., Mattusch, P., Johnston, T. D., Dixon, G. R. and Hobolth, L. A. (1975). *Trans. Br. mycol. Soc.* **65**, 295–303.

Buczacki, S. T. and White, J. G. (1979). *Plant Pathol.* **28**, 36–39.

Buddenhagen, I. W. (1958). *Am. J. Bot.* **45**, 355–365.

Budzier, H. H. (1956). *NachrBl. dt. PflSchdienst, Berl.* **10**, 33–35.

Buisman, C. J. (1927). *Meded. phytopath. Lab. Willie Commelin Scholten* **11**, 1–51.

Bumbieris, M. (1972). *Aust. J. agric. Res.* **23**, 651–657.

Bumbieris, M. (1974). *Aust. J. Bot.* **22**, 655–660.

Burgess, A. H. (1964). "Hops: Botany, cultivation, and utilization". Leonard Hill, London, and Interscience Publ., New York.

Burnett, J. H. (1975). "Mycogenetics". John Wiley, London, New York, Sydney, Toronto.

Burrell, R. G., Clayton, C. W., Gallegly, M. E. and Lilly, V. G. (1966). *Phytopathology* **56**, 422–426.

Butcher, D. N., El-Tigani, S. and Ingram, D. S. (1974). *Physiol. Plant Pathol.* **4**, 127–140.

Byford, W. J. (1976). *Ann. appl. Biol.* **83**, 69–77.

Caetano, V. R., Kitajima, E. W. and Costa, A. S. (1978). *Fitopatologica Brasileira* **3**, 39–46.

Calvert, E. L. (1976). *Rep. Res. Tech. Work. Min. Agr. Northern Ireland*, pp. 84–85.

Cameron, J. N. and Carlile, M. J. (1977). *J. gen. Microbiol.* **98**, 599–602.

Cameron, J. N. and Carlile, M. J. (1978). *Nature, London* **271**, 448–449.

Cameron, J. N. and Carlile, M. J. (1980). *J. gen. Microbiol.* **120**, 347–353.

Campbell, R. N. (1962). *Nature* **195**, 675–677.

Campbell, R. N. and Fry, P. R. (1966). *Virology* **29**, 222–233.

Campbell, R. N., Greathead, A. S. and Westerlund, F. V. (1980). *Phytopathology* **70**, 741–746.

Campbell, R. N. and Grogan, R. G. (1964). *Phytopathology* **54**, 681–690.

Campbell, R. N. and Lin, M. T. (1976). *Am. J. Bot.* **63**, 826–832.

Campbell, W. A. (1949). *Plant Dis. Reptr.* **3**, 134–135.

Campbell, W. A. and Copeland, Jr., O. L. (1954). *USDA Circular* 940.

Campbell, W. A. and Gallegly, M. E. (1965). *Plant Dis. Reptr.* **49**, 233–234.

Campbell, W. A. and Hendrix, F. F. (1967). *Mycologia* **59**, 274–278.

Campbell, W. A. and Presley, J. T. (1946). *USDA Circ.* 749.

Canova, A. (1966). *Phytopathol. Mediter.* **5**, 53–58.

Cantino, E. C. (1950). *Quart. Rev. Biol.* **25**, 269–277.

Cantino, E. C. (1955). *Quart. Rev. Biol.* **30**, 138–149.

Cantino, E. C. and Turian, G. F. (1959). *Ann. Rev. Microbiol.* **13**, 97–124.

Carruthers, W. (1875). *Jl. R. agric. Soc.* **6**, 396–399.

Castro, J. and Zentmyer, G. A. (1969). *Phytopathology* **59**, 10.

Caten, C. E. (1970). *Can. J. Bot.* **48**, 897–905.

Caten, C. E. (1971). *Trans. Br. mycol. Soc.* **56**, 1–7.

Caten, C. E. (1974). *Ann. appl. Biol.* **77**, 259–270.

Caten, C. E. and Day, A. W. (1977). *A. Rev. Phytopathol.* **15**, 295–318.

Caten, C. E. and Jinks, J. L. (1968). *Can. J. Bot.* **46**, 329–348.

Catherall, P. L. and Boulton, R. E. (1979). *Pl. Path.* **28**, 57–60.

Cavalier-Smith, T. (1974). *J. Cell. Sci.* **16**, 529–556.

Chambers, A. Y., Hadden, C. H. and Merrill, S. (1974). *Tenn. Farm Home Sci. Prog. Rep.* **90**, 30–31.

Chod, J., Polák, J., Kůdela, V. and Jokeš, M. (1976). *Biol. Plant.* **18**, 63–66.

Chong, J. and Barr, D. J. S. (1973). *Can. J. Bot.* **51**, 1411–1420.

Chupp, C. (1917). *N.Y. State Agr. Expt. Sta. Bull.* 387, 451–452.

Clare, B. G. and Zentmyer, G. A. (1966). *Phytopathology* **56**, 1334–1335.

Clare, B. G., Flentje, N. T. and Atkinson, M. R. (1968). *Aust. J. biol. Sci.* **21**, 275–295.

Clark-Walker, G. D. and Gleason, F. H. (1973). *Arch. Mikrobiol.* **92**, 209–216.

Claussen, P. (1908). *Ber. dt. bot. Ges.* **26**, 144–161.

Clayton, E. E., Gaines, J. G., Shaw, K. J., Smith, T. E., Foster, M. H., Lunn, W. M. and Graham, T. W. (1942). *U.S.D.A. Tech. Bull.* 799.

Cohen, E. and Schiffmann-Nadel, M. (1978a). *Pl. Dis. Reptr.* **62**, 386–388.

Cohen, E. and Schiffmann-Nadel, M. (1978b). *Pl. Dis. Reptr.* **62**, 388–389.

Cohen, Y., Reuveni, M. and Eyal, H. (1979). *Phytopathology* **69**, 645–649.

Cohn, F. (1854). *Nova Acta Acad. Leop.-Carol.* **24**, 101–256.

Coker, W. C. (1923). "The Saprolegniaceae, with notes on other water molds". Chapel Hill.

Colhoun, J. (1958). *Phytopathological Paper* No. 3, C.M.I., 108 pp.

Colhoun, J. (1966). *In* "The Fungus Spore" (M. F. Madelin ed.), pp. 85–94. Butterworths, London.

Commonwealth Mycological Institute (1974). Distribution Maps of Plant Diseases. Map No. 34. *Spongospora subterranea* (Wallr.) Lagerh. var. *subterranea.*

Commonwealth Mycological Institute (1978). Distribution Maps of Plant Diseases. Map No. 522. *Polymyxa betae* Keskin.

Commonwealth Mycological Institute (1979). Distribution Maps of Plant Diseases. Map No. 430. *Olpidium brassicae* (Woron.) Dang.

Constantinescu, O. (1979). *Trans. Br. mycol. Soc.* **72**, 510–515.

Cook, W. R. I. (1926). *Trans. Br. mycol. Soc.* **11**, 196–213.

Cook, W. R. I. and Nicholson, W. H. (1933). *Ann. Bot.* **47**, 851–859.

Cook, W. R. I. and Schwartz, E. J. (1930). *Phil. Trans. R. Soc., B.* **218**, 282–314.

Cooper, B. A. and Aronson, J. M. (1967). *Mycologia* **59**, 658–670.

Cooper, J. I. and Harrison, B. D. (1973). *Pl. Path.* **22**, 73–78.

Cooper, J. I., Jones, R. A. C. and Harrison, B. D. (1976). *Ann. appl. Biol.* **83**, 215–230.

Copeland, H. F. (1956). "The Classification of Lower Organisms." Pacific Books, Palo Alta, California.

Copeland, R. B. and Logan, C. (1977). *Pl. Path.* **26**, 175–179.

Corbett, M. K. and Styer, E. L. (1976). *Proc. Am. Phytopath. Soc.* **3**, 332.

Cornu, M. (1872). *Annls Sci. nat.* (Bot.) **15**, 112–198.

Cottam, C. (1945). *Plant Dis. Reptr.* **29**, 302–310.

Couch, J. N., Leitner, J. and Whiffen, A. (1939). *J. Elisha Mitchell scient. Soc.* **55**, 399–408.

Coulombe, L. J. (1974). *Phytoprotection* **55**, 45–54.

Crandall, B. S. (1947). *Phytopathology* **37**, 928–929.

Crandall, B. S., Gravatt, G. F. and Ryan, M. M. (1945). *Phytopathology* **35**, 162–180.

Crawford, I. P. (1975). *Bacteriol. Rev.* **39**, 87–120.

Crute, I. R. (1980). *Tests Agrochem. Cult.* (*Ann. appl. Biol.* **94**, *suppl.*) No. 1, 24–35.

Crute, I. R. (1981). *In* "The Downy Mildews" (D. M. Spencer, ed.), pp. 237–253. Academic Press, New York and London.

Crute, I. R., Buczacki, S. T., Crisp, P., Stevenson, K. and James, H. M. (1981). *Rep. natn. Veg. Res. Stn* for 1980, pp. 75–76.

Crute, I. R. and Dickinson, C. H. (1976). *Ann. appl. Biol.* **82**, 433–450.

Crute, I. R. and Dixon, G. R. (1981). *In* "The Downy Mildews" (D. M. Spencer, ed.), pp. 421–460. Academic Press, New York and London.

Crute, I. R. and Gordon, P. L. (1980a). *Rep. natn. Veg. Res. Stn.* for 1979, p. 75.

Crute, I. R. and Gordon, P. L. (1980b). *Tests Agrochem. Cult.* (*Ann. appl. Biol.* **94**, *suppl.*) No. 1, 32–33.

Crute, I. R., Gray, A. R., Crisp, P. and Buczacki, S. T. (1980a). *Pl. Breed. Abstr.* **50**, 91–104.

Crute, I. R., Halstead, A. L. and Gordon, P. L. (1980b). *Rep. natn. Veg. Res. Stn.* for 1979, pp. 73–75.

Crute, I. R. and Jagger, B. M. (1979). *Rep. natn. Veg. Res. Stn.* for 1978, pp. 78–79.

Crute, I. R. and Johnson, A. G. (1976). *Ann. appl. Biol.* **83**, 125–137.

Crute, I. R. and Norwood, J. M. (1977). *Rep. natn. Veg. Res. Stn.* for 1976, p. 102.

Crute, I. R. and Norwood, J. M. (1978). *Rep. natn. Veg. Res. Stn.* for 1977, p. 103.

Crute, I. R. and Norwood, J. M. (1980). *Ann. appl. Biol.* **94**, 275–278.

Csinos, A. S. (1979). *Can. J. Bot.* **57**, 2059–2063.

Csinos, A. and Hendrix, J. W. (1977a). *Phytopathology* **67**, 434–438.

Csinos, A. and Hendrix, J. W. (1977b). *Can. J. Bot.* **55**, 26–29.

Cutter, V. M. (1951). *A. Rev. Microbiol.* **5**, 17–34.

D'Ambra, V. and Mutto, S. (1975). *Riv. Patol. veg.*, **11**, 115–124.

D'Ambra, V. and Mutto, S. (1977). *Can. J. Bot.* **55**, 831–839.

Damle, V. P. (1943). *J. Indian bot. Soc.* **22**, 137–158.

Dangeard, P. (1901). *Le Botaniste* **8**, 1–370.

Davey, C. B. and Papavizas, G. C. (1961). *Phytopathology* **51**, 131–133.

Davidse, L. C. (1975). *In* "Microtubules and Microtubule Inhibitors" (M. Borgers and M. de Brabander, eds), pp. 483–495. North Holland Publ. Corp., Amsterdam.

Davidse, L. C. (1980). *Acta Bot. Neerl.* **29**, 216.

Davidse, L. C. (1981). *Neth. J. Pl. Path.* **87**, 17–31.

Davidse, L. C. and Flach, W. (1977). *J. Cell Biol.* **72**, 174–193.

Davidse, L. C., Looijen, D., Turkensteen, L. J. and Van der Wal, D. (1981). *Neth. J. Pl. Path.* **87**, 65–68.

Day, P. R. (1974). "Genetics of Host–Parasite Interaction". Freeman, San Francisco.

Dayhoff, M. O., McLaughlin, P. J., Barker, W. C. and Hunt, L. T. (1975). *Naturwissenschaften* **62**, 154–161.

De Bary, A. (1852). *Bot. Z.* **10**, 473–479, 486–496, 505–511.

De Bary, A. (1876). *Jl. R. agric. Soc.* **12**, 240.

De Bary, A. (1884). "Vergleichende Morphologie und Biologie der Pilze, Mycetozoen und Bacterien." Leipzig.

de Bruyn, H. L. G. (1937). *Genetica* **19**, 553–558.
Dekker, J. (1969). *World Rev. Pest Control* **8**, 79–85.
Dekker, J. (1976). *A. Rev. Phytopathol.* **14**, 405–428.
Dekker, J. (1982). *In* "Fungicide Resistance in Crop Protection" (J. Dekker and S. G. Georgopoulos, eds), pp. 128–138. Pudoc, Wageningen.
Dekhuijzen, H. M. (1975). *Physiol. Plant Pathol.* **6**, 187–192.
Dekhuijzen, H. M. (1979). *Phytopath. Z.* **87**, 171–192.
Dekhuijzen, H. M. and Overeem, J. C. (1971). *Physiol. Plant Pathol.* **1**, 151–161.
Delp, C. (1980). *Plant Disease* **64**, 652–657.
Dennett, C. W. and Stanghellini, M. E. (1977). *Phytopathology* **67**, 1134–1141.
Denward, T. (1970). *Hereditas* **66**, 35–48.
Diaconu, V. (1979). *An. Inst. Cerc. Prot. Plant.* **15**, 285–294.
Dias, H. F. and McKeen, C. D. (1972). Cucumber necrosis virus. *C.M.I./A.A.B. Descriptions of Plant Viruses* No. 82.
Dick, M. W. (1968). *Veröff. Inst. Meeresforsch. Bremerhaven* **3**, 27–38.
Dick, M. W. (1969). *New Phytol.* **68**, 751–775.
Dick, M. W. (1972). *New Phytol.* **71**, 1151–1159.
Dick, M. W. (1973a). *In* "The Fungi" Vol. IVB (G. C. Ainsworth, F. K. Sparrow and A. S. Sussman, eds), pp. 113–144. Academic Press, New York and London.
Dick, M. W. (1973b). *In* "The Fungi" Vol. IVB (G. C. Ainsworth, F. K. Sparrow and A. S. Sussman, eds), pp. 145–163. Academic Press, New York and London.
Dick, M. W. (1976). *In* "Recent Advances in Aquatic Mycology" (E. B. G. Jones, ed.), pp. 513–542. Elek, London.
Dick, M. W. and Win-Tin (1973). *Biol. Rev.* **48**, 133–158.
Dickinson, G. H. and Greenhalgh, J. R. (1977). *Trans. Br. mycol. Soc.* **69**, 111–116.
Didier, P., Puytorac, P. de., Wilbert, N. and Detcheva, R. (1980). *J. Protozool.* **27**, 72–79.
Diriwächter, G., Herter, G. and Gindrat, D. (1979). *Ber. schweiz. bot. Ges.* **89**, 105–113.
Dixon, G. R. (1976). *Pl. Path.* **25**, 129–134.
Dodge, J. D. (1973). "The Fine Structure of Algal Cells". Academic Press, London and New York.
Dodge, J. D. (1979). *In* "Biochemistry and Physiology of Protozoa" 2nd edn., Vol. 1 (M. Levandowsky and S. H. Hunter, eds), pp. 7–57. Academic Press, New York and London.
Dogma, I. J. (1975). *Kalikasan* **4**, 69–105.
Dorst, H. J. M. van (1975). *Neth. J. Plant. Pathol.* **81**, 45–48.
Douchet, J. P., Absi, M., Hay, S. J. B., Muntan, L. and Villani, A. (1977). *Proc. Br. Crop Prot. Conf.*, pp. 535–540.
Drechsler, C. (1930). *J. Wash. Acad. Sci.* **20**, 398–418.
Drinkall, M. J. and Price, T. V. (1979). *Trans. Br. mycol. Soc.* **72**, 91–98.
Dueck, J. and Stone, J. R. (1979). *Can. J. Pl. Sci.* **59**, 423–427.
Duniway, J. M. (1979). *A. Rev. Phytopath.* **17**, 431–460.
Dyakov, Yu. T. and Kulish, V. B. (1979). *Genetika* **15**, 49–56.
Dyakov, Yu. T. and Kuzovnikova, T. A. (1974). *Mikol. i Fitopatol.* **8**, 81–89.
Dylewski, D. P., Braselton, J. P. and Miller, C. E. (1978). *Am. J. Bot.* **65**, 258–267.
Eckert, J. W. (1967). *In* "Fungicides" (D. C. Torgeson, ed.), Vol. I, pp. 287–378. Academic Press, New York and London.
Eckert, J. W. and Tsao, P. H. (1962). *Phytopathology* **52**, 771–777.
Edgington, L. V., Khew, K. L. and Barron, G. L. (1971). *Phytopathology* **61**, 42–44.
Edgington, L. V., Martin, R. A., Bruin, G. C. and Parsons, I. M. (1980). *Plant Disease* **64**, 19–23.

Edgington, L. V. and Peterson, C. A. (1977). *In* "Antifungal Compounds" (M. R. Siegel and H. D. Sisler, eds), Vol. II, pp. 51–89. Marcel Dekker, New York.

Edney, K. L. (1976). *Rep. E. Malling Res. Stn.* for 1975, p. 116.

Edney, K. L. and Chambers, D. A. (1981). *Ann. appl. Biol.* **97**, 237–241.

Edney, K. L., Chambers, D. A. and Beecham, A. (1979). *Rep. E. Malling Res. Stn.* for 1978, pp. 92–93.

El Fahl, A. M. and Calvert, E. L. (1976). *Record Agr. Res.* **24**, 7–12.

Elliott, C. G. and MacIntyre, D. (1973). *Trans. Br. mycol. Soc.* **60**, 311–316.

Ellzey, J. T. and Huizar, E. (1977). *Ark. Mikrobiol.* **112**, 311–313.

Emerson, R. (1941). *Lloydia* **4**, 77–144.

Emerson, R. (1950). *A. Rev. Microbiol.* **4**, 169–200.

Emerson, R. (1958). *Mycologia* **50**, 589–621.

Erselius, L. J. and Shaw, D. S. (1982). *Trans. Br. mycol. Soc.* **78**, 227–238.

Erwin, D. C., Zentmyer, G. A., Galindo, J. and Niederhauser, J. S. (1963). *A. Rev. Phytopathol.* **1**, 375–396.

Erwin, J. A. (1973). *In* "Lipids and Biomembranes of Eukaryotic Microorganisms" (J. A. Erwin, ed.), pp. 41–143. Academic Press, New York and London.

Esmarch, F. (1927). *Angew. Bot.* **9**, 88–124.

Evans, E. (1968). "Plant Diseases and Their Chemical Control". Blackwell, Oxford and Edinburgh.

Exconde, O. R. (1970a). *Indian Phytopath.* **23**, 275–284.

Exconde, O. R. (1970b). *Indian Phytopath.* **23**, 389–395.

Exconde, O. R. (1975). *Trop. Agr. Res.* **8**, 157–163.

Fehrmann, H. (1976). *Phytopath. Z.* **86**, 144–185.

Fedorintschik, N. S. (1935). *Zashch. Rast. Vredit.* **5**, 87–95.

Fielding, M. J. and Rhodes, R. C. (1967). *Cotton Dis. Council Proc.* **27**, 56–58.

Finlayson, D. G. and Campbell, C. J. (1971). *Can. Pl. Dis. Survey* **51**, 122–126.

Firman, I. D. (1974). *In* "Phytophthora Diseases of Cocoa" (P. H. Gregory, ed.), pp. 131–139. Longman, New York.

Fischer, A. (1882). *Jb. wiss. Bot.* **13**, 286–371.

Fischer, A. (1892). *In* "Die Pilze Deutschlands Oesterreichs und der Schweiz (Rabenhorst's Kryptogamen-Flora)". Vol. 1, pp. 1–490.

Fitch, W. M. (1976). *J. Mol. Evol.* **8**, 13–40.

Fitzpatrick, H. M. (1923). *Mycologia* **15**, 166–173.

Fitzpatrick, H. M. (1930). "The Lower Fungi–Phycomycetes". McGraw-Hill, New York.

Flanagan, P. W. (1970). *Can. J. Bot.* **9**, 1673–1675.

Flentje, N. T. (1964). *Aust. J. Biol. Sci.* **17**, 643–650.

Flor, H. H. (1956). *Adv. Genet.* **8**, 29–59.

Flowers, R. A. and Hendrix, J. W. (1972). *Phytopathology* **62**, 474–477.

Flowers, R. A. and Hendrix, J. W. (1974). *Phytopathology* **74**, 718–720.

Frank, Z. R. (1968). *Phytopathology* **58**, 542–543.

Frank, Z. R. (1972). *Phytopathology* **62**, 1331–1334.

Fraser, R. S. S. and Buczacki, S. T. (1983). *Trans. Br. mycol. Soc.* **80**, 107–112.

Frederiksen, R. A. and Odvodi, G. (1979). *Sorghum Newsl.* **22**, 129.

Frederiksen, R. A. and Renfro, B. L. (1977). *A. Rev. Phytopathol.* **15**, 249–275.

Frezzi, M. J. (1956). *Rev. Invest. Agr.*, T.X. No. 2, Buenos Aires. 241 pp.

Frossard, P., Haury, A. and Laville, E. (1977). *Phytiat. Phytopharm.* **26**, 55–61.

Fry, P. R. (1958). *N.Z. J. Agric. Res.* **1**, 301–304.

Fry, P. R. and Campbell, R. N. (1966). *Virology* **30**, 517–527.

Fuchs, A., De Ruig, S. P., Van Tuyl, J. M. and De Vries, F. W. (1977). *Neth. J. Pl. Path.* **83** (*Suppl.*), 189–205.

Fujisawa, I. and Sugimoto, T. (1976). *Res. Bull. Hokkaido Natl. Agric. Exp. Stn.* **115**, 123–131.

Fuller, M. S. (1966). *In* "The Fungus Spore" (M. F. Madelin, ed.), pp. 67–84. Butterworths, London.

Fuller, M. S. (1976). *Int. Rev. Cytol.* **45**, 113–153.

Fuller, M. S. and Barshad, I. (1960). *Am. J. Bot.* **47**, 105–109.

Fuller, M. S., Lewis, B. and Cook, P. (1966). *Mycologia* **58**, 313–318.

Fuller, M. S. and Reichle, R. E. (1965). *Mycologia* **57**, 946–961.

Galindo, J. A. and Gallegly, M. E. (1960). *Phytopathology* **50**, 123–128.

Galindo, J. A. and Zentmyer, G. A. (1964). *Phytopathology* **54**, 238–239.

Galindo, J. and Zentmyer, G. A. (1967). *Phytopathology* **57**, 1300–1304.

Gallegly, M. E. (1968). *A. Rev. Phytopathol.* **6**, 375–396.

Gallegly, M. E. (1970). *Phytopathology* **60**, 1135–1141.

Gallegly, M. E. and Galindo, J. (1958). *Phytopathology* **48**, 274–277.

Gallegly, M. E. and Niederhauser, J. S. (1959). *In* "Plant Pathology Problems Progress 1908–1958", pp. 168–182. University of Wisconsin Press, Madison.

Garber, R. C. and Aist, J. R. (1979a). *J. Cell. Sci.* **40**, 89–110.

Garber, R. C. and Aist, J. R. (1979b). *Can. J. Bot.* **57**, 2509–2518.

Gardner, D. E. and Hendrix, F. F. (1973). *Can. J. Bot.* **51**, 1593–1598.

Garren, K. H. (1970). *Plant Dis. Reptr.* **54**, 840–843.

Garrett, R. G. and Tomlinson, J. A. (1967). *Trans. Br. mycol. Soc.* **50**, 429–435.

Gates, L. F. (1975). *Can. J. Pl. Sci.* **55**, 891–895.

Gäumann, E. A. (1964). "Die Pilze", 2nd ed. Birkhäuser, Basel.

Gay, J. L. and Greenwood, A. D. (1966). *In* "The Fungus Spore" (M. F. Madelin, ed.), p. 95. Butterworth, London.

Gay, J. D. and McCarter, S. M. (1968). *Plant Dis. Reptr.* **52**, 416.

Georgopoulos, S. G. (1977). *In* "Antifungal Compounds" (M. R. Siegel and H. D. Sisler, eds), Vol. II, pp. 439–495. Marcel Dekker, New York.

Georgopoulos, S. G. and Zaracovitis, C. (1967). *A. Rev. Phytopathol.* **5**, 109–130.

Georgopoulos, S. G., Sarris, M. and Ziogas, B. N. (1979). *Pestic. Sci.* **10**, 389–392.

Gibbs, A. J. (1977). Tobamovirus group. *C.M.I./A.A.B. Descriptions of Plant Viruses* No. 184.

Giddings, N. J. and Berg, A. A. (1919). *Phytopathology* **9**, 209–211.

Gill, D. L. (1970). *Plant Dis. Reptr.* **54**, 1077–1079.

Gill, H. S. and Powell, D. (1968a). *Phytopath. Z.* **63**, 23–29.

Gill, H. S. and Powell, D. (1968b). *Phytopathology* **58**, 722–723.

Gill, H. S. and Zentmyer, G. A. (1978). *Phytopathology* **68**, 163–167.

Gindrat, D. (1976). *Phytopathology* **66**, 312–316.

Gleason, F. H. (1972). *Mycologia* **64**, 663–666.

Gleason, F. H. (1976). *In* "Recent Advances in Aquatic Mycology" (E. B. G. Jones, ed.), pp. 543–572. Elek Science, London.

Gleason, F. H. and Unestam, T. (1968). *J. Bacteriol.* **95**, 1599–1603.

Goldie-Smith, E. K. (1951). *J. Elisha Mitchell scient. Soc.* **67**, 108–121.

Goldie-Smith, E. K. (1954). *Amer. J. Bot.* **41**, 441–448.

Goldie-Smith, E. K. (1956). *J. Elisha Mitchell scient. Soc.* **72**, 348–356.

Gooday, G. W. and Trinci, A. P. T. (1980). *In* "The Eukaryotic Microbial Cell" (G. W. Gooday, D. Lloyd and A. P. J. Trinci, eds), 33rd. Sym. Soc. Gen. Microbiol., 1980, Cambridge Univ. Press, Cambridge, London, New York, New Rochelle, Melbourne, Sydney.

Gorenz, A. M. (1974). *In* "Phytophthora Diseases of Cocoa" (P. H. Gregory, ed.), pp. 235–257. Longman, New York.

Gotelli, D. (1974). *Mycologia* **66**, 846–858.

Graham, K. M. (1954). *Phytopathology* **44**, 490.

Graham, K. M. (1955). *Am. Potato J.* **32**, 277–282.

Grau, C. R. and Reiling, T. P. (1977). *Phytopathology* **67**, 273–276.

Green, B. R. and Dick, M. W. (1972). *Can. J. Microbiol.* **18**, 963–968.

Gregory, C. T. (1912). *Phytopathology* **2**, 235–249.

Griffin, D. M. (1972). "Ecology of Soil Fungi". Chapman and Hall, London.

Griffin, D. M. (1978). *In* "Water Deficits and Plant Growth". (T. T. Kozlowski, ed.), Vol. 5, pp. 175–197, N.Y. Acad.

Griffin, M. J. (1972). *Pl. Path.* **21**, 95.

Grogan, R. G., Zink, F. W., Hewitt, W. B. and Kimble, K. A. (1958). *Phytopathology* **48**, 292–297.

Grosso, J. J. (1954). *Pl. Dis. Reptr.* **38**, 333.

Grove, S. N. and Bracker, C. E. (1978). *Expl. Mycol.* **2**, 51–98.

Haas, J. H. (1964). *Phytopathology* **54**, 894.

Haasis, F. A., Nelson, R. R. and Marx, D. H. (1964). *Phytopathology* **54**, 1146–1151.

Haenseler, C. M. (1929). *N.J. Agr. Exp. Sta. Ann. Rpt.* **50**, 262–270.

Haeske, E. (1958). Untersuchungen über das Blattnervenmosaik des Salats. Dissertation L 470, University Giessen.

Hall, R., Zentmyer, G. A. and Erwin, D. C. (1969). *Phytopathology* **59**, 770–774.

Halos, P. M. and Huisman, O. C. (1976a). *Phytopathology* **66**, 152–157.

Halos, P. M. and Huisman, O. C. (1976b). *Phytopathology* **66**, 158–164.

Halpin, J. E., Hanson, E. W. and Dickson, J. G. (1952). *Phytopathology* **42**, 245–249.

Halsall, D. M. (1976). *J. gen. Microbiol.* **94**, 149–158.

Hampson, M. C. (1977). *Can. Pl. Dis. Survey* **57**, 75–78.

Hampson, M. C. (1979). *Can. Pl. Dis. Survey* **59**, 7–14.

Hampton, R. O. and Buchholtz, W. F. (1959). *Iowa St. Coll. J. Sci.* **33**, 489–495.

Harris, D. C. (1979). *Ann. appl. Biol.* **91**, 331–336.

Harris, D. C. and Gibbs, J. A. (1980). *Rep. E. Malling Res. Stn.* for 1979, p. 96.

Harris, S. E., Braselton, J. P. and Miller, C. E. (1980). *Mycologia* **72**, 916–925.

Harrison, B. D. (1956). *Ann. appl. Biol.* **44**, 215–226.

Harrison, B. D. (1974). *Proc. 1st Intersectional Congress of the International Association of Microbiological Societies* **3**, 303–312.

Harrison, B. D. (1977). *Ann. Rev. Phytopathol.* **15**, 331–360.

Hartley, C. (1921). "Damping-off in Forest Nurseries". USDA Bull. 934.

Hartley, C. and Merrill, T. C. (1914). *Phytopathology* **4**, 89–92

Hartman, R. E. (1955). *Am. Pot. J.* **32**, 317–326.

Hartog, M. (1896). *Ann. Bot.* **10**, 98–100.

Harvey, R. G., Hagedorn, D. J. and DeLoughery, R. L. (1975). *Crop Sci.* **15**, 67–71.

Hashioka, Y. (1941). *Formosan Agr. Rev.* **27**, 329–345.

Haskins, E. F. (1978). *Protoplasma* **94**, 193–206.

Haskins, R. H. (1963). *Can. J. Microbiol.* **9**, 451–457.

Hawker, L. E. (1962). *Trans. Brit. mycol. Soc.* **45**, 190–199.

Heath, I. B. (1976). *In* "Recent Advances in Aquatic Mycology" (E. B. G. Jones, ed.), pp. 603–650. Elek, London.

Heath, I. B. (1978). *In* "Nuclear Division in the Fungi" (I. B. Heath, ed.), pp. 89–176. Academic Press, New York and London.

Heath, I. B. (1980). *In* "International Review of Cytology", Vol. 64 (G. H. Bourne and J. F. Danielli, eds), pp. 1–80. Academic Press, New York and London.

Heath, I. B. and Greenwood, A. D. (1971). *Z. Zellforsch* **112**, 371–389.

Heath, I. B., Greenwood, A. D. and Griffiths, H. B. (1970). *J. Cell. Sci.* **7**, 445–461.

Hebert, T. T. and Panizo, C. H. (1975). Oat mosaic virus. *C.M.I./A.A.B. Descriptions of Plant Viruses* No. 145.

Hegnauer, H. and Hohl, H. R. (1978). *Expl. Mycol.* **2**, 216–233.

Heim, P. (1955). *Revue Mycol.* **20**, 131–157.

Held, A. A. (1972). *Mycologia* **64**, 871–886.

Held, A. A. (1973). *Bull. Torrey bot. Club* **100**, 203–216.

Held, A. A. (1975). *Can. J. Bot.* **53**, 2212–2232.

Helling, C. S., Dennison, D. G. and Kaufman, D. D. (1974). *Phytopathology* **64**, 1091–1100.

Hendrix, F. F. and Campbell, W. A. (1966). *Plant Dis. Reptr.* **50**, 393–395.

Hendrix, F. F. and Campbell, W. A. (1968a). *Mycologia* **60**, 802–805.

Hendrix, F. F. and Campbell, W. A. (1968b). *Forest Sci.* **14**, 292–297.

Hendrix, F. F. and Campbell, W. A. (1969). *Mycologia* **61**, 397–391.

Hendrix, F. F. and Campbell, W. A. (1970). *Can. J. Bot.* **48**, 377–384.

Hendrix, F. F. and Campbell, W. A. (1973). *A. Rev. Phytopathol.* **11**, 77–98.

Hendrix, F. F. and Campbell, W. A. (1974). *Mycologia* **66**, 681–684.

Hendrix, F. F., Campbell, W. A. and Moncrief, J. B. (1970). *Plant Dis. Reptr.* **54**, 419–421.

Hendrix, F. F., Campbell, W. A. and Chien, C. Y. (1971). *Mycologia* **63**, 283–289.

Hendrix, F. F. and Kuhlman, E. G. (1965). *Phytopathology* **55**, 1183–1187.

Hendrix, F. F. and Papa, K. E. (1974). *Proc. Am. Phytopath. Soc.* **1**, 200–207.

Hendrix, F. F. and Powell, W. M. (1970). *Phytopathology* **60**, 16–19.

Hendrix, J. W. (1970). *A. Rev. Phytopathol.* **8**, 111–130.

Hesse, R. (1874). Inaugural Dissertation, Gottingen.

Hibberd, D. J. (1976a). *Bot. J. Linnean Soc.* **72**, 55–80.

Hibberd, D. J. (1976b). *J. Protozool.* **23**, 374–385.

Hibberd, D. J. (1979). *BioSystems* **11**, 243–261.

Hibberd, D. J. and Leedale, G. F. (1972). *Ann. Bot.* **36**, 49–71.

Hickman, C. J. (1958). *Trans. Brit. mycol. Soc.* **41**, 1–13.

Hickman, C. J. (1970). *Phytopathology* **60**, 1128–1135.

Hickman, C. J. and Ho, H. H. (1966). *A. Rev. Phytopathol.* **4**, 195–220.

Hickey, E. L. and Coffey, M. D. (1980). *Physiol. Plant Path.* **17**, 199–204.

Hidaka, Z., Uozumi, T. and Shimizu, T. (1956). *Bull. Hatano Tobacco Expt. Sta.* **40**, 31–35.

Hills, F. J. and Leach, L. D. (1962). *Phytopathology* **52**, 51–56.

Hiruki, C. (1975). *Can. J. Bot.* **53**, 2425–2434.

Hiruki, C. and Alderson, P. G. (1976). *Can. J. Bot.* **54**, 2820–2826.

Hiruki, C., Alderson, P. G., Kobayashi, N. and Furusawa, I. (1975). *Proc. 1st Intersectional Congress of the International Association of Microbiological Societies* **3**, 297–302.

Ho, H. H. (1981). *Mycologia* **73**, 705–714.

Ho, H. H. and Foster, B. (1972). *Mycopath. Mycol. appl.* **46**, 335–339.

Hoch, H. C. and Abawi, A. (1979). *Phytopathology* **69**, 417–419.

Hoch, H. C. and Mitchell, J. E. (1972). *Protoplasma* **75**, 113–138.

Hock, W. K. and Sisler, H. D. (1969). *Phytopathology* **59**, 627–632.

Hodges, C. S. (1962). *Mycologia* **54**, 221–229.

Hodgson, W. A. (1963). *Am. Pot. J.* **40**, 143–148.

Hoitink, H. A. J., Herr, L. J. and Schmitthenner, A. F. (1976). *Phytopathology* **66**, 1369–1372.

Holliday, P. (1970). *Commonw. Mycol. Inst., Descriptions of Pathogenic Fungi and Bacteria*, No. 229.

Holloway, S. A. and Heath, I. B. (1977a). *Expl. Mycol.* **1**, 9–29.

Holloway, S. A. and Heath, I. B. (1977b). *Can. J. Bot.* **55**, 1328–1339.

Hooley, P., Fyfe, A. M., Evola Maltese, C. and Shaw, D. S. (1982). *Trans. Br. mycol. Soc.* **79**, 563–566.

Hoppe, P. E. (1949). *Phytopathology* **39**, 77–84.

Horak, I. and Schlösser, E. (1978). *Meded. Fac. Landbouww. Rijksuniv. Gent.* **43**, 979–987.

Hori, M. (1935). *Ann. Phytopath. Soc. Japan* **5**, 10–22.

Horiuchi, S. and Hori, M. (1970). *Bull. Chugoku natn. agric. Expt. Stn.* E **17**, 35–55.

Horne, A. S. (1930). *Ann. Bot.* **44**, 199–225.

Horner, C. E. (1963). *Phytopathology* **53**, 472–474.

Horsfall, J. G. (1945). "Fungicides and Their Action". Chronica Botanica, Waltham, Massachusetts.

Horsfall, J. G. (1956). "Principles of Fungicidal Action". Chronica Botanica, Waltham, Massachusetts.

Howard, F. L., Rowell, J. B. and Keil, H. L. (1951). *Univ. Rhode Island Agr. Expt. Sta. Bull.* 308.

Howard, K. L. and Bryant, T. R. (1971). *Mycologia* **63**, 58–68.

Howard, K. L. and Moore, R. T. (1970). *Bot. Gaz.* **131**, 311–336.

Howatt, J. L. and Grainger, P. N. (1955). *Am. Potato J.* **32**, 180–188.

Hsu, D. and Hendrix, F. F. (1973). *Can. J. Bot.* **51**, 1421–1424.

Hudspeth, M. E. S., Timberlake, W. E. and Goldberg, R. B. (1977). *Proc. natn. Acad. Sci. USA* **74**, 4332–4336.

Huguenin, B. (1973). *Cah. O.R.S.T.O.M. Ser. Biol.* **20**, 59–61.

Huguenin, B. and Boccas, B. (1970). *C.r. hebd. Séanc. Acad. Sci., Paris* D **271**, 660–663.

Hunt, N. R., O'Donnell, F. G. and Marshall, R. P. (1925). *J. Agr. Res.* **31**, 301–363.

Huth, W. (1979). *NachrBl. Dt. PflSchutzdienst. Berl.* **31**, 53–55.

Hütter, R. and DeMoss, J. A. (1967). *J. Bacteriol.* **94**, 1896–1907.

Ichida, A. A. and Fuller, M. S. (1968). *Mycologia* **60**, 141–155.

Il'ina, M. N. (1979). *Probl. obshch. i chastn. fitotoksikol.* (Leningrad), pp. 117–121.

Ingram, D. S. (1969a). *J. gen. Microbiol.* **55**, 9–18.

Ingram, D. S. (1969b). *J. gen. Microbiol.* **56**, 55–67.

Ingram, D. S. and Tommerup, I. C. (1972). *Proc. Roy. Soc.,* B **180**, 103–112.

Inouye, I. and Fujii, S. (1977). Rice necrosis mosaic virus. *C.M.I./A.A.B. Descriptions of Plant Viruses* No. 172.

Inouye, T. and Saito, Y. (1975). Barley yellow mosaic virus. *C.M.I./A.A.B. Descriptions of Plant Viruses* No. 143.

Ito, S. and Tokunaga, Y. (1933). *J. Faculty Agr. Hokkaido Imp. Univ.* **32**, 201–227.

Iwaki, M., Nakano, A., Iemura, H. and Tochihara, H. (1978). *Ann. Phytopath. Soc. Japan* **44**, 578–584.

Iwata, Y. (1957). *Ann. Phytopath. Soc. Japan* **22**, 108–110.

Jacobsen, B. J. and Williams, R. H. (1970). *Pl. Dis. Reptr.* **54**, 456–460.

Jagger, I. C. (1940). *Phytopathology* **30**, 53–64.

Janssen, B. D. (1973). Studies of host-specific variations in the late blight fungus of potato. Ph.D. Thesis, University of Wales.

Jeffrey, S. I. B., Jinks, J. L. and Grindle, M. (1962). *Genetica* **32**, 323–338.

Jinks, J. L. and Grindle, M. (1963). *Heredity* **18**, 245–264.

Johnson, F. and Chambers, Y. (1973). *Plant Dis. Reptr.* **57**, 848–852.

Johnson, G. F. (1935). *Agr. Hist.* **9**, 67–79.

Johnson, G. I., Davis, R. D. and O'Brien, R. G. (1979). *Pl. Dis. Reptr.* **63**, 212–215.

Johnson, T. W., Jr. (1956). "The Genus Achlya: Morphology and Taxonomy". University of Michigan Press, Ann Arbor.

Johnson, T. W. (1957). *Am. J. Bot.* **44**, 875–878.

Johnson, T. W. (1983). "The Saprolegniaceae". University of North Carolina Press, Chapel Hill.

Jones, J. P. (1978). *Pl. Dis. Reptr.* **62**, 798–802.

Jones, P. M. (1928). *Arch. Protistenk.* **62**, 313–327.

Jones, R. A. C. and Harrison, B. D. (1969). *Ann. appl. Biol.* **63**, 1–17.
Joubert, J. J. and Rijkenberg, F. H. J. (1971). *A. Rev. Phytopathol.* **9**, 499–507.
Kaars Sijpesteijn, A. (1977). *In* "Systemic Fungicides" 2nd edn. (R. W. Marsh, ed.), pp. 131–159. Longman, London.
Kaars Sijpesteijn, A. (1982). *In* "Fungicide resistance in crop protection" (J. Dekker and S. C. Georgopoulos, eds), pp. 32–45. Pudoc, Wageningen.
Kaars Sijpesteijn, A., Kerkenaar, A. and Overeem, J. C. (1974). *Meded. Fac. Landbouww. Rijksuniv. Gent.* **39**, 1027–1034.
Kable, P. F. and Jeffrey, H. (1980). *Phytopathology* **70**, 8–12.
Kaji, H. and Tanaka, Y. (1968). *J. mol. Biol.* **32**, 221–230.
Kajiwara, T., Kobayashi, T., Inaka, T., Sudjadi, M. and Otjim, S. (1979). *J. Pestic. Sci.* **4**, 425–430.
Kamimura, S., Akutsa, M. and Takahi, Y. (1976a). *Ann. Phytopath. Soc. Japan* **42**, 204–215.
Kamimura, S., Nishikawa, M. and Takahi, Y. (1976b). *Ann. Phytopath. Soc. Japan* **42**, 242–252.
Kanlong, S. and Hendrix, J. W. (1977). *Can. J. Bot.* **55**, 17–22.
Kaosiri, T. and Zentmyer, G. A. (1980). *Mycologia* **72**, 988–1000.
Kaosiri, T., Zentmyer, G. A. and Erwin, D. C. (1978). *Can. J. Bot.* **56**, 1730–1738.
Karling, J. S. (1942). "The Plasmodiophorales". Author, New York.
Karling, J. S. (1943). "The Simple Holocarpic Biflagellate Phycomycetes". Author, New York.
Karling, J. S. (1945). *Amer. J. Bot.* **30**, 637–648.
Karling, J. S. (1950). *Lloydia* **13**, 1–71.
Karling, J. S. (1964). "Synchytrium". Academic Press, New York and London.
Karling, J. S. (1968). "The Plasmodiophorales". 2nd. edn. Hafner Publishing Co., New York and London.
Karling, J. S. (1977). "Chytridiomycetarum Iconographia". Lubrecht and Cramer, Monticello, New York.
Kassanis, B. (1970a), Tobacco necrosis virus. *C.M.I./A.A.B. Descriptions of Plant Viruses* No. 14.
Kassanis, B. (1970b). Satellite virus. *C.M.I./A.A.B. Descriptions of Plant Viruses* No. 15.
Katsura, K. (1971). *Rev. Plant Prot. Res.* **4**, 58–70.
Kataria, H. R. and Grover, R. K. (1975). *Ind. J. Exp. Biol.* **13**, 281–285.
Kazama, F. Y. (1972). *J. gen. Microbiol.* **71**, 555–566.
Kazama, F. Y. (1980). *Can. J. Bot.* **58**, 2434–2446.
Keeling, B. L. (1974). *Phytopathology* **64**, 1445–1447.
Keen, N. T., Reddy, M. N. and Williams, P. H. (1969). *Phytopathology* **59**, 637–644.
Kenneth, R. (1957a). *Commonw. Mycol. Inst., Descriptions of Pathogenic Fungi and Bacteria*, No. 451.
Kenneth, R. (1975b). *Commonw. Mycol. Inst., Descriptions of Pathogenic Fungi and Bacteria*, No. 452.
Kerkenaar, A. and Kaars Sijpesteijn, A. (1977). *Neth. J. Pl. Path.* **83** (Suppl.), 145–152.
Kerr, A. (1963). *Aust. J. Biol. Sci.* **16**, 55–69.
Keskin, B. (1964). *Arch. Mikrobiol.* **49**, 348–374.
Keskin, B. (1971). *Arch. Mikrobiol.* **77**, 344–348.
Keskin, B. and Fuchs, W. H. (1969). *Arch. Mikrobiol.* **68**, 218–226.
Khaki, I. A. and Shaw, D. S. (1974). *Genet. Res.* **23**, 75–86.
Khan, S. R. (1977). *Can. J. Bot.* **55**, 730–739.
Kim, S. H., Kantzes, J. G. and Weaver, L. O. (1974). *Phytopathology* **64**, 373–380.

Klaveriappa, K. M. and Safeeulla, K. M. (1978). *Proc. Ind. Acad. Sci.* B **87**. 303–308.

Klein, R. M. and Cronquist, A. (1967). *Q. Rev. Biol.* **42**, 105–296.

Klotz, L. J., Calavan, E. C. and De Wolfe, T. A. (1972). *Calif. Citrogr.* **57**, 267–268.

Knights, B. A. (1970). *Phytochemistry* **9**, 701–704.

Ko, W. H. (1978). *J. gen. Microbiol.* **107**, 15–18.

Ko, W. H., Chang, H. S. and Su, H. J. (1978a). *Trans. Br. mycol. Soc.* **71**, 496–499.

Ko, W. H., Chang, H. S., Su. H. J., Chen, C. C. and Leu, L. S. (1978b). *Mycologia* **70**, 380–384.

Koch, W. J. (1956). *Am. J. Bot.* **43**, 811–819.

Koch, W. J. (1968). *Am. J. Bot.* **55**, 841–859.

Koike, H. and Yang, S. (1971). *Phytopathology* **61**, 1090–1092.

Kole, A. P. (1954). *Tijskr. PlZiekt.* **60**, 1–65.

Kole, A. P. and Gielink, A. J. (1961). *Proc. k. ned. Akad. Wet. C.* **64**, 157–161.

Kole, A. P. and Gielink, A. J. (1962). *Proc. k. ned. Akad. Wet. C.* **65**, 117–121.

Kole, A. P. and Gielink, A. J. (1963). *Neth. J. Plant Pathol.* **69**, 258–262.

Komuro, Y. (1971). *JARQ* **6**, 41–45.

Kotova, V. V. and Tsvetkova, N. A. (1979). *Khimiya v Sel'skom Khozyaïstve* **27**, 37–39.

Kraft, J. M. and Erwin, D. C. (1967). *Phytopathology* **57**, 374–376.

Kreutzer, W. A. (1963). *A. Rev. Phytopathol.* **1**, 101–126.

Krywienczyk, J. and Dorworth, C. E. (1980). *Can. J. Bot.* **58**, 1412–1417.

Kubai, D. F. (1978). *In* "Nuclear Division in the Fungi" (I. B. Heath, ed.), pp. 177–229. Academic Press, New York, San Francisco and London.

Kuhlman, E. G. and Hendrix, F. F. (1963). *Plant Dis. Reptr.* **47**, 552–553.

Kulish, V. B. and Dyakov, Yu. T. (1979). *Dokl. Akad. Nauk SSSR* **244**, 735–738.

Kusano, S. (1912). *J. Coll. Agric. Imp. Univ. Tokyo*, **4**, 141–199.

Kusano, S. (1930). *Japan J. Bot.* **5**, 35–132.

Kusano, S. (1932). *J. Coll. Agric. Imp. Univ. Tokyo* **11**, 359–426.

Kuwata, S. and Kubo, S. (1981). *Ann. Phytopath. Soc. Japan* **47**, 264–268.

Lafon, R., Bugaret, Y. and Bulit, J. (1977). *Phytiat. Phytopharm.* **26**, 19–40.

Lal, B. B. and Chakravarti, B. P. (1977). *Pl. Dis. Reptr.* **61**, 334–336.

Lal, S., Nath. K. and Saxena, S. C. (1977). *Indian Phytopathol.* **30**, 143–144.

Lange, L. (1976). *T. Planteavl.* **80**, 153–169.

Lange, L. (1978). *Phytopath. Z.* **92**, 132–142.

Lange, L. and Insunza, V. (1977). *Trans. Br. mycol. Soc.* **69**, 377–384.

Lange, L. and Olson, L. W. (1976a). *Protoplasma* **89**, 339–351.

Lange, L. and Olson, L. W. (1976b). *Protoplasma* **90**, 33–45.

Lange, L. and Olson, L. W. (1977). *Protoplasma* **93**, 27–43.

Lange, L. and Olson, L. W. (1978a). *Trans. Br. mycol. Soc.* **71**, 377–384.

Lange, L. and Olson, L. W. (1978b). *Can. J. Bot.* **56**, 1229–1239.

Lange, L. and Olson, L. W. (1979). *Dansk. Bot. Arkiv* **33**, 1–95.

Lange, L. and Olson, L. W. (1980a). *Protoplasma* **102**, 323–342.

Lange, L. and Olson, L. W. (1980b). *Trans. Br. mycol. Soc.* **74**, 449–457.

Lasure, L. L. and Griffin, D. H. (1974). *Mycologia* **66**, 391–396.

Laville, E. (1979). *Fruits* **34**, 35–41.

Laviola, C. (1968). Studies on the genetics of *Phytophthora infestans*. Ph.D. Thesis, West Virginia University, Morgantown.

Laviola, C. and Portacci, M. (1974). *Phytopathol. Mediterr.* **15**, 87–92.

Leach, L. D. (1947). *J. agric. Res.* **75**, 161–179.

Leach, L. D., Garber, R. H. and Tolmsoff, W. J. (1960). *Phytopathology* **50**, 643–644.

Leach, S. S. and Rich, A. E. (1969). *Phytopathology* **59**, 1360–1365.

Leclerc, J. C. and Couté, A. (1976). *C.r. hebd. Séanc. Acad. Sci. Paris* **282**, Série D, 2067–2070.
Ledingham, G. A. (1936). *Can. J. Res., C.* **14**, 117–121.
Ledingham, G. A. (1939). *Can. J. Res.* **17**, 38–51.
Leedale, G. F. (1974). *Taxon* **23**, 261–270.
LéJohn, H. B. (1971). *Nature* **231**, 164–168.
LéJohn, H. B. (1974). *Evol. Biol.* **7**, 79–125.
Lemmerzahl, J. (1930). *Phytopath. Z.* **2**, 257–320.
Leppik, E. E. (1962). *FAO Pl. Prot. Bull.* **10**, 126–129.
Lessie, P. E. and Lovett, J. S. (1968). *Am. J. Bot.* **55**, 220–236.
Levine, N. D. (Chairman) *et al.* (1980). *J. Protozool.* **27**, 37–58.
Lin, C. C., Sicher, R. C. and Aronson, J. M. (1976). *Arch. Microbiol.* **108**, 85–91.
Lin, M. T. (1979). *Pl. Dis. Reptr.* **63**, 10–12.
Linford, M. B. and McKinney, H. H. (1954). *Plant Dis. Reptr.* **38**, 711–713.
Lingappa, B. T. (1956). *Mycologia* **48**, 427–432.
Lingappa, Y. (1959a). *Am. J. Bot.* **46**, 145–150.
Lingappa, Y. (1959b). *Am. J. Bot.* **46**, 233–240.
Lipps, P. E. and Bruehl, G. W. (1978). *Phytopathology* **68**, 1120–1127.
Lister, R. M. and Murant, A. F. (1967). *Ann. appl. Biol.* **54**, 167–176.
Litvinov, M. A. (1959). *Acta Inst. Bot. Acad. Sci. U.R.S.S., Ser. II, Plantae Cryptogamae* **12**, 188–212.
Llanos, M. C. and Lockwood, J. L. (1960). *Phytopathology* **50**, 826–830.
Locke, J. C., Papavizas, G. C. and Rubel, M. K. (1979). *Phytopathology* **69**, 536.
Lockwood, J. L. and Ballard, J. C. (1959). *Phytopathology* **49**, 406–410.
Loening, U. E. (1968). *J. molec. Biol.* **38**, 355–365.
Loening, U. E. (1969). *Biochem. J.* **113**, 131.
Loening, U. E. (1973). *Biochem. Soc. Symp.* **37**, 95–104.
Long, M. and Keen, N. T. (1977a). *Phytopathology* **67**, 670–674.
Long, M. and Keen, N. T. (1977b). *Phytopathology* **67**, 675–677.
Long, M., Keen, N. T., Ribeiro, O. K., Leary, J. V., Erwin, D. C. and Zentmyer, G. A. (1975). *Phytopathology* **65**, 592–597.
López, B. P. and López, G. M. (1977). *Fitopatologia* **12**, 20–23.
Lotsy, J. P. (1907). "Algae und Pilze; Vorträge über botanische Stammesgeschichte." Jena.
Lovett, J. S. (1975). *Bact. Rev.* **39**, 345–404.
Lovett, J. S. and Haselby, J. A. (1971). *Arch. Mikrobiol.* **80**, 191–204.
Lucas, G. B. (1965). "Diseases of Tobacco". Scarecrow Press, New York.
Lukens, R. J. (1971). "Chemistry of Fungicidal Action". Springer Verlag, New York.
Lukens, R. J. (1978). *Phytopath. News* **12**, 142.
Luna, L. V. and Hine, R. B. (1964). *Phytopathology* **54**, 955–959.
Lunney, C. Z. and Bland, C. E. (1976). *Protoplasma* **90**, 119–137.
Lynn, D. H. (1981). *Biol. Rev.* **56**, 243–292.
Lyr, H. (1977). *In* "Antifungal Compounds" (M. R. Siegel and H. D. Sisler, eds), Vol. II, pp. 301–332. Marcel Dekker, New York.
Lyr, H., Casperson, G. and Laussmann, B. (1977). *Z. allg. Mikrobiol.* **17**, 117–129.
McCarter, S. M. and Roncadori, R. W. (1972). *Ga. agric. Res.* **13**, 3–5.
McCorkindale, N. J., Hutchinson, S. A., Pursey, B. A., Scott, W. T. and Wheeler, R. (1969). *Phytochemistry* **8**, 861–867.
McDonough, E. S. (1937). *Mycologia* **29**, 151–173.
Macfarlane, I. (1970). *Trans. Br. mycol. Soc.* **55**, 113–116.
McGhee, R. B. and Hanson, W. L. (1964). *J. Protozool.* **11**, 555–562.

McGhee, R. R. and McGhee, A. H. (1979). *J. Protozool.* **26**, 348–351.
McGrath, H. and Miller, P. R. (1958). *Pl. Dis. Reptr.* **250**, *Suppl.*, 3–31.
MacIntyre, D. and Elliott, C. G. (1974). *Genet. Res.* **24**, 295–309.
McIntyre, J. L. and Hankin, L. (1978). *Can. J. Microbiol.* **24**, 75–78.
Mackie, W. and Preston, R. D. (1974). *In* "Algal Physiology and Biochemistry" (W. D. P. Stewart, ed.), pp. 40–85. Blackwell Scientific, Oxford, London, Edinburgh and Melbourne.
McKinney, H. H. (1937). Mosaic diseases of wheat and related cereals. *USDA Circular* 442.
McKinney, H. H. (1948). *Phytopathology* **38**, 1003–1013.
McKinney, H. H., Paden, W. R. and Koehler, B. (1957). *Pl. Dis. Reptr.* **41**, 256–266.
McKeen, W. E. (1977). *Can. J. Bot.* **55**, 44–47.
McMeekin, D. (1960). *Phytopathology* **50**, 93–97.
Maia, N., Venard, P. and Lavrut, F. (1976). *Ann. Phytopathol.* **8**, 141–146.
Maire, R. and Tison, A. (1911). *Annls. Mycol.* **9**, 226–246.
Malcolmson, J. F. (1969). *Trans. Br. mycol. Soc.* **53**, 417–423.
Malcolmson, J. F. (1970). *Nature* **225**, 971–972.
Mandel, M. (1968). *In* "Handbook of Biochemistry" (H. A. Sober, ed.), pp. H-27–H-29. CRC Press, Cleveland, Ohio.
Manocha, M. S. and Colvin, J. R. (1968). *J. Bacteriol.* **95**, 1140–1152.
Manton, I. (1959). *J. exp. Bot.* **10**, 448–461.
Manton, I. (1965). *In* "Advances in Botanical Research", Vol. II (R. D. Preston, ed.), pp. 1–34. Academic Press, New York and London.
Margulis, L. and Schwartz, K. (1982). "Phyla of the Five Kingdoms". W. H. Freeman, San Francisco.
Markowitz, M. M. (1978). *J. Phycol.* **14**, 289–302.
Marlatt, R. B. and McKittrick, R. T. (1963). *Phytopathology* **53**, 597–599.
Marshall, W. (1787). "The Rural Economy of Norfolk". Vol. 2. London.
Marshall, W. D. and Jarvis, W. R. (1979). *J. agric. Fd Chem.* **27**, 766–769.
Martin, H. and Worthing, C. R. (1976). "Insecticide and Fungicide Handbook for Crop Protection". Blackwell, Oxford, London, Edinburgh and Melbourne.
Martin, W. W. (1971). *J. Elisha Mitchell scient. Soc.* **87**, 209–221.
Marx, D. H. (1967). *XIV. IUFRO-Kongress* **5**, 172–181.
Marx, D. H. (1969a). *Phytopathology* **59**, 153–163.
Marx, D. H. (1969b). *Phytopathology* **59**, 411–417.
Marx, D. H. (1973). *Phytopathology* **63**, 18–23.
Matocha, P., Frederiksen, R. A. and Reyes, L. (1974). *Ind. Phytopath.* **27**, 322–324.
Mastenbroek, D. and de Bruin, T. (1955). *Tijdschr. PlZiekt.* **61**, 88–92.
Mathre, D. E. and Otta, J. D. (1967). *Plant Dis. Reptr.* **51**, 864–866.
Matthews, R. E. F. (1979). *Intervirology* **12**, 129–296.
Matthews, V. D. (1931). "Studies on the Genus Pythium". University of North Carolina Press, Chapel Hill.
Mattusch, P. (1978). *In* "Woronin + 100 International Conference on Clubroot" (S. T. Buczacki and P. H. Williams, eds), pp. 24–28a. Department of Plant Pathology, University of Wisconsin, Madison.
Maxwell, D. P., Maxwell, M. D., Hänssler, G., Armentrout, V. N., Murray, G. M. and Hoch, H. C. (1975). *Planta* **124**, 109–123.
Maynard Smith, J. (1978). "The Evolution of Sex". Cambridge University Press, Cambridge.
Melhus, I. E., Rosenbaum, J. and Schultz, E. S. (1916). *J. agric. Res.* **7**, 213–254.
Melkonian, M. (1980). *BioSystems* **12**, 85–104.
Melville, S. C. and Hawken, R. H. (1967). *Pl. Path.* **16**, 145–147.

Merz, W. G., Burrell, R. G. and Gallegly, M. E. (1969). *Phytopathology* **59**, 367–370.
Meurs, A. L. (1934). *Phytopath. Z.* **7**, 170–185.
Meyer, D. and Schönbech, F. (1975). *Z. Pflanzenkr. Pflanzenschutz* **82**, 337–354.
Mez, C. (1929). *Schaftliche der Königsberger Gelehrten Gesellschaft-Naturwissenschaftliche Klasse* **6**, 1–58.
Michelmore, R. W. and Ingram, D. S. (1980). *Trans. Br. mycol. Soc.* **75**, 47–56.
Michelmore, R. W. and Ingram, D. S. (1981). *Trans. Br. mycol. Soc.* **77**, 131–137.
Michelmore, R. W. and Ingram, D. S. (1982). *Trans. Br. mycol. Soc.* **78**, 1–9.
Michelmore, R. W., Pawar, M. N. and Williams, R. J. (1982). *Phytopathology* **72**, 1368–1372.
Michelmore, R. W. and Sansome, E. (1982). *Trans. Br. mycol. Soc.* **79**, 291–297.
Middleton, J. T. (1943). *Mem. Torrey Bot. Club.* **20**, 1–171.
Miele, W. H. and Linkins, A. E. (1978). *Can. J. Bot.* **56**, 1974–1981.
Milbrath, D. G. (1923). *J. agric. Res.* **23**, 989–994.
Mildenhall, J. P., Pratt, R. G., Williams, P. H. and Mitchell, J. E. (1971). *Plant Dis. Reptr.* **55**, 536–540.
Millardet, A. (1885a). *J. Agr. Prat.* (Paris) **49**, 801–805 [English translation in *Phytopathol. Classics* **3**, 18–25 (1933)].
Millardet, A. (1885b). *J. Agr. Prat.* (Paris) **49**, 513–516 [English translation in *Phytopathol. Classics* **3**, 7–11 (1933)].
Millardet, A. and Gayon, U. (1885). *J. Agr. Prat.* (Paris) **49**, 707–710 [English translation in *Phytopathol. Classics* **3**, 12–17 (1933)].
Miller, C. E. (1958). *J. Elisha Mitchell scient. Soc.* **74**, 49–64.
Miller, C. E. and Dylewski, D. P. (1981). *Am. J. Bot.* **68**, 342–349.
Miller, C. E. and Ristanović, B. (1969). *Ohio J. Sci.* **69**, 105–109.
Miller, C. R., Dowler, W. M., Petersen, D. H. and Ashworth, R. P. (1966). *Phytopathology* **56**, 46–49.
Mills, W. R. (1940a). *Phytopathology* **30**, 17.
Mills, W. R. (1940b). *Phytopathology* **30**, 830–839.
Milovidov, P. F. (1931). *Arch. Protistenk.* **73**, 1–46.
Mitchell, D. J. (1975). *Phytopathology* **65**, 570–575.
Mitchell, J. E. and Hagedorn, D. J. (1966). *Pl. Dis. Reptr.* **50**, 91–95.
Mitchell, J. E. and Hagedorn, D. J. (1969). *Pl. Dis. Reptr.* **53**, 697–701.
Mitchell, J. E. and Hagedorn, D. J. (1971). *Phytopathology* **61**, 978–983.
Moestrup, Ø. (1978). *BioSystems* **10**, 117–144.
Moestrup, Ø. and Thomsen, H. A. (1974). *Protoplasma* **81**, 247–269.
Montecillo, C. M., Bracker, C. E. and Powell, M. J. (1980). *Can. J. Bot.* **58**, 1885–1897.
Montgomerie, I. G. and Kennedy, D. M. (1977). *Pl. Path.* **26**, 139–143.
Moore, E. D. and Miller, C. E. (1973). *Mycologia* **65**, 145–154.
Moore, L. D. and Couch, H. B. (1961). *Plant Dis. Reptr.* **45**, 616–619.
Mortimer, A. M. and Shaw, D. S. (1975). *Genet. Res.* **25**, 201–205.
Mortimer, A. M., Shaw, D. S. and Sansome, E. R. (1977). *Ark. Microbiol.* **111**, 255–259.
Mowat, W. P. (1970). *Ann. appl. Biol.* **66**, 17–28.
Moxham, S. E. and Buczacki, S. T. (1983). *Trans. Br. mycol. Soc.* **80**, 297–304.
Moxham, S. E., Fraser, R. S. S. and Buczacki, S. T. (1983). *Trans. Br. mycol. Soc.* **80**, 497–505.
Mukerji, K. G. (1975). *Commonw. Mycol. Inst., Descriptions of Pathogenic Fungi and Bacteria* No. 460.
Mukerji, K. G. and Holliday, P. (1975). *Commonw. Mycol. Inst., Descriptions of Pathogenic Fungi and Bacteria*, No. 453.
Mulder, D. (1969). *Neth. J. Pl. Path.* **75**, 178–181.

Muller, G. J., Linn, M. B. and Sinclair, J. B. (1972). *Pl. Dis. Reptr.* **56**, 1054–1057.
Muller, H. J. (1964). *Mutation Res.* **1**, 2–9.
Müller, K. O. (1933). *Bot. Abstr.* **15**, 84–96.
Müller, K. O., Mackay, J. H. E. and Friend, J. N. (1954). *Nature* **174**, 878–879.
Mullins, J. T. (1961). *Am. J. Bot.* **48**, 377–387.
Mullins, J. T. and Barksdale, A. W. (1965). *Mycologia* **67**, 352–359.
Munnecke, D. E., Bricker, J. L. and Kolbezen, M. J. (1978). *Phytopathology* **68**, 1210–1216.
Murphy, M. N. and Lovett, J. S. (1966). *Devel. Biol.* **14**, 68–95.
Murphy, P. A. (1918). *Ann. Bot.* **32**, 115–153.
Murray, G. M. and Hoch, H. C. (1975). *Planta (Berl.)* **124**, 109–123.
Muse, R. R., Schmitthenner, A. F. and Partyka, R. E. (1974). *Phytopathology* **64**, 252–253.
Nawaschin, S. (1899). *Flora* **86**, 404–427.
Neish, G. A. and Green, B. R. (1976). *J. gen. Microbiol.* **96**, 215–219.
Neish, G. A. and Green, B. R. (1977). *Biochim. Biophys. Acta* **479**, 411–415.
Nemec, S. (1970). *Plant Dis. Reptr.* **54**, 416–418.
Nemec, S. (1971). *Phytopathology* **61**, 711–714.
Nemec, S. (1972). *Can. J. Bot.* **50**, 1091–1096.
Nemec, S. (1974). *Mycopath. Mycol. Appl.* **52**, 283–289.
Newhook, F. J. (1959). *N.Z. J. agric. Res.* **2**, 808–843.
Newhook, F. J. and Podger, F. D. (1972). *A. Rev. Phytopathol.* **10**, 299–326.
Newhook, F. J., Waterhouse, G. M. and Stamps, D. J. (1978). *Commonw. Mycol. Inst., Mycol. Papers* No. 143.
Niederhauser, J. S. (1961). *In* "Recent Advances in Botany", pp. 491–497. University of Toronto Press.
Nienhaus, F. and Stille, B. (1965). *Phytopath. Z.* **54**, 335–337.
Noon, J. P. and Hickman, C. J. (1974). *Can. J. Bot.* **52**, 1591–1595.
Noviello, C. and Snyder, W. C. (1962). *Phytopath. Z.* **46**, 139–163.
Oakley, B. R. (1978). *BioSystems* **10**, 59–64.
O'Brien, R. G. (1978). *Pl. Dis. Reptr.* **62**, 277–279.
Ohh, S. H., King, T. H. and Kommedahl, T. (1978). *Phytopathology* **68**, 1644–1649.
Ojha, M., Dutta, S. K. and Turian, G. (1973). *J. gen. Microbiol.* **77**, v–vi.
Olive, L. S. (1953). *Bot. Rev.* **19**, 439–586.
Olive, L. S. (1969). *Science* **164**, 857.
Olive, L. S. (1975). "The Mycetozoans". Academic Press, New York, San Francisco and London.
Olsen, O. A. (1966). *Can. Pl. Dis. Survey* **46**, 1–4.
Olson, L. W. (1973). *Protoplasma* **78**, 113–127.
Olson, L. W. and Borkhardt, B. (1978). *Trans. Br. mycol. Soc.* **71**, 65–76.
Olson, L. W., Edén, U. M. and Lange, L. (1980). *Protoplasma* **103**, 1–16.
Olson, L. W., Edén, U. M. and Lange, L. (1981). *Proc. 3rd Int. Spore Symp.* pp. 43–70. Academic Press, N.Y. and London.
Olson, L. W. and Fuller, M. S. (1968). *Arch. Mikrobiol.* **62**, 237–250.
Olson, L. W. and Lange, L. (1978). *Protoplasma* **97**, 275–290.
Olson, L. W. and Lange, L. (1983). *Nordic J. Bot.* (In press).
Olson, L. W. and Reichle, R. (1978). *Trans. Br. mycol. Soc.* **70**, 423–437.
Ooka, J. J. and Yamamoto, B. (1979). *Phytopathology*, **69**, 918.
Oren, Y. and Solel, Z. (1978). *Phytoparasitica* **6**, 65–70.
Palti, J. (1975). *Commonw. Mycol. Inst., Descriptions of Pathogenic Fungi and Bacteria* No. 457.
Palzer, C. (1975). *Phytophthora Newsletter* **3**, 9–11.

Papa, K. E., Campbell, W. A. and Hendrix, F. F. (1967). *Mycologia* **59**, 589–595.

Papavizas, G. C. (1964). *Pl. Dis. Reptr.* **48**, 537–541.

Papavizas, G. C. and Ayers, W. A. (1974). *Tech. Bull. Agr. Res. Serv. USDA*, **1485**, 118–130.

Papavizas, G. C. and Davey, C. B. (1962a). *Pl. Dis. Reptr.* **46**, 646–650.

Papavizas, G. C. and Davey, C. B. (1962b). *Phytopathology* **52**, 24.

Papavizas, G. C. and Davey, C. B. (1963). *Phytopathology* **53**, 116–122.

Papavizas, G. C. and Lewis, J. A. (1976). *Pl. Dis. Reptr.* **60**, 484–488.

Papavizas, G. C., Lewis, J. A., Lumsden, R. D., Adams, P. B., Ayers, W. A. and Kantzes, J. G. (1977). *Phytopathology* **67**, 1293–1299.

Papavizas, G. C., O'Neill, N. R. and Lewis, J. A. (1978). *Phytopathology* **68**, 1667–1671.

Papavizas, G. C., Schwenk, F. W., Locke, J. C. and Lewis, J. A. (1979). *Pl. Dis. Reptr.* **63**, 708–712.

Park, D. (1975). *Trans. Br. mycol. Soc.* **65**, 249–257.

Parthasarathy, M. V., Van Slobbe, W. G. and Soudant, C. (1976). *Science* **192**, 1346–1348.

Patterson, G. W. (1971). *Lipids* **6**, 120–127.

Paulus, A. O., Nelson, J., Gafney, J. and Snyder, M. (1977). *Proc. Br. Crop Prot. Conf.* pp. 929–935.

Payak, M. M., Renfro, B. L. and Lal, S. (1970). *Indian Phytopathol.* **23**, 183–193.

Pearlmutter, N. L. and Lembi, C. A. (1978). *J. Hist. Cyto.* **26**, 782–791.

Pendergrass, W. R. (1950). *Mycologia* **42**, 279–289.

Perkins, F. O. (1974). *Veröff. Inst. Meeresforsch. Bremerh. Suppl.* **5**, 45–63.

Perkins, F. O. (1976). *In* "Recent Advances in Aquatic Mycology" (E. B. G. Jones, ed.), pp. 279–312. Elek, London.

Perry, D. A. (1973). *Trans. Br. mycol. Soc.* **61**, 135–144.

Petersen, H. E. (1910). *Annls mycol.* **8**, 494–560.

Peterson, C. A. and Edgington, L. V. (1971). *Phytopathology* **61**, 91–92.

Petri, L. (1925). *Rev. Appl. Mycol.* **10**, 122.

Pfender, W. F., Hine, R. B. and Stanghellini, M. E. (1977). *Phytopathology* **67**, 657–663.

Pieczarka, D. J. and Abawi, G. S. (1978). *Phytopathology* **68**, 403–408.

Pickett-Heaps, J. D. (1969). *Cytobios* **3**, 257–280.

Plaats-Niterink, A. J. van der (1968). *Acta Bot. Neer.* **17**, 320–329.

Plaats-Niterink, A. J. van der (1969). *Acta Bot. Neer.* **18**, 489–495.

Plaats-Niterink, A. J. van der (1981). "Monograph of the genus *Pythium.*" Centralbureau voor Schimmelcultures, Baarn. Studies in Mycol. No. 21.

Plaats-Niterink, A. J. van der, Samson, R. A., Stalpers, J. A. and Weijman, A. C. M. (1976). *Persoonia* **9**, 85–93.

Plumb, R. T. and Macfarlane, I. (1977). *Rothamsted Experimental Station Report for 1976*, Part 1, pp. 256–257.

Podger, F. D. (1972). *Phytopathology* **62**, 972–981.

Pokorny, K. S. (1967). *J. Protozool.* **14**, 697–708.

Pont, W. (1959). *Queensl. J. agr. Sci.* **16**, 299–327.

Porter, D. and Smiley, R. (1979). *Expl. Mycol.* **3**, 188–193.

Porter, J. R. and Shaw, D. S. (1978). *Trans. Br. mycol. Soc.* **71**, 515–518.

Powell, M. J. (1978). *BioSystems* **10**, 167–181.

Powell, M. J. (1980). *Am. J. Bot.* **67**, 839–853.

Powell, M. J. and Kock, W. J. (1977). *Can. J. Bot.* **55**, 1686–1695.

Powell, N. T., Melendez, P. L. and Batten, C. K. (1971). *Phytopathology* **61**, 1332–1337.

Pratt, B. H. and Heather, W. A. (1972). *Trans. Br. mycol. Soc.* **59**, 87–96.
Pratt, B. H. and Heather, W. A. (1973). *Aust. J. Biol. Sci.* **26**, 559–573.
Pratt, B. H. and Mitchell, D. J. (1975). *Can. J. Bot.* **51**, 333–339.
Pratt, B. H., Sedgley, J. H., Heather, W. A. and Shepherd, C. J. (1972). *Aust. J. Biol. Sci.* **25**, 861–863.
Pratt, R. G. and Green, R. J. (1973). *Can. J. Bot.* **51**, 429–436.
Prévost, B. (1807). "Memoire sur la cause immédiate de la carie ou charbon des bles, et de plusieurs autres maladies des plantes, et sur les préservatifs de la carie". Paris.
Price, D. (1981). *Rep. Glasshouse Crops Res. Inst.* 1979, pp. 137–138.
Pringsheim, N. (1858). *Jahrb. wiss. Bot.* **1**, 284–304.
Prowazek, S. (1905). *Arb. K. GesundhAmt.* **22**, 396–410.
Pryor, D. E. (1946). *Phytopathology* **36**, 264–272.
Putz, C. (1977). *J. gen. Virol.* **35**, 397–401.
Ragan, M. A. and Chapman, D. J. (1978). "A Biochemical Phylogeny of the Protists." Academic Press, New York, San Francisco and London.
Ragozzino, A. and Furia, A. (1974). *Annali. Fac. Sci. Agr. Univ. Napoli Portici S. IV* **8**, 265–273.
Ramachandraiah, M., Venkataratham, P. and Sulochana, C. B. (1979). *Plant Dis. Reptr.* **63**, 949–951.
Rands, R. D. (1922). *Dept. Landb. Inst. Plantenziekten Nijv. Handel, Meded.* **54**, 53.
Rands, R. D. and Dopp, E. (1938). *USDA Tech. Bull.* 666.
Rao, A. S. (1968). *Phytopathology* **58**, 1516–1521.
Raper, J. R. (1957). *Symp. Soc. Exp. Biol. XI*, 143–165.
Raper, J. R. (1966). *In* "The Fungi" (G. C. Ainsworth and A. S. Sussman, eds), Vol. 2, pp. 473–511. Academic Press, New York and London.
Reddick, D. and Mills, W. (1938). *Am. Pot. J.* **15**, 29–34.
Reeves, R. J. and Jackson, R. M. (1974). *J. gen. Microbiol.* **84**, 303–310.
Reichhard, T. (1976). *Pflanzenschutzberichte* **45**, 57–69.
Reichle, R. E. (1969). *Mycologia* **61**, 30–51.
Reisert, P. S. and Fuller, M. S. (1962). *Mycologia* **54**, 647–657.
Reuveni, M., Eyal, H. and Cohen, Y. (1980). *Pl. Disease* **64**, 1108–1109.
Reyes, A. A., Davidson, T. R. and Marks, C. F. (1974). *Phytopathology* **64**, 173–177.
Ribeiro, O. K. (1978). "A Source Book of the Genus *Phytophthora*." J. Cramer, Hirschberg.
Ribeiro, O. K., Zentmyer, G. A. and Erwin, D. C. (1976). *Phytopathology* **66**, 172–174.
Rich, S. (1950). *Pl. Dis. Reptr.* **34**, 253–255.
Rich, S. (1960). *Pl. Dis. Reptr.* **44**, 352–353.
Rich, S. and Horsfall, J. G. (1952). *Phytopathology* **42**, 457–460.
Richardson, L. T. (1976). *Can. J. Plant Sci.* **56**, 365–369.
Richardson, L. T. and Munnecke, D. E. (1964). *Can. J. Bot.* **42**, 301–306.
Ridings, W. H., Gallegly, M. E. and Lilly, V. G. (1969). *Phytopathology* **59**, 737–742.
Ridings, W. H. and Zettler, F. W. (1973). *Phytopathology* **63**, 289–295.
Rines, H. W., Case, M. E. and Giles, H. H. (1969). *Genetics* **61**, 789–800.
Ringo, D. L. (1967). *J. Cell Biol.* **33**, 543–571.
Robak, J. and Dobrzanski, A. (1978). *Acta Agrobot.* **31**, 151–158.
Robertson, G. I. (1973). *N.Z. J. Agric. Res.* **16**, 357–365.
Robertson, G. I. (1980). *New Zealand J. Bot.* **18**, 73–102.
Robertson, J. A. (1972). *Arch. Mikrobiol.* **85**, 259–266.
Rodriguez-Kanana, R., Backman, P. A. and Curl, E. A. (1977). *In* "Antifungal Compounds" (M. R. Siegel and H. D. Sisler, eds), Vol. I, pp. 117–161. Marcel Dekker, New York.

Romero, S. (1967). "Effects of genetic recombination on the pathogenicity of *Phytophthora infestans*". Doct. thesis, Univ. California, Riverside, California.

Romero, S. and Erwin, D. C. (1969). *Phytopathology* **59**, 1310–1317.

Roncadori, R. W. (1965). *Phytopathology* **55**, 595–599.

Roncadori, R. W., Lehman, P. S. and McCarter, S. M. (1974). *Phytopathology* **64**, 1303–1306.

Roncadori, R. W. and McCarter, S. M. (1972). *Phytopathology* **62**, 373–376.

Rosek, C. E. and Timberlake, W. E. (1979). *Nucleic Acid Research* **7**, 1567–1578.

Rosenburg, O. (1903). *Bih. k. Svenska Vetensk-Akad. Handl.* **28**, 1–20.

Rosinski, M. A. and Campana, R. J. (1964). *Mycologia* **56**, 738–744.

Rotem, J., Palti, J. and Rawitz, E. (1962). *Pl. Dis. Reptr.* **46**, 145–149.

Royle, D. J. and Hickman, C. J. (1964). *Can. J. Microbiol.* **10**, 151–162.

Ruhland, W. (1902). *Hedwigia* **41**, 179–180.

Sachs, J. (1874). "Lehrbuch der Botanik", Leipzig.

Sacristan, M. D. and Hoffman, F. (1979). *Theor. Appl. Genet.* **54**, 129–132.

Safeeulla, K. M. (1976). "The Biology and Control of the Downy Mildew of Pearl Millet, Sorghum and Finger Millet". Mysore University, Mysore.

Sahtiyanci, S. (1962). *Arch. Mikrobiol.* **41**, 187–228.

Saksena, M. R. K. (1936a). *Rev. gén. de Botanique* **48**, 156–188.

Saksena, M. R. K. (1936b). *Rev. gén. de Botanique* **48**, 215–52.

Saksena, M. R. K. (1936c). *Rev. gén. de Botanique* **48**, 273–313.

Salmon, S. C. (1951). *Pl. Dis. Reptr.* **35**, 251–254.

Samuel, G. and Garrett, S. D. (1945). *Ann. appl. Biol.* **32**, 96–101.

Sansome, E. (1961). *Nature* **191**, 827–828.

Sansome, E. (1963a). *Trans. Br. mycol. Soc.* **46**, 63–72.

Sansome, E. (1963b). *Proc. Sci. Assoc. Nigeria* **6**, 48–52.

Sansome, E. (1965). *Cytologia* **30**, 103–117.

Sansome, E. (1966). *In* "Chromosomes Today, Volume One" (C. D. Darlington and K. R. Lewis, eds), pp. 77–83. Oliver and Boyd, Edinburgh.

Sansome, E. (1970). *Trans. Br. mycol. Soc.* **54**, 101–107.

Sansome, E. (1976). *Can. J. Bot.* **54**, 1535–1545.

Sansome, E. (1977). *J. gen. Microbiol.* **99**, 311–316.

Sansome, E. (1980). *Trans. Br. mycol. Soc.* **74**, 175–185.

Sansome, E. and Brasier, C. M. (1973). *Nature* **241**, 344–345.

Sansome, E. and Brasier, C. M. (1974). *Trans. Br. mycol. Soc.* **63**, 461–467.

Sansome, E. and Harris, B. J. (1962). *Nature* **196**, 291–292.

Sansome, E. and Sansome, F. W. (1974). *Trans. Br. mycol. Soc.* **62**, 323–332.

Sansome, E., Brasier, C. M. and Sansome, F. W. (1979). *Trans. Br. mycol. Soc.* **73**, 293–302.

Sargent, J. A., Ingram, D. S. and Tommerup, I. C. (1977). *Proc. Roy. Soc. Lond. B* **198**, 129–138.

Satour, M. M. and Butler, E. E. (1968). *Phytopathology* **58**, 183–192.

Savage, E. J., Clayton, C. W., Hunter, J. H., Brenneman, J. A., Laviola, C. and Gallegly, M. E. (1968). *Phytopathology* **58**, 1004–1021.

Savage, E. J. and Gallegly, M. E. (1960). *Phytopathology* **50**, 573.

Savory, B. M. (1966). *Res. Rev., Commonw. Bur. Hort. Plantation Crops, E. Malling*, 1.

Savile, D. B. O. (1968). *In* "The Fungi, An Advanced Treatise", Vol. III (G. C. Ainsworth and A. S. Sussman, eds), pp. 649–675. Academic Press, New York and London.

Savile, D. B. O. (1979). *Bot. Rev.* **45**, 377–503.

Scherffel, A. (1925). *Arch. Protistenk.* **52**, 1–141.

Schick, R. (1932). *Züchter* **4**, 233–236.

Schick, R., Schick, E. and Haussdorfer, M. (1958). *Phytopath. Z.* **31**, 225–236.
Schlub, R. L. and Schmitthenner, A. F. (1978). *Phytopathology* **68**, 1186–1191.
Schnepf, E., Deichgräber, G. and Drebes, G. (1978a). *Arch. Microbiol.* **116**, 141–150.
Schnepf, E., Deichgräber, G. and Drebes, G. (1978b). *Protoplasma* **94**, 263–280.
Schnepf, E., Deichgräber, Röderer, G. and Herth, W. (1977). *Protoplasma* **92**, 87–107.
Schwinn, F. J. (1979). *Proc. Br. Crop Prot. Conf.*, pp. 791–802.
Schwinn, F. J., Staub, T. and Urech, P. A. (1977). *Meded. Fac. Landbouww. Rijksuniv. Gent.* **42**, 1181–1188.
Scott, W. W. (1961). *Virginia Agr. Expl. Stat. Tech. Bull.* No. 151.
Sewell, G. W. F. (1981). *Ann. appl. Biol.* **97**, 31–42.
Sharma, B. B. and Wahab, S. (1976). *Indian Phytopathol.* **29**, 81–83.
Shattock, R. C. and Shaw, D. S. (1975). *Trans. Br. mycol. Soc.* **64**, 29–41.
Shattock, R. C. and Shaw, D. S. (1976). *Trans. Br. mycol. Soc.* **67**, 201–206.
Shattock, R. C., Janssen, B. D., Whitbread, R. and Shaw, D. S. (1977). *Ann. appl. Biol.* **86**, 249–260.
Shaw, C. G. (1970). *Indian Phytopath.* **23**, 364–370.
Shaw, C. G. (1978). *Mycologia* **70**, 594–604.
Shaw, D. S. (1983). *In* "*Phytophthora*: its Biology, Taxonomy, Ecology and Pathology" (D. C. Erwin, S. Bartnicki-Garcia and P. H. Tsao, eds). American Phytopathological Society, St. Paul, Minnesota. In press.
Shaw, D. S. and Elliott, C. G. (1968). *J. gen. Microbiol.* **51**, 75–84.
Shaw, D. S. and Khaki, I. A. (1971). *Genet. Res.* **17**, 165–167.
Shaw, R. (1965). *Biochim. Biophys. Acta* **98**, 230–237.
Shear, C. L. and Dodge, B. O. (1927). *J. agric. Res.* **34**, 1019–1042.
Shepard, C. J. (1975). *Search* **6**, 484–490.
Shepard, J. F., Bidney, D. and Shahin, E. (1980). *Science* **208**, 17–24.
Shew, H. D. and Beute, M. K. (1979). *Phytopathology* **69**, 204–207.
Shokes, F. M. and McCarter, S. M. (1979). *Phytopathology* **69**, 510–516.
Short, G. E. and Lacy, M. L. (1976). *Phytopathology* **66**, 188–192.
Shukla, D. D., Shanks, G. J., Teakle, D. S. and Behncken, G. M. (1979). *Aust. J. Biol. Sci.* **32**, 267–276.
Shurtleff, M. C. (1980). "Compendium of Corn Diseases". Am. Phytopath. Soc., Minnesota.
Sideris, C. P. (1931). *Mycologia* **23**, 252–295.
Sideris, C. P. (1932). *Mycologia* **24**, 14–61.
Siegel, M. R. (1977). *In* "Antifungal Compounds" (M. R. Siegel and H. D. Sisler, eds), Vol. II, pp. 399–438. Marcel Dekker, New York.
Siegel, M. R. and Sisler, H. D. (eds) (1977). "Antifungal Compounds", Vol. I and Vol. II. Marcel Dekker, New York.
Sietsma, J. H., Child, J. J., Nesbitt, L. R. and Haskins, R. H. (1975). *J. gen. Microbiol.* **86**, 29–38.
Signoret, P. A., Alliott, B. and Poinso, B. (1977). *Ann. Phytopathol.* **9**, 377–379.
Sinclair, J. B. and Darrag, I. E. M. (1966). *Cotton Dis. Council Proc.* **26**, 108.
Singh, R. S. and Mitchell, J. E. (1961). *Phytopathology* **51**, 440–444.
Singh, V. O. and Bartnicki-Garcia, S. (1975). *J. Cell. Sci.* **18**, 123–132.
Sleigh, M. A. (1982). *BioSystems* **14**, 423–424.
Slifkin, M. K. (1961). *Mycologia* **53**, 183–193.
Slifkin, M. K. (1962). *Mycologia* **54**, 105–106.
Slykhuis, J. T. (1970). *Phytopathology* **60**, 319–331.
Slykhuis, J. T. and Barr, D. J. S. (1978). *Phytopathology* **68**, 639–643.

Smith, J. E. (1955). Host–parasite physiology in relation to club-root disease of crucifers. Doctoral Diss., Purdue University.

Smith, J. M. (1979). *Proc. Br. Crop Prot. Conf.*, pp. 331–339.

Smith, J. .M., Cartwright, J. H. M. and Smith, E. G. (1977). *Proc. Br. Crop Prot. Conf.*, pp. 633–640.

Smith, P. C. and Walker, J. C. (1941). *J. agr. Res.* **63**, 1–20.

Smith, P. M. (1978). *Rep. Glasshouse Crops Res. Inst.* for 1977, pp. 120–121.

Smoot, J. J., Gough, F. J., Lamey, H. A., Eichenmuller, J. J. and Gallegly, M. E. (1958). *Phytopathology* **48**, 165–171.

Sneh, B. (1972). *Can. J. Microbiol.* **18**, 1389–1392.

Sneh, B., Humble, S. J. and Lockwood, J. L. (1977). *Phytopathology* **67**, 622–628.

Southall, M. A., Motta, J. J. and Patterson, G. W. (1977). *Am. J. Bot.* **64**, 246–252.

Southern, J. W., Schenck, N. C. and Mitchell, D. J. (1976). *Phytopathology* **66**, 1380–1385.

Sparrow, F. K. (1931). *Ann. Bot.* **45**, 257–275.

Sparrow, F. K. (1935). *In* "Proceedings of the Zesde International Botany Congress", Vol. 2, pp. 181–185. E. J. Brill, Leiden.

Sparrow, F. K. (1943). "The Aquatic Phycomycetes, Exclusive of the Saprolegniaceae and *Pythium*". University of Michigan Press, Ann Arbor.

Sparrow, F. K. (1947). *Am. J. Bot.* **34**, 94–97.

Sparrow, F. K. (1958). *Mycologia* **50**, 797–813.

Sparrow, F. K. (1960). "The Aquatic Phycomycetes", 2nd edn. Revised. University of Michigan Press, Ann Arbor.

Sparrow, F. K. (1973a). *In* "The Fungi," Vol. IVB (G. C. Ainsworth, F. K. Sparrow and A. S. Sussman, eds), pp. 61–73. Academic Press, New York and London.

Sparrow, F. K. (1973b). *In* "The Fungi," Vol. IVB (G. C. Ainsworth, F. K. Sparrow and A. S. Sussman, eds), pp. 85–110. Academic Press, New York and London.

Spieckermann, A. and Kotthoff, P. (1924). *Deutsche Landw. Pr.* **51**, 114–115.

Spurr, A. R. (1969). *J. Ultrastruct. Res.* **26**, 31–43.

Stamps, D. J. (1953). *Trans. Br. mycol. Soc.* **36**, 255–259.

Stanghellini, M. E. and Hancock, J. G. (1971). *Phytopathology* **61**, 165–168.

Staub, T., Dahmen, H. and Schwinn, F. J. (1978a). *Z. Pflkrankh. Pflschutz.* **85**, 162–168.

Staub, T., Dahmen, H. and Schwinn, F. J. (1978b). *Proc. 3rd Int. Congr. Plant Pathol.*, München, 16–23 August, 1978, p. 366.

Staub, T., Dahmen, H., Urech, P. and Schwinn, F. (1979). *Pl. Dis. Reptr.* **63**, 385–389.

Stephenson, L. W., Erwin, D. C. and Leary, J. V. (1974a). *Phytopathology* **64**, 149–150.

Stephenson, L. W., Erwin, D. C. and Leary, J. V. (1974b). *Can. J. Bot.* **52**, 2141–2143.

Stevens, F. L. (1899a). *Bot. Gaz.* **28**, 149–172.

Stevens, F. L. (1899b). *Bot. Gaz.* **28**, 225–245.

Stevens, R. B. (1974). "Mycology Guidebook". Washington Press, Washington.

Stewart, K. D. and Mattox, K. (1980). *In* "Phytoflagellates" (E. R. Cox, ed.), pp. 433–462. Elsevier North Holland, Amsterdam.

Storck, R. and Alexopoulos, C. J. (1970). *Bacteriol. Rev.* **34**, 126–154.

Strandberg, J. O., Williams, P. H. and Yukawa, Y. (1966). *Phytopathology* **56**, 903.

Strugger, S. (1949). "Fluorescence Microscopy and Microbiology". M. & H. Schapter, Hanover.

Sundheim, L. (1972). *Physiol. Plant Pathol.* **2**, 301–306.

Sutton, J. C. (1975). *Can. J. Plant Sci.* **55**, 139–143.

Takahi, Y., Nakanishi, T. and Kamimura, S. (1974). *Ann. Phytopath. Soc. Japan* **40**, 362–367.

Talley, M. R., Miller, C. E. and Braselton, J. P. (1978). *Mycologia* **70**, 1241–1247.
Tamada, T. (1975). Beet necrotic yellow vein virus. *C.M.I./A.A.B. Descriptions of Plant Viruses* No. 144.
Tamada, T., Abe, H. and Baba, T. (1974). *Proc. 1st Intersectional Congress of the International Association of Microbiological Societies* **3**, 313–320.
Taylor, F. J. R. (1978). *BioSystems* **10**, 67–89.
Taylor, J., Biesbrock, J. A., Hendrix, Jr., F. F., Powell, W. M., Daniell, J. W. and Crosby, F. L. (1970). *Univ. of Ga. Agric. Exp. Stn. Res. Bull.* **77**.
Teakle, D. S. (1962). *Virology* **18**, 224–231.
Teakle, D. S. (1969). *In* "Viruses, Vectors and Vegetation", pp. 23–54. Interscience, New York.
Teakle, D. S. and Gold, A. H. (1963). *Virology* **19**, 310–315.
Teakle, D. S. and Hiruki, C. (1964). *Virology* **24**, 539–544.
Teasdale, J. R., Harvey, R. G. and Hagedorn, D. J. (1979). *Weed Sci.* **27**, 195–201.
Temmink, J. H. M. (1971). *Meded. Landbouwhoogeschool Wageningen* **71**, 1–135.
Temmink, J. H. M. and Campbell, R. N. (1968). *Can. J. Bot.* **46**, 951–956.
Temmink, J. H. M. and Campbell, R. N. (1969). *Can. J. Bot.* **47**, 227–231.
Temmink, J. H. M., Campbell, R. N. and Smith, P. R. (1970). *J. gen. Virol.* **9**, 201–213.
Thakur, S. B. (1977). *Mycologia* **69**, 637–641.
Thirumalachar, M. J. (1969). *Indian Phytopathol.* **22**, 155.
Thirumalachar, M. J. and Whitehead, M. D. (1952). *Am. J. Bot.* **39**, 416–418.
Thirumalachar, M. J., Whitehead, M. D. and Boyle, J. C. (1949). *Bot. Gaz.* **110**, 487–491.
Thomas, D. L., McCoy, R. E., Norris, R. C. and Espinoza, A. S. (1979). *Phytopathology* **69**, 222–226.
Thomas, W. (1973). *N.Z. J. Agric. Res.* **16**, 150–154.
Thomas, W. and Fry, P. R. (1972). *N.Z. J. Agric. Res.* **15**, 857–866.
Thorn, W. A., Zentmyer, G. A. and Wong, P. O. (1959). *Yb. Calif. Avocado Soc.* **43**, 104–106.
Thorold, C. A. (1955). *Trans. Br. mycol. Soc.* **38**, 435–452.
Thouvenel, J. C., Dollet, M. and Fauquet, C. (1976). *Ann. appl. Biol.* **84**, 311–320.
Thouvenel, J. C. and Fauquet, C. (1980). *Plant Disease* **64**, 957–958.
Thouvenel, J. C. and Fauquet, C. (1981). *Ann. appl. Biol.* **97**, 99–107.
Tillman, R. W. and Sisler, H. D. (1973). *Phytopathology* **63**, 219–225.
Timmer, L. W. (1977). *Phytopathology* **67**, 1149–1154.
Timmer, L. W., Castro, J., Erwin, D. C., Belser, W. L. and Zentmyer, G. A. (1970). *Am. J. Bot.* **57**, 1211–1218.
Tinggal, S. H. and Webster, J. (1981). *Trans. Br. mycol. Soc.* **76**, 191–195.
Tisdale, W. H. (1919). *J. Agric. Res.* **16**, 137–154.
Tocchetto, A. (1974). *Agron. Subriograndense* **10**, 227–231.
Todd, F. A. (1955). *N.C. Agr. Exp. Sta. Tech. Bull.* **111**, 16.
Tokunaga, J. and Bartnicki-Garcia, S. (1971). *Arch. Mikrobiol.* **79**, 293–310.
Tokunaga, Y. (1933). *Trans. Sapporo nat. Hist. Soc.* **13**, 25–32.
Tomlinson, J. A. (1958a). *Trans. Br. mycol. Soc.* **41**, 491–498.
Tomlinson, J. A. (1958b). *Ann. appl. Biol.* **46**, 608–621.
Tomlinson, J. A. and Faithfull, E. M. (1979a). *Rep. natn. Veg. Res. Stn.* for 1978, pp. 80–82.
Tomlinson, J. A. and Faithfull, E. M. (1979b). *Ann. appl. Biol.* **93**, 13–19.
Tomlinson, J. A. and Faithfull, E. M. (1979c). *Proc. Br. Crop Prot. Conf.*, pp. 341–346.
Tomlinson, J. A. and Faithfull, E. M. (1980). *Acta Horticulturae* **98**, 325–331.

Tommerup, I. C. and Ingram, D. S. (1971). *New Phytol.* **70**, 327–332.
Tommerup, I. C., Ingram, D. S. and Sargent, J. A. (1974). *Trans. Br. mycol. Soc.* **62**, 145–150.
Tompkins, C. M. (1950). *Hilgardia* **20**, 183–190.
Tompkins, C. M. and Middleton, J. T. (1950). *Hilgardia* **20**, 171–182.
Torgeson, D. C. (ed.) (1967). "Fungicides. An Advanced Treatise". Vol. I and II. Academic Press, New York and London.
Toxopeus, H. J. (1956). *Euphytica* **5**, 221–237.
Trow, A. H. (1899). *Ann. Bot.* **13**, 131–179.
Tsakiridis, J. P., Vasilakakis, Ch. B. and Chrisochou, A. P. (1979). *Pl. Dis. Reptr.* **63**, 63–66.
Tsang, K. C. (1929). *Botaniste* **21**, 1–128.
Tsao, P. H. (1970). *A. Rev. Phytopathol.* **8**, 157–186.
Tsao, P. H. and Guy, S. O. (1977). *Phytopathology* **67**, 796–801.
Tsuchizaki, T., Hibino, H. and Saito, Y. (1975). *Phytopathology* **65**, 523–532.
Tucker, C. M. (1928). *Puerto Rico Ag. Exp. Sta. Rept.* 25–27.
Tucker, C. M. (1931). *Missouri Agri. Expt. Sta. Bull.* 184.
Turner, P. D. (1963). *Phytopathology* **53**, 1337–1339.
Turner, P. D. (1965). *Pl. Dis. Reptr.* **49**, 135–137.
Uesugi, Y. (1978). *Jap. Pestic. Inf.* No. 35, 5–9.
Ulland, B. M., Weisburger, J. H., Weisburger, E. K., Rice, J. M. and Cypher, R. (1972). *J. natl. Cancer Inst.* **49**, 583–584.
Unestam, T. (1966). *Physiologia Pl.* **19**, 15–30.
Upshall, A. (1969). *Can. J. Bot.* **47**, 863–867.
Upstone, M. E. (1976). *Ann. appl. Biol.* **84**, 103–106.
Upstone, M. E. (1977). *Proc. Br. Crop Prot. Conf.*, pp. 197–202.
Urech, P. A., Schwinn, F. J. and Staub, T. (1977). *Proc. Br. Crop Prot. Conf.*, pp. 623–631.
Usugi, T. and Saito, Y. (1976). *Ann. Phytopathol. Soc. Japan* **42**, 12–20.
Uyemoto, J. K., Grogan, R. G. and Wakeman, J. R. (1968). *Virology*, **34**, 410–418.
Uyemoto, J. K. and Gilmer, R. M. (1972). *Phytopathology* **62**, 478–481.
Vaartaja, O. (1965). *Mycologia* **57**, 417–430.
Van der Plank, J. E. (1963). "Plant Diseases: Epidemics and Control". Academic Press, New York and London.
Van der Zaag, D. E. (1956). *Tijdschr. Plantenziekten* **62**, 89–156.
Vanterpool, T. C. and Ledingham, G. A. (1930). *Can. J. Res.* **2**, 171–194.
Vaziri-Tehrani, B. and Dick, M. W. (1980a). *Biochem. System. Ecol.* **8**, 105–108.
Vaziri-Tehrani, B. and Dick, M. W. (1980b). *Trans. Br. mycol. Soc.* **74**, 225–230.
Vaziri-Tehrani, B. and Dick, M. W. (1980c). *Trans. Br. mycol. Soc.* **74**, 231–238.
Vera-Chaston, H. P. de (1977). *Kalikasan* **6**, 183–198.
Vermeulen, H. (1963). *J. Protozool.* **10**, 216–222.
Vickerman, K. (1962). *J. Protozool.* **9**, 26–33.
Vickerman, K. (1976). *In* "Biology of the Kinetoplastida", Vol. 1 (W. H. R. Lumsden and D. A. Evans, eds), pp. 1–34. Academic Press, London, New York and San Francisco.
Vidhyasekaran, P. and Parambaramani, C. (1971). *Indian Phytopathol.* **24**, 305–309.
Vigil, E. L. (1973). *Cell Biochem.* **2**, 237–285.
Vogel, H. J. (1960). *Biochim. Biophys. Acta* **41**, 172–173.
Vogel, H. J. (1964). *Am. Nat.* **98**, 435–446.
Vogel, H. J. (1965). *In* "Evolving Genes and Proteins" (V. Bryson and H. J. Vogel, eds), pp. 25–40. Academic Press, New York.

328 REFERENCES

Vörös, J. (1965). *Phytopath. Z.* **54**, 249–257.
Vo-Thi-Hai, Bompeix, G. and Ravisé, A. (1979). *C.r. hebd. Séanc. Acad. Sci. Paris* **288**, Série D, 1171–1174.
Waard, M. A. de and Gieskes, S. A. (1977). *Neth. J. Pl. Path.* **83** (*Suppl.*), 177–188.
Wager, H. (1900). *Ann. Bot.* **14**, 263.
Wager, V. A. (1931). *Africa Dept. Agric. Bull.* 105.
Wager, W. A. (1940). *Yb. Calif. Avocado Soc.* 30–43.
Walker, J. C. and Larson, R. H. (1960). *Wisc. Agr. Exp. Stn. Bull.* **547**, 12–16.
Wang, H. S. and LéJohn, H. B. (1974a). *Can. J. Microbiol.* **20**, 567–574.
Wang, H. S. and LéJohn, H. B. (1974b). *Can. J. Microbiol.* **20**, 575–580.
Ward, E. W. B., Lazarovits, G., Stössel, P., Barrie, S. D. and Unwin, C. H. (1980). *Phytopathology* **70**, 738–740.
Ware, W. M. (1926). *Trans. Br. mycol. Soc.* **11**, 91–107.
Warren, C. G., Sanders, P. and Cole, Jr., H. (1975). *Phytopathology* **65**, 836.
Watanabe, T., Hashimoto, K. and Sata, M. (1977). *Phytopathology* **67**, 1324–1332.
Waterhouse, G. M. (1956). *Commonw. Mycol. Inst., Misc. Publ.* No. 12.
Waterhouse, G. M. (1967). *Commonw. Mycol. Inst., Mycol. Papers* No. 109.
Waterhouse, G. M. (1968a). *Mycologia* **60**, 976–978.
Waterhouse, G. M. (1968b). *Commonw. Mycol. Inst., Mycol. Papers* No. 110.
Waterhouse, G. M. (1973). *In* "The Fungi", Vol. IVB (G. C. Ainsworth, F. K. Sparrow and A. S. Sussman, eds), pp. 165–183. Academic Press, New York and London.
Waters, H. (1978). *Ann. Appl. Biol.* **90**, 293–302.
Webb, P. C. R. (1949). *Nature* **163**, 608.
Webster, J. (1980). "Introduction to Fungi". 2nd edn. Cambridge University Press, Cambridge.
Webster, R. K. (1974). *A. Rev. Phytopathol.* **12**, 331–353.
Wechmar, M. B. von (1980). *Phytopath. Z.* **99**, 289–293.
Weete, J. D. (1974). "Fungal Lipid Biochemistry." Plenum Press, New York and London.
Wei, C. T., Kung, H. and Pan, R. S. (1955). *Acta Phytopath. Sinica* **1**, 127–140.
Weisaeth, G. (1977). *Gartneryrket* **67**, 937–939.
Weste, G. (1973). *Aust. J. Bot.* **21**, 50–51.
Weste, G. (1978). *Phytopath. Z.* **93**, 41–55.
Westerlund, F. V., Campbell, R. N., Grogan, R. G. and Duniway, J. M. (1978). *Phytopathology* **68**, 927–935.
Wettstein, F., von (1921). *Sitzber. Akad. Wiss. Wien. Math.-Naturw. Kl., Abt.* 1, **130**, 3–20.
Whipps, J. M. and Cooke, R. C. (1978). *Trans. Br. mycol. Soc.* **70**, 285–287.
White, D. G. (1976). *Phytopathology* **66**, 523–525.
White, J. G. (1980). *Pl. Path.* **29**, 124–130.
White, J. G. and Buczacki, S. T. (1977). *Ann. appl. Biol.* **85**, 287–300.
White, J. G. and Buczacki, S. T. (1979). *Trans. Br. mycol. Soc.* **73**, 271–275.
White, R. P. (1930). *Phytopathology* **20**, 131.
Whitehouse, H. L. K. (1949). *New Phytol.* **48**, 212–244.
Whitney, E. D. (1974). *Phytopathology* **64**, 380–383.
Whittaker, R. H. (1969). *Science* **163**, 150–160.
Wiertsema, W. P. and Wissink, G. H. (1977). *Meded. Fac. Landbouww. Rijksuniv. Gent.* **42**, 1189–1194.
Wilde, P. (1961). *Ark. Microbiol.* **40**, 163–195.
Wilhelm, S. (1965). *Phytopathology* **55**, 1016–1020.

Williams, D. J., Beach, B. G. W., Horriére, D. and Marechal, G. (1977). *Proc. Br. Crop Prot. Conf.*, pp. 565–573.

Williams, K. L. (1976). *Nature* **260**, 785–786.

Williams, P. H. and McNabola, S. S. (1967). *Can. J. Bot.* **45**, 1665–1669.

Williams, P. H. and McNabola, S. S. (1970). *Phytopathology* **60**, 1557–1561.

Williams, P. H., Reddy, M. N. and Strandberg, J. O. (1969). *Can. J. Bot.* **47**, 1217–1221.

Williams, P. H. and Yukawa, Y. B. (1967). *Phytopathology* **57**, 682–687.

Wilson, G. W. (1907). *Bull. Torrey Bot. Club.* **34**, 387–416.

Win-Tin and Dick, M. W. (1975). *Arch. Microbiol.* **105**, 283–293.

Wisselingh, C. van (1898). *Jb. wiss. Bot.* **31**, 619–687.

Wolf, F. A. and Wolf, F. T. (1947). "The Fungi". John Wiley and Sons, New York.

Wolfe, J. (1972). *In* "Advances in Cell and Molecular Biology", Vol. 2 (E. J. DuPraw, ed.), pp. 151–192. Academic Press, New York and London.

Wolfe, M. S. (1978). *In* "Plant Disease Epidemiology" (Scott, P. R. and Bainbridge, A. S., eds), pp. 201–207. Blackwell, Oxford.

Wolfe, M. S., Barrett, J. A., Shattock, R. C., Shaw, D. S. and Whitbread, R. (1976). *Ann. appl. Biol.* **82**, 369–374.

Woronin, M. S. (1877). *Trudy S-peterb. Obshch. Estest.* **8**, 169–201.

Worthing, C. R. (ed.) (1979). "The Pesticide Manual: A World Compendium" 6th Ed. Br. Crop. Prot. Council, London.

Wynn, A. R. and Epton, H. A. S. (1979). *Trans. Br. mycol. Soc.* **73**, 255–259.

Yarwood, C. E. (1971). *Pl Dis. Reptr.* **55**, 342–344.

Yarwood, C. E., and Hecht-Poinar, E. (1973). *Phytopathology* **63**, 1111–1115.

Yerkes, W. D. and Shaw, C. G. (1959). *Phytopathology* **49**, 499–507.

Young, P. A. and Morris, H. E. (1927). *Am. J. Bot.* **14**, 551–552.

Yukawa, Y. (1957). *Bull. Fac. Agric. Yamaguchi Univ.* **8**, 649–664.

Yusa, A. (1963). *J. Protozool.* **10**, 253–262.

Yusa, A. (1965). *J. Protozool.* **12**, 51–60.

Zan. K. (1962). *Trans. Br. mycol. Soc.* **45**, 205–221.

Zaumeyer, W. J. (1958). *A. Rev. Microbiol.* **12**, 415–440.

Ziegler, A. W. (1953). *Am. J. Bot.* **40**, 60–66.

Zentmyer, G. A. (1961). *Science* **133**, 1595–1596.

Zentmyer, G. A. (1963). *Phytopathology* **53**, 1383–1386.

Zentmyer, G. A. (1973). *Phytopathology* **63**, 267–272.

Zentmyer, G. A. (1979). *Phytopathology* **69**, 1129–1131.

Zentmyer, G. A. (1980). *Am. Phytopathological Society Monograph* No. 10.

Zentmyer, G. A., Kaosiri, T. and Idosu, G. O. (1977). *Trans. Br. mycol. Soc.* **69**, 329–332.

Zentmyer, G. A. and Klotz, L. J. (1947). *Phytopathology* **37**, 25.

Zentmyer, G. A., Klure, L. J. and Pond, E. C. (1978). *Pl. Dis. Reptr.* **62**, 918–922.

Zentmyer, G. A. and Ohr, H. D. (1978). *Phytopathol. News* **12**, 142–143.

Zentmyer, G. A. and Richards, S. J. (1952). *Phytopathology* **42**, 35–37.

Zoeten, G. A. de, Arny, D. C., Grau, C. R., Saad, S. M. and Gaard, G. (1980). *Phytopathology* **70**, 1019–1022.

Zopf, W. (1894). *Beitr. Physiol. Morph. nied. Organismen* **4**, 43–68.

Index

A

Acaricides, 190
Acetone, fungicide solvent, 206
Achlya, 35, 60–61, 73, 87–88, 90, 104, 119, 222, 253–254, 256, 280, 293
  ambisexualis, 90, 119
  bisexualis, 61, 104
  flagellata, 253–254, 256
  klebsiana, 293
Acquisition
  of virus by thallus, 240–241
  of virus by zoospores, 239–240
Acrasiales, relationships of, based on rRNA, 166–167
Acrasiomycetes, relationships to Plasmodiophoromycetes, 165
Acridine orange *see* Fluorochromes
Acylalanine fungicides, 194–195, 197–198, 200–202, 204, 209, 218, 222–230 *also see* individual fungicide names
Akaryote nucleus, 183 *also see* Meiosis
Albuginaceae, 72
Albugo, 87, 89, 108–109, 209, 223, 293
  candida, 89, 108, 209, 223, 293
  ipomoeae-panduratae, 209
  tragopogonis, 293
Algae
  centrioles, 49
  flagella, 49, 71
  infection by Chytridiomycetes, 78–79
  infection by Plasmodiophoromycetes, 79
  infection by Pythiaceae, 135
  lysine synthesis, 67
  mastigonemes, 48
  oogamous reproduction, 57
  phytopathogenic, 80–81
  relationship to aquatic fungi, 45–47, 82
  relationship to Chytridiomycetes, 168
  sterols, 64
  trichocysts, 79, 168
Allomyces, 34, 64–65, 68, 252, 254, 256
  macrogynous, 64, 68
  reticulatus, 34

Amino acids
  analogues, 119
  control of Pythiaceae by, 208
  ratios between, as taxonomic criterion, 63, 166
Aminoadipic acid (AAA), lysine biosynthesis pathway, 67–68
Ammonium
  as nitrogen source, 60–61
  nitrate, 212
  thiocyanate, 214
Anastomosis, 229
Anisochytridiales, 285
Anisolpidium, 54, 57, 76
  ectocarpii, 57
Antagonism, to Pythiaceae by other organisms, 157, 159
Antheridium
  Peronosporales, 74
  Pythiaceae, as criterion for subdividing, 124–125, 130
  reduction in number of, in Oomycetes, 73
  in selfings, 101–102
Antibiotics
  control of Peronosporaceae by, 199–200, 202, 218–220, 227
  Oomycetes, insensitivity to, 194
  in selective media, 132–133, 252
Aphanomyces, 10, 14, 20, 30, 40, 60–61, 72–73, 208–209, 216–217, 220, 222, 280, 290, 293, 295
  brassicae, 293
  camptostylus, 293
  cladogamus, 293
  cochlioides, 293
  euteiches, 14, 20, 30, 208, 290, 293, 296
  pisci, 296
  raphani, 72, 208, 293
Aplanopsis terrestris, 254
Apodachlya, 62, 90
  brachynema, 90
Appressorium
  formation of, and fungicidal action, 223
Aquatic fungi
  definition, difficulties with, 44

Ascomycetes *also see* Ascomycotina
    possible derivation from algae, 46
    possible derivation from Phyco-
        mycetes, 47
    relationships of, based on rRNA, 166–
        167
Ascomycotina, 44 *also see* Ascomycetes
    cytochromes, 70
    evolution of, based on fatty acid syn-
        thesis, 64
    lysine synthesis, 67
    origin of, based on cell-wall analysis,
        62
    origin of, based on rRNA, 65
    sterols in, 64
    tryptophan synthesis, 68
Asexual reproduction
    advantage of, 55
Aspergillus, 86, 119
ATP *see* Energy
Autotaxis, 35, 37
Auxin *see* Growth regulators
Axoneme, 12
    terminal plate, 21

B

Bacitracin, 213 *also see* Antibiotics
Bacteria
    lysine synthesis, 67
    tryptophan synthesis, 68
Bactericides *see* individual names and
        Antibiotics
Baits
    to isolate Pythiaceae, 132
Basal body *see* Kinetosome
Basidiomycetes *also see* Basidiomycotina
    cytochrome, 70
    possible derivation from algae, 46
    relationships of, based on rRNA, 166–
        167
Basidiomycotina, 44 also see Basidio-
        mycetes
    based on rRNA, 65
    evolution of, based on fatty acid,
        synthesis, 64
    lysine synthesis, 67
    origin of, based on cell-wall analysis,
        62
    sterols, 64
    tryptophan synthesis, 68

Basidiophora, 296
    butleri, 296
    entospora, 296
Beet necrotic yellow-vein virus, 241,
        248
Benfluralin *see* Dinitroaniline herbicides
Benomyl, 159, 189, 208–214, 218
Benzimidazole fungicides, 194, 218,
        227 *also see* individual fungicide
        names
Benzol, as fungicide, 200
Binary fission, 76
Black fruit (pepper), 80
Black pod (cocoa), 66, 205
Black rot (radish), 208
Black shank (tobacco), 140, 154
Blastocladiales
    centriole angle, 49
    cytoplasmic microtubules, 5
    DNA–RNA hybridisation, 65–66
    life cycles, 58
    meiosis, 58
    mitochondria and lipid body, 16
    mitosis, 54–55
    nitrogen requirement, 61
    nuclear cap and ribosomes, 14
    nuclear shape, 10–11
    parasitised by other zoosporic fungi,
        252
    ploidy, 58
    requirement for sulphur, 60
    resting spore development, 58
    rootlets, 51
    zoospore shape, 5
    types 24–25
Blastocladiella, 14, 65
Blepharoplast *see* Kinetosome
Bordeaux mixture, 194–196, 199, 202,
        204–205, 209, 216 *also see* Copper
Boric acid, 210
Botrytis, 197, 199
    squamosa, 197
Bremia, 86, 89, 99, 104, 106, 108–110,
        116–118, 197, 223–224, 232, 296
    lactucae, 89, 99, 104, 106, 108–110,
        116–118, 197, 223–224, 232, 296
Bremiella, 296
Brevilegnia, 280
Brome mosaic virus, 234
Brown spot (maize), 214
Burgundy mixture, 194–195 *also see*
        Copper

C

Calcium cyanamide
  as fungicide, 163, 210, 214
Calomel *see* Mercury salts
Cankers, caused by Pythiaceae, 205
Captafol, 204–206, 208–209
Captan, 196, 199, 203, 205, 207–211, 214, 217
Carbamide, 214
Carbathiin, 214
Carbendazim, 210, 213–214, 218
Carbon disulphide, 211–212
Carboxin, 214
Catenaria, 55
  anqillulae, 55
Cell wall
  algal, 62–63
  cellulose, 62–63
  chitin, 62–63, 165–166
  composition, taxonomy based on, 62–63
  fungal, 62
  host, in relation to infection, 33, 268
  thickening, in response to fungicide, 220
  sporangial, 267–268, 279
Cellulase, 61
Cellulose
  absent from walls of Chytridiomycetes, 63, 168
  of Plasmodiophoromycetes, 165–166
  degradation, 61
  in walls of Ascomycetes, 62–63
    Hyphochytriomycetes, 62
    Oomycetes, 61–63, 73
Centrioles, 48–50
  absent from resting spores, 177, 278
  angle of alignment, 49–50
  function of, 48–49
  non-functional *see* Kinetosome, vestigial
Cephaleuros virescens 80
Ceratocystis, 62–63
  ulmi 62
CGA 29212, 209, 228
Chemotaxis, 35, 37
  of zoospores, to artificial attractants, 35
    hormonal, 34–35
    to plants, 35, 37
  in relation to classification, 38–40

Chemotaxonomy
  zoosporic fungi, 59–70
Chitin
  decomposition, 61
  enhancing seed infection by Pythiaceae, 135–136
  synthesis, evolution of, 63
  synthetase and fungicidal activity, 218
  in walls of algae, 62
    fungi, 61
    Hyphochytriomycetes, 62–63
    Oomycetes, 62–63, 73
    Plasmodiophoromycetes, 165–166, 174
Chitinase, production of, by fungi, 61
Chitosan, absent from walls of Plasmodiophoromycetes, 166
Chlamydomonas, 76
Chlamydospores, in Pythiaceae, 125, 206
Chloramphenicol, 227 *also see* Antibiotics
Chloranil, 203, 208
Chlorochytrium, 80
Chloroneb, 195, 201, 210, 212, 220–221, 227
Chloropicrin, 139, 203, 210–212, 214, 246
Chlorophyta
  fatty acid synthesis and evolution, 64
  phytopathogenic, 81
  rootlets, 50
  transition zone, 50
Chlorothalonil, 195, 196, 198, 204, 209–210, 217
Chorismate *see* Tryptophan
Chromosomes *also see* Meiosis and Mitosis
  alignment in mitosis, 55
  rearrangement and tetraploidy in Peronosporales, 99–101
Chrysophyceae, rootlets, 50
Chrysophyta, fatty acid synthesis and evolution, 64
Chytrid *see* Chytridiales
  Definition, 44
Chytridiaceae
  relationship to Plasmodiophoromycetes, 164
  ribosomes, 14
Chytridiales *also see* Chytridiomycetes
  centriole angle, 49
  chitin degradation, 61

Chytridiales—*cont.*
  cytoplasmic microtubules, 5, 17–18
  mitochrondria and lipid bodies, 16
  nitrogen requirement, 61
  nuclear shape, 11
  nucleus and kinetosomes, 11–13
  parasitic on other fungi, 249, 252–253
  ploidy, 58
  possible origin of, 46–47
  relationship to Oomycetes, 47
    to Plasmodiophorales, 46
    to Zygomycetes, 47
  rhizoplast and rootlet structure, 21
  ribosome arrangement, 14
  rootlets, 51
  virus vectors, 253
  zoospore shape, 5
    size, 5
    types, 22–23
Chytridiomyces, 14, 58, 61
  hyalinus, 58
Chytridiomycetes *also see* individual orders and families
  ability to synthesise vitamins, 60
  asexual reproduction, 55–56
  cellulosic substances degraded by, 61
  centrioles, 49
  chemical control, 212–214, 222
  convergent evolution, 53–54
  cytochromes, 69
  dehydrogenases, 68–69
  evolution, based on fatty acid synthesis, 64
  flagellation, 47–48, 167–168
  groups derived from, based on cell-wall analysis, 62
  lysine synthesis, 67–68
  mitosis, 54–55
  oogamous reproduction, 56
  origin, 65, 76–78, 166–168
  phylogeny and classification, 76–78, 170
  reclassification, 26, 52
  relationship to Plasmodiophoromycetes, 164–168
  resting spores, 58
  rootlets, 51
  sexual reproduction, 58
  sterols, 64
  transition zone, 50
  tryptophan synthesis, 68

  types of thallus development, 53–54
  virus vectors, 212–213, 234–235, 242
  zoospore release, 56
    shape, 5
Chytridium, 14, 285
Chytriomyces hyalinus, 282
Ciliates
  affinity of Oomycetes with, 88
Citric acid cycle, 68–69
Cladochytrium, 14
Classification
  Chytridiomycetes, 76–78
  Hyphochytriomycetes, 75–76
  Oomycetes, 70–75
  phytopathogenic Myxomycota, 80
  Plasmodiophoromycetes, 78–80
  Pythiaceae, 124–131
  zoosporic plant pathogens, 38–40, 81–82
Cleavage
  of plasmodia, 183, 259, 267, 276–279
  in zoosporangia, 174–175, 259
Clubroot (Brassicas), 80, 161–191, 210, 257, 299
Codium, 46
Coelomomyces punctatus, 34
Conidia
  and infection by Peronosporaceae, 200
  Oomycetes, 73
  -producing compared with zoospore-producing organisms, 28–30
  resistance to adverse conditions, 200
Conjugatae, Zygomycetes possibly derived from, 46
Control
  biological
    of Plasmodiophoromycetes, 190–191
    of Pythiaceae, 156–157
  chemical *see* Fungicides and individual fungicide names
  cultural, of Pythiaceae, 155–156
Copper, salts as fungicides, 194–196, 198–199, 202–205, 207, 209, 211, 214, 216, 221
Corticium, 157
Cotton-seed oil, as fungicide amendment, 199
Cotton stunt, 154
Cottony blight (turf), 154
Cottony leak (cucurbits), 154
Coumarin, in selective media, 132

Criconemoides, 138, 144
    quadricornis, 144
Cristae, in mitochondria, types of 15, 77, 276
Cronartium quercuum f.sp. *fusiforme*, 154
Crook-root (watercress), 234, 257, 301
Crop loss, caused by Pythiaceae, 152–154
Cruciform nuclear division, 55, 79, 163, 168, 183 *also see* Mitosis
Cryptophyceae, 71
Cucumber necrosis virus, 236, 239–240, 242
Cyclohexamide, 206
Cymoxanil, 199–200, 204, 209, 218, 221, 226–228
Cyrtolophosis, 51
Cyst *see* Resting spore
Cytochrome
    affected by fungicide, 220
    and taxonomy of lower fungi, 69–70
Cytology, of Peronosporales, 86–87
Cytoplasmically controlled variation, 229

D

Damping-off *see* Rots and Seed and seedling diseases
Dazomet, 210–211
Decline diseases, caused by Pythiaceae, 136–139
22-Dehydrocholesterol *see* Sterols
Dehydrogenases, in relation to fungal phylogeny, 68–69
Deuteromycotina, 44
Dexon *see* *p*-Dimethylaminobenzene-diazo sodium sulphonate
Dialkyldithiocarbamates *see* Dithiocar-bamates and individual fungicide names
Diaminopimelic acid (DAP), lysine bio-synthesis pathway, 67–68
1,2-Dibromo-3-chloropropane, 212
Dicarboximides *see* individual fungicide names
Dichlone, 203
p-Dichlorobenzene, 221
Dichloromethane, as fungicide solvent, 206

Dichloropropane-dichloropropene (D-D), 203, 210–212, 246
Dichlozolin, 210
Dicksonomyces, 297
Dicloran, 221
Dictyomorpha dioica, 253, 256
Dictyosomes, 276–278
Dictyuchus, 280
Difolatan, 209
Dikaryosis, in Plasmodiophoromycetes, 173, 182–183
Dimethirimol, 227
p-Dimethylaminobenzenediazo sodium sulphonate, 158
Dinoflagellates, affinities of Oomycetes with, 88
Dinitramine *see* Dinitroaniline herbi-cides
Dinitroaniline herbicides, 209–210
Dinoseb, 214
Diphenyl, 221
Diplanetism, 71
Diploidy *see* Ploidy
Diplomastigomycotina, 165
Disease prediction, 231
Dithiocarbamate fungicides, 195, 216–217 *also see* individual fungicide names
    control of Chytridiales by, 214
        Peronosporaceae by, 196–202
        Pythiaceae by, 204–205
DNA
    measurements in studies of ploidy, 90–91, 97, 119–120
    in recombinants, 107
    in taxonomic studies, 65–66
Dodine, 209
Downy mildews *see* Peronosporaceae and host plant names
Drainage *also see* Water
    root rot, 206
    to control lettuce big-vein disease, 244
Drazoxolon, 203, 217
Drugs, resistance to in Peronosporales, 96–99, 104, 107, 119

E

Ecology, of Pythiaceae, 49–52
Electron microscope *see* Ultrastructure

Electrophoresis
  for distinguishing species of fungi, 66–
    67
  use of in crosses, 119
Endobiotic growth, 250
Endochitinase *see* Chitinase
Endodesmidium, 78
Endogone, 41
Endoplasmic reticulum, 15, 268
Endothia parasitica, 145
Energy, and zoospores, 31
Entophlyctis, 53, 55, 78
  graminis, 53
  variabilis, 78
Environment
  influence on disease caused by
    Pythiaceae, 143–144
  on zoospores, 25
Enzymes *also see* individual enzyme
    names
  extracellular, test for, 62
  and fungicide activity, 218
  and fungicide insensitivity in
    Oomycetes, 194
  and metabolic pathways, evolutionary
    significance of 67–69
  peculiar to Oomycetes and Hypho-
    chytriomycetes, 194
  treatment of resting spores with, 191
Epibiotic growth, 250
Epidemiology
  zoosporic organisms, 31–34
  zoospore-transmitted viruses, 241–246
L-Ethionine, and fungicide action, 222
Ethirimol, 227
Ethyl alcohol, 211
Ethyl mercaptan, and fungicide activity,
    222
Ethyl phosphonate fungicides, 209, 222,
    226 *also see* individual fungicide
    names
Ethylene dibromide, 212
Ethylene thiourea, degradation product
    of dithiocarbamate fungicides,
    204, 216–217
Etridiazole, 195, 203, 206–208, 213, 220,
    227
Eucarpic development, 250
Euglena, 67–68
Evolution *also see* Phylogeny
  convergent, 53–54
  trends in Oomycetes, 72–73

Exit tubes, zoosporangial, 259, 266
Eye-spot, -like organelle, 25

F

Fatty acids, in cells, as basis of evol-
    utionary scheme, 63–64
Fenaminosulf, 203, 206, 208–211, 213,
    217, 227
Fenarimol, 229
Ferbam, 155, 195, 199, 203, 214, 216
Fertilisers, 208, 211
Fields, artificial infestation of, 188
Flagellar apparatus, 47–48
Flagellates, rootlet system in, 50–51
Flagellum, 3, 47, 268
  length, 5
Flagellation
  difference between organisms with dif-
    ferent types, 26
  types, for separation of groups, 5, 48,
    167–168
  relationship between mono- and uni-
    flagellate groups, 46
  whiplash (smooth or naked) and tinsel
    (flimmer), differences, 48
Fluorochromes
  to assess spore viability, 188
  to induce mutation, 227
  in studies of ploidy, 90–91, 119–120
Folpet, 196, 204–205, 221
Formaldehyde, 203, 210–211, 214, 247
Fosetyl-Al, 195, 198–200, 204, 206–209,
    219, 222, 227, 229–230
Freesia leaf necrosis virus, 239, 247
Frontonia, 79
Fruit rots, caused by Pythiaceae, 204–
    205
Fucosterol *see* Sterols
Fungicides *also see* individual fungicide
    names
  bulb treatment, 203
  to control Albuginaceae, 209
    Chytridiales, 212–214
    Leptomitales, 222
    Peronosporaceae, 196–202, 221–223
    Plasmodiophoromycetes, 163, 189–
      190, 209–212
    Pythiaceae, 154, 158–159, 202–208,
      221
    Saprolegniaceae, 208–209, 222

Fungicides—*cont.*
  curative, 198–199, 223
  dusts, 197, 199, 203
  economic considerations, 196
  effect on non-target organisms, 159
  eradicant, 200–201, 204–205, 207
  furrow application, 201, 203, 223
  granules, 223, 231
  log-P values, 224–225
  phytotoxicity, 196–199, 204, 206, 216, 221, 225
  protectant, 195, 199–202, 204–205, 207, 215, 221, 225, 228, 230
    resistance to, 215–217, 226–227
  residues, 196, 199–200, 204, 216
  resistance to, in
    fungi, 215, 220–221
    general, 226–230
    Peronosporaceae, 199–200, 223
    Pythiaceae, 159, 220, 223
  root-dip, 206, 210, 212
  seed treatments, 195, 201, 203, 206, 208, 212, 217, 221, 223, 231
  side-effects, 225–226
  soil application, 197–198, 202–203, 205–208, 210–211, 214, 217, 220–221, 223, 231
  solubility, 195, 216–217, 222, 224–225
  spray application, 196–199, 201–202, 205–207, 214, 217, 221–222, 231
  sterol-inhibiting, 194, 229
  systemic, 195, 198–201, 203–205, 209–211, 214–215, 220–221, 225, 230
    combined with host plant resistance, 115, 201, 222, 224
    resistance to, 215, 218–224, 226–230
  toxicity, 217
  tuber treatment, 211
  vapour, 200, 223
Furalaxyl, 195, 198–199, 202–203, 206–208, 222–223, 228
Fusarium, 140, 144, 220
  oxysporum, 144
  solani, 140, 144

G

Galben, 222
Gall, biochemistry of development, 187
Gallic acid, in selective media, 133

Gametangium, 3
  definition, 3
  fusion in selfings, 102
  heterozygous, fusion of, 57
  Peronosporales, early ideas on nuclear division in, 87–88
Gametes, 3
  definition, 3
  infection by, 3
Gemma, in Oomycetes, 73
Gene for gene reaction, 92, 111–112, 230
Genetical analysis, Peronosporales, 95–96, 102–103
Germ-tube
  in Peronosporales, 74, 95
  and sporangium development, 52–53
β-Glucan, in cell walls of Plasmodiophoromycetes, 165
Glyceollin *see* Phytoalexin
Glycine, 226
Golgi apparatus, in Oomycete zoospores, 71
Grape downy mildew, 194–195, 199
Growth regulators, produced by Pythiaceae, 142

H

Haploidy *see* Ploidy
Haplomastigomycotina, 165
Harpochytriales, mitochondria and lipid bodies, 16
Hartrot disease (cocoa), 81
Haustoria
  and fungicidal action, 223
  in Oomycetes, 73
  in Peronosporales, 74
Heat
  for fruit treatment, 207
  for soil sterilisation, 203
Herbicides, control of zoosporic pathogens by, 209–210, 214
Heterodera schactii, 144
Heterokaryons, 86, 194, 229
Heteroplasmons, 86
Heterothallism
  evolution, 57
  in Oomycetes, 73
  in Peronosporales, 74
  in Pythiaceae, 126–127, 130–131
Hexachlorobenzene, 221

Holocarpic development, 250
  significance of in Oomycetes, 73
Homothallism
  evolution, 57
  in Peronosporales, 74
  in Pythiaceae, 126–127
  suppressed in Oomycetes, 73
Hop downy mildew, 202
Hormones *also see* Chemotaxis
  produced by zoosporic organisms, 34–35
Host plants
  Algae, 78–79, 249–250
  Anemone, 28
  Apple, 139, 142, 152, 207–208
  Aster, 296
  Avocado, 136, 138, 141, 146, 157–158, 206
  Banana, 302
  Barley, 136, 139 *also see* Cereals
  Bean, 140, 144, 154, 203, 236, 244–245, 302 *also see* Soybean
  Beet, 243, 248 *also see* Sugar beet
  Blueberry, 136
  Brassica, 80, 161, 190 *also see* Crucifers and individual plant names
  Cabbage, 161–162, 224, 242, 257 *also see Brassica* and Crucifers
  Caladium, 136
  Calla, 140
  Camellia, 146
  Carrot, 140, 242
  Cereals, 211, 289–290 *also see* individual plant names
  Chamaecyparis, 206
  Chenopodium, 80, 243
  Chestnut, 145, 148
  Chinese cabbage, 162, 190 *also see Brassica* and Crucifers
  Chrysanthemum, 242
  Cinnamon, 145
  Citrus, 142, 152, 205, 207
  Clover, 136
  Cocoa, 60, 81, 205, 302
  Coconut, 205
  Coffee, 81, 302
  Compositae, 293, 296, 299 *also see* individual plant names
  Corn *see* Maize and Cereals
  Cotton, 136, 139, 150, 154, 157, 220
  Crucifers, 80, 161, 209–210, 212, 257,
    293, 298–299 (non-crucifers, 186)
    also see individual plant names
  Cucumber, 140, 221, 224, 236, 244–245 *also see* Cucurbits
  Cucurbits, 154, 242, 300 *also see* individual plant names
  Echinodorus, 26
  Eucalyptus, 148
  Fir, 147
  Foliage plants, 136
  Fruit, 220 *also see* individual plant names
  Fungi, 249, 250
  Geranium, 140
  Gramineae, 296, 298, 301 *also see* Cereals and individual plant names
  Grapevine, 136, 138, 194, 199, 223, 228
  Grasses, 140, 220, 244 *also see* Gramineae and individual plant names
  Groundnut, 140, 220, 244
  Guayule, 150
  Hop, 136, 202
  Hyacinth, 136
  Ipomoea, 136
  Iris, 136
  Jarrah *see Eucalyptus*
  Juncus, 297
  Lettuce, 27, 30, 104, 116–120, 197–198, 212, 221, 242, 244–246, 296
  Lower plants, 249–283
  Maize, 136, 196–197, 200–201, 214, 289, 298
  Melon, 150
  Millet, finger, 28
    pearl, 30
  Muskmelon, 136
  Oat, 257 *also see* Cereals
  Oil palm, 81
  Onion, 198–199, 298
  Ornamental plants (general), 220
  Papaya, 205
  Pea, 136, 144, 157, 159, 203, 208–209, 296
  Peach, 136, 138, 152, 154
  Peanut *see* Groundnut
  Pecan, 136, 144, 152
  Pelargonium, 140, 158
  Pepper, 80, 149
  Pine, 35, 76, 131, 136–138, 146–147, 150, 152, 154, 157
  Poinsettia, 140, 155

Host plants—*cont.*
Potato, 29, 32, 80, 91–93, 111–116, 129, 140, 195, 204, 211, 224, 227, 234, 246, 257, 288, 301–302
Pythium, 250, 280
Radish, 208, 296
Ranunculaceae, 299
Rape, 136 *also see Brassica* and Crucifers
Rhododendron, 141, 145, 151, 153, 206
Rice, 293 *also see* Cereals
Safflower, 136
Sorghum, 200–201, 244, 298 *also see* Cereals
Soybean, 153, 206
Spinach, 234, 243
Spruce, 136
Strawberry, 136, 143, 206
Sugar beet, 80, 144, 203, 208, 211–212, 257, 296 *also see* Beet
Sugar-cane, 136, 154, 298
Sunflower, 202
Sweet potato, 209
Taro, 140
Tea, 80
Tobacco, 140, 142, 144, 154–156, 199–200, 223, 236, 245–246, 298
Tomato, 136, 149–150, 204, 220, 222–223, 228, 244, 289
Tulip, 236, 244
Turf, 141, 152, 154, 220 *also see* Grasses and Gramineae
Turnip, 161, 190 *also see Brassica* and Crucifers
Umbelliferae, 299
Vegetables (general), 220
Watercress, 211, 234, 243, 257, 301 *also see* Crucifers
Watermelon, 140 *also see* Cucurbits
Wheat, 136, 139, 141, 244, 246, 257 *also see* Cereals
Zostera, 80
Host range, specialisation in, Oomycetes, 73
Hydromyxomycetes, relationship to Plasmodiophoromycetes, 165
Hydroponic culture, significance of Pythiaceae in, 203
Hydroxypyrimidine fungicides, 194 *also see* individual fungicide names
Hymexazol, 195, 208–209, 220, 227
Hyphae, Oomycetes, 72–74

Hyphochytriaceae, 285
Hyphochytriales, 47 *also see* Hypho-chytriomycetes
nuclear shape, 11
nucleus and kinetosomes, 11–13
parasitic on other fungi, 250
Hypochytridiomycetes *see* Hyphochy-triomycetes
Hyphochytriomycetes, 48
asexual reproduction, 55–56
cellulose degradation, 61
centrioles, 49
chemical control, 223, 226
convergent evolution with Oomycetes and Chytridiomycetes, 53–54
dehydrogenases, 69
flagellation, 47
mitosis, 54–55
name, validated, 285
nutrition, 60
origin and relationships of, based on rRNA, 65, 166–167
as parasites of Oomycetes, 76, 226
phylogeny and classification, 45, 75–76
relationship to Plasmodiophoro-mycetes, 164–165
role of, in soil, 45
rootlets, 51
sexual reproduction, lack of conclus-ive evidence for, 57–58
transition zone, 50
zoospore release, 56
Hyphochytrium, 21, 45, 60–61, 76, 285
catenoides, 21, 45, 60–61, 76

I

IAA *see* Growth regulators
Immunodiffusion *see* Serology
Immunofluorescence *see* Serology
Infection/penetration, 33–34
artificial, obtaining, 254–256
host resistance to, 33–34
types of host response to, 39
in relation to available water, 29
seed-borne, 202
systemic, 201–202
types, in relation to classification, 38–40
by zoospores compared with conidia or hyphae, 200, 202

Inoculum
  longevity of, 33
  methods for maximising production, 32
  relationship between, and amount of fungicide to control Pythiaceae, 158
Insecticides, 190
Interactions, between zoosporic and other organisms, 144
Interface, host-parasite, 174–175, 179–180, 266–267
Iprodione, 221
Isolation, of Pythiaceae, 131–133
Isopropalin *see* Dinitroaniline herbicides

J

Java downy mildew, 201

K

Karlingia, 78
Karyogamy
  in Peronosporales, 98
  in Plasmodiophoromycetes, 59, 183
Kasugamycin, 227 *also see* Antibiotics
Kinetosomes, 47–50, 268
  angle between, 21, 71, 268
  characters associated with, 20–21
  location in relation to nucleus, 10, 12–13
  and microtubules, 17
  props, 26, 50
  vestigial, 26, 49, 167
    in Chytridiomycetes, 167
    in Hyphochytriomycetes, 75, 167
Kitazin, 214 *also see* Antibiotics
Kochiomyces, 77

L

Labyrinthula, 80
Lagena, 57, 74, 297
  radicicola, 57, 297
Lagenidiales, 72
  derivation of, 74
  meiosis, 57
  parasitic on other fungi, 250, 252
  relationship to Plasmodiophoromycetes, 164
  zoospore discharge in, 74

Lagenidium, 10, 14, 39–40, 57, 72
  callinectes, 57
Lagenisma, 10, 14, 57
  coscinodisci, 57
Lagenocystis, 297
Lateral groove, 5
Leptomitaceae, parasitised by zoosporic fungi, 251
Leptomitales, 44, 72
  chitin in cell walls, 73–74
  evolution of, 73–74
Leptomitus, 62
Lettuce big vein disease, 27, 30, 212, 239, 241–242, 244–247
Lettuce downy mildew, 116–120, 197–198
Ligniera, 27, 79, 297, 301
  junci, 297
  pilorum, 79, 301
Lime, as fungicide, 194–195, 210–211
α- and γ-Linolenic acid *see* Fatty acids
Lipid, in walls of Plasmodiophoromycetes, 165, 174
Lipid body
  function of, 16–17
  location of, 13, 73
Lipoid globules, 258–259, 268, 276
Lower fungi, definition, 44
LS 73–1038 *see* Sodium ethyl phosphonate
Lysine, biosynthesis of, evolutionary significance, 67–69, 166

M

Maize downy mildew, 200–201
Maize white line mosaic virus, 247
Mancozeb, 195, 204, 207, 209, 217, 221, 224
Maneb, 196–197, 209, 216
Manganese, salts as fungicides, 195
Manure, 212
Mastigomycotina, 44
  dehydrogenases in, 68
Mastigomycota, 165
Mastigonemes, 48
Mating type
  inheritance of, in Peronosporales, 107–111
  in Pythiaceae, 130–131, 146–147
ME135, 208

Meiosis, 3
  in Oomycetes, 57
  in Peronosporales, 88–91, 97–101, 108
  in Plasmodiophoromycetes, 183, 186, 276–277, 279
Meiosporangium, 3
Meiospores, 3
Meloidogyne, 144
  incognita, 144
Membranosorus, 297
Mercury salts, as fungicides, 163, 203, 210–212, 215, 227, 247
Metalaxyl, 195, 197–201, 206–209, 213, 222–225, 228–229, 232, 246
Metham sodium, 203, 210
L-Methionine, antagonising fungicides, 222
Methyl benzimadazole-2-yl-carbamate, 163
Methyl bromide, 153, 159, 163, 203, 210, 212, 246
24-Methylene cholesterol *see* Sterols
Methyl isothiocyanate, 210, 212, 246
N-Methyl-*N*′-nitro-*N*-nitrosoguanidine, 227
Metiram, 195, 209, 217
Microbody, and microbody–lipid body complex, 16–17
Micromyces, 78
Microtubules, 17–20, 268
  axonemal, 17
  cytoplasmic, 5, 12, 17–18, 20
  in flagella and cilia, 47–48
  and fungicide activity, 218
  kinetosomal, 17
  protein synthesis, 25
  rootlet systems,
    in Oomycetes, 71
    in Plasmodiophoromycetes, 168
    in protozoans, 168
  spindle, 54
Milfuram, 195, 199, 222, 228
Mites
  associated with infection by Pythiaceae, 140
  as virus vectors, 236
Mitochondria
  effect of fungicide on, 220–221
  formation of, 15
  location of, 13, 15, 268
  types of, 15, 77

Mitosis *also see* Cruciform nuclear division
  crossing over at, in Peronosporales, 98, 103–105, 108, 110
  effect of fungicides on, 218, 221
  in Oomycetes, 72
  in Plasmodiophoromycetes, 183
  in zoosporic fungi, 54–55
Moisture *see* Water
Monoalkyldithiocarbamates *see* Dithiocarbamates and individual fungicide names
Monoblepharidales
  cytoplasmic microtubules, 17
  mitochondria and lipid bodies, 16
  nitrogen requirement, 61
  nuclear shape, 11
  nucleus and kinetosomes, 11–13
  oogamous reproduction, 57
  rhizoplast and rootlet structures, 21
  ribosome arrangement, 14
  rootlets, 51
  zoospore types, 24–25
Monoblepharis, 35, 46
Monocentric development, 250
Mucor, 220–221
  mucedo, 220–221
Mutants
  characteristics of, resistant, 229–230
  and heterozygosity in Peronosporales, 98–99
  induced in Peronosporales, 96–98
  as origin of asexual variation, 229
  production of in Peronosporales, 96–98
Mycetozoa
  relationship to Oomycetes, 88
  relationship to Plasmodiophoromycetes, 170
Mycorrhiza
  endotrophic, developed by zoosporic parasites, 41
  preventing infection by zoospores, 151, 157
Myxamoebae, in Plasmodiophoromycetes, 164, 182
Myxomycetes
  relationships of, based on rRNA, 166–167
  relationship to Plasmodiophoromycetes, 46, 163–168

Myxomycota
  classification and phylogeny of phyto-
    pathogenic forms, 80
  dehydrogenases in, 68
  Plasmodiophoromycetes classified in,
    165

N

Nabam, 195, 216
NAD/NADP *see* Dehydrogenases
Naphthalene, 221
Napropamid *see* Herbicides
Neem oil, as fungicide, 201
Nematicides, 156, 163, 190
Nematodes, 138, 144, 156, 242
Nepoviruses, 242
Neurospora, 86, 88, 91, 119
Nitralin *see* Dinitroaniline herbicides
Nitraphen, 214
Nitrate, as nitrogen source, 60
Nitrogen
  from ammonium and nitrate, 60
  related to little-leaf disease, 138
  requirement for in different groups,
    60–61, 73
4-Nitrosopyrazoles, 225
Non-parasitic pathogens, 142–143
Nowakowskiella, 14
Nuclear cap
  function of, 25
  ribosomes in, 14
Nuclear envelope, in mitosis, 54
Nucleic acids
  inhibition by fungicides, 220–221, 223
  taxonomic value of, in various groups,
    65–66
  in viruses, 236, 238–239
Nucleolus, 276
  disappearance of at onset of sporo-
    genesis, 177, 182–183, 259, 276
  in mitosis, 54–55
Nucleus
  in cystosoral plasmodia, 174, 183,
    258–259
  location in zoospores, 9–13
  migration into germ tube, 52–53
  shape, 9–13, 276
  volume, in cystosoral plasmodia, 183
Nutrition, absorptive and holozoic, sig-
    nificance of, 168–171, 174
Nystatin *see* Antibiotics

O

Ochromonas, 51, 71
Octomyxa, 79, 165, 171, 280
  achlyae, 79, 280
Oedogonium, 46
Olpidiaceae, 13, 46
Olpidiopsaceae, 46
Olpidiopsis, 54, 252, 254, 256, 282
  incrassata, 252
  varians, 254
Olpidium, 3, 9, 12–13, 15, 18, 21, 26–27,
    30, 32, 34, 35, 39, 40, 53–54, 58, 62,
    78, 175, 210, 212–213, 216, 218,
    234–236, 239–242, 244, 246–247,
    288, 297
  brassicae, 9, 12, 15, 18, 21, 26–27, 30,
    32, 40, 53, 58, 62, 78, 175, 212–213,
    216, 234–236, 239, 242, 244, 246–
    247, 288, 297
  cucurbitacearum, 58, 234, 242
  pendulum, 26
  radicale, 9, 12, 18, 26, 32, 58, 78, 234–
    235, 242, 247, 297
  trifolii, 32, 35
  viciae, 32, 35, 58
Onion downy mildew, 197–198
Onion leaf blight, 197
Oogamous reproduction
  in Chytridiomycetes, 56
  in Oomycetes, 56
Oogonium, 56
  abortion, 99
  as criterion for subdivisions to
    Pythiaceae, 125–130
  in selfings, 101–102
Oomycetes, 44, 165
  ability to synthesise vitamins, 60
  cell walls, 72
    in relation to chemical control,
      194
  cellulose in walls of, 61
  centrioles, 49
  chemical control of (general), 193–209,
    218–221, 230–231 *also see*
    Fungicides and individual fungi-
    cide names
  convergent evolution with
    Hypochytriomycetes and
    Chytridiomycetes, 53–54
  cytochromes, 69
  cytoplasmic microtubules, 18–20

Oomycetes—*cont.*
  dehydrogenases, 69
  ecological origin of, 72
  evolution based on fatty acid syn-
      thesis, 64
    from heterokont algae, 72
  evolutionary adaptation, 53
    trends, 73
  flagellation, 47–48
  hyphae, 72–74
  meiosis, 57
  mitosis, 54–55, 72
  nitrogen requirement, 60
  nucleus, 9–11
  nutrition, 72
  oogamous reproduction, 56–57, 72
  origin and relationships, based on
      rRNA, 65, 166–167
  phylogeny and classification, 70–75
  possible origin of, 46
  relationship to Chytridiomycetes, 47
      Hypochytriomycetes, 45
  ribosomes, 14
  rootlets, 51
  sexual reproduction, 56–57, 72
  transition zone, 50
  zoospore size, 9
    types, 25
Oosphere, 56
Oosporae (Oedogoniales), Oomycetes
    possibly derived from, 46
Oospore, 57, 206 *also see* Oomycetes and
    sub-groups
  abortion, 99
  as criterion for subdivision in
      Pythiaceae, 125–130
  decrease in number in Oomycetes, 73
  in diseased tissue, 139
  parasitised by Hypochytriomycetes,
      226
Oosporogenesis, in Oomycetes, 73
Operculum, in Chytridiales, importance
    of, 55–56
Organotin fungicides, 204
Osmotic pressure, of zoospores, 22
Oxathiin fungicides, 194, 218 *also see*
    individual fungicide names
Oxycarboxin, 214, 227

P

Paramecium, 79
Parasexual cycle, 105–107
Parasitism
  change from saprophytism in
      Oomycetes, 73
  importance in classification of
      Plasmodiophoromycetes, 164, 171
PCNB (pentachloronitrobenzene) *see*
    Quintozene
Peanut clump virus, 242, 244, 246
Penetration, into host, 22
Peronophythora, 296, 298
  litchii, 298
Peronoplasmopara, 298
Peronosclerospora, 45, 74–75, 196–197,
    200–202, 297–298, 301
  maydis, 201, 298
  philippensis, 197
  sacchari, 196, 201, 298
  sorghi, 201–202, 297–298
Peronospora, 27, 45, 75, 86–87, 89, 108–
    109, 129, 197–200, 224, 229–230,
    298
  destructor, 197–198, 298
  parasitica, 89, 108–109, 218, 224, 298
  pisi, 27
  tabacina, 199–200, 218, 229–230, 298
Peronosporaceae, 72
  chemical control, 196–202, 216–217,
      222–223
  differentiated from Pythiaceae, 74–75
  environmental factors affecting, 196
  meiosis, 57
  relationship to Plasmodiophoro-
      mycetes, 164
Peronosporales *also see* Albuginaceae,
    Peronosporaceae, Pythiaceae
  abortion in oogonia and oospores, 99
  antheridia, 74
  chromosome rearrangements and
      tetraploidy, 99–101
  cytological difficulties, 86–87
  dehydrogenases, 69
  drug resistance, 96–99, 104, 107, 119
  evolution, 73–74
  gametangial nuclear division, early
      opinions on, 87–88
  genetics, 91–121
  karyogamy, 98
  mastigonemes, 48

Peronosporales—*cont.*
  mating type, inheritance, 107–111
  meiosis, 3, 88–91, 94–101
  mitosis, crossing over at, 103–105, 108, 110
  mutants, 96–98, 227–230
  physiologic race, 92–96, 106
  ploidy, 3, 88–91, 94–101
  problems of working with, 86
  selfing, 101–103
  somatic recombination, 105–107
  sporangia, 74
  zoospores, 74, 86, 109–110
pH
  effect on Plasmodiophoromycetes, 163, 211, 246
  possible effect on Pythiaceae, 155–156
  influence on fungicides, 222
Phaeophyceae
  sterols in, 64
  transition zone, 50
Phenolics, 222
M-1-phenylthiosemicarbazide
  and virus inoculation, 238
Philippine downy mildew, 197
Phlyctidiaceae *see* Spizellomycetaceae
Phlyctochytrium, 12, 21, 25, 55
  arcticum, 12, 21
  irregulare, 55
Phospholipases
  in mitochondria, 220
Phototaxis, 25, 34, 37 *also see* Rumposome
Phycomycetes
  Ascomycetes possibly derived from, 47
  biflagellate origin of, 26
  lateral groove, 5
  monophyletic origin of, 46–47
  Plasmodiophorales in, 46
  polyphyletic origin of, 27, 46–47
  relationship to Plasmodiophoromycetes, 164
  taxonomic status of name, 44
  zoospore shape, 5
Phyllosiphon, 81
Phylogeny
  of Chytridiomycetes, 76–78
  of Hyphochytriomycetes, 75–76
  of Oomycetes, 70–75
  of phytopathogenic Myxomycota, 80
  of Plasmodiophoromycetes, 78–80
  of zoosporic fungi, 25–26

Physiologic race
  Chytridiales, 242
  Peronosporales, 92–96, 106, 111–118, 229
  Plasmodiophoromycetes, 189, 243, 280–281
  Pythiaceae, 139
Physiology, of zoosporic fungi, in relation to phylogeny, 60–62
Physoderma, 3, 11, 14, 17, 25, 29, 32, 39–40, 53, 58, 77, 210, 212, 214, 218, 289, 298–299, 302
  maydis, 3, 25, 29, 58, 214, 289, 299
Phytoalexins, 222, 224
Phytomonas, 81
  elmassiana, 81
Phytopathogen, most primitive, 73
Phytophthora, 9–10, 14, 20–22, 28, 30, 33–35, 38, 40–41, 56–57, 60–67, 72, 74, 87–99, 101–112, 114–115, 117–119, 124–125, 129–134, 136–138, 140–152, 155–160, 193, 204–208, 216–218, 220–224, 227–230, 280, 288, 290, 296, 299, 301
  cactorum, 33, 57, 66–67, 94, 97–98, 134, 141–142, 151, 205, 227
  cambivora, 34, 67
  capsici, 89, 94, 96, 98, 227–229
  cinnamomi, 35, 61, 66–67, 90, 130–131, 134, 137–138, 141, 145–148, 150–151, 155, 206, 221
  citricola, 141–142, 151
  citrophthora, 35, 66, 207
  cryptogea, 67, 142–143
  drechsleri, 57, 67, 89–91, 94, 96–97, 99, 101–102, 104, 106, 108–109, 112, 114, 119, 150, 227
  erythroseptica, 67, 99
  fragariae, 60, 206
  heveae, 151
  hibernalis, 207
  infestans, 30, 35, 74, 89–90, 92–96, 101, 103, 105–107, 111–112, 114–115, 117–118, 129–130, 193, 204, 218, 223–224, 229–230
  macrospora, 301
  megasperma, 30, 66, 101
    f.sp. *medicaginis*, 99, 224, 228
    var. *megasperma*, 90, 101
    var. *sojae*, 90, 98, 101, 105, 206, 224, 227

Phytophthora—cont.
  nicotianae var. *parasitica*, 14, 20–21, 89, 134, 140–141, 151, 155–156, 205, 207
  palmivora, 14, 22, 33, 35, 66–67, 112, 205
  syringae, 35, 97, 207
Pine little-leaf disease, 131, 137–138, 146–147, 150, 154, 156
Plasmalemma, 17–18, 22, 175, 240
Plasmodia
  cleavage, 183, 276–279
  cystosoral, 59, 174, 266, 276–279
  encystment, 183
  fusion, 183
  ingestion by, 169–171
  isolation of, 187
  sporangial, 79, 179–180, 258–259, 266–268
  spread through host tissues, 183
Plasmodiophora brassicae, 3, 39–40, 49, 54, 59, 64–65, 79–80, 161–166, 168–171, 173–175, 179, 182–183, 186–191, 209–210, 213, 216, 218, 227, 257, 266–268, 281–282, 290, 299
Plasmodiophorales *also see* Plasmodio-phoromycetes
  cytology of sporangial development, 53
  parasitic on other fungi, 250, 253–283
  relationship to other organisms, 26, 46–47, 163–164
Plasmodiophorid *see* Plasmodiophoro-mycetes
  definition, 164
Plasmodiophoromycetes *also see* Plasmo-diophorales
  akaryote state, 183
  as possible ancestors of Chytridio-mycetes, 168
  biological control, 189–190
  cell wall composition, 165–166
  centrioles, 49, 177
  chemical control, 163, 189–190, 209–212
  chromosomes, 183, 186
    number, 186
  culture, on agar, 171, 187
    in host callus, 187
  difficulties of studying, 187–189
  host plant resistance, 162, 189–190
  importance as plant pathogens, 161–162
  karyogamy, 59, 183
  life history, 171–186
  lysine synthesis, 166
  mitosis, 54–55
  meiosis, 59, 183
  nutrition, type of, significance, 169–171
  parasitic on other fungi, 250, 253–283
  effect of pH on, 163
  phylogeny and classification, 78–80, 163–171
  plasmodium, cystosoral, 59, 169–171, 183, 186
    development and structure, 173–178
    zoosporangial, 179–182
  ploidy, 59, 171, 173, 183
  relationship to Protozoa, 165, 167
  rRNA molecular weight, 166–167
  resting spore, 59
    counting, 188
    development and structure, 173–178
    germination, 177–178
    wall composition, 165–166
    viability, 188
  sexual reproduction, 59
  sterols, 64
  as virus vectors, 233–235, 243, 246
  zoospores
    centrioles, 49
    encystment, 178–179
    flagellation, 47
    function of secondary, 182
    rootlets, 51–52
    transition zone, 50
Plasmogamy, possible, in Plasmodio-phoromycetes, 182
Plasmopara, 27, 32, 35, 40, 75, 87, 194, 199, 202, 223, 228, 289, 296, 299–300
  halstedii, 202, 299
  viticola, 35, 194, 199, 223, 228, 299
Ploidy
  in Blastocladiales, 58
  in Chytridiales, 58
  in Lagenidiales, 57
  in Oomycetes, 194
  in Peronosporales, 88–91, 94–101
  in Plasmodiophoromycetes, 59, 171, 173, 183
  in Spizellomycetales, 58
  in zoosporic organisms, 3

Polycentric development, 53, 250
Polymyxa, 27, 39, 49, 50–51, 59, 79–80,
    168, 179, 209, 211–212, 234–236,
    241–248, 257, 266–268, 279, 282,
    297, 300
    betae, 39, 59, 79–80, 179, 212, 234–
        235, 241, 245–246, 248, 257, 266–
        268, 279, 282, 300
      f.sp. *amaranthi*, 243
    graminis, 49–51, 59, 234, 242, 244–246,
        257, 300
Polyoxin, 227
Polyphagus euglenae, 34
Polyram, 209
Post-harvest diseases, caused by Pythi-
    aceae, 207–208
Potassium azide, 201
Potato blight, 91–93, 111–116, 129, 204
Potato mop-top virus, 211, 238, 240, 242,
    244, 246
Potato powdery scab, 80, 211, 234, 257,
    301
Potato virus X, 213, 234
Potato wart disease, 32, 213–214, 288,
    302
Poterioochromonas, 51
Potyviruses, 238
Powdery mildew
    possibly increased by acylalanine
      fungicides, 225
    spores as virus vectors, 234
Procymidone, 221
Proline *see* Amino acids
Propamocarb, 195, 198, 200, 204, 206,
    209, 218, 221, 227
Protein
    in cell walls of Plasmodiophoro-
      mycetes, 165, 174, 191
    and fungicide sensitivity in Oomycetes,
      194, 220, 222
    'pin-wheel' inclusion bodies, 238
    synthesis in zoospores, 31
Prothiocarb, 195, 198, 203–204, 206, 209,
    212, 218, 221, 227–228
Protista, 170
Protococcideae, Oomycetes possibly de-
    rived from, 46
Protozoa
    fungi possibly derived from, 46
    lack of cruciform nuclear division in,
      168
    phytopathogenic, 81

relationships of, based on rRNA, 167
relationship to Plasmodiophoro-
    mycetes, 165
rootlet system in, 51
Pseudomonas, 138
    syringae, 138
Pseudoperonospora, 27, 32, 35, 40, 202,
    218, 228, 230, 289, 298, 300
    cubensis, 35, 228, 230, 300
    humuli, 35, 202, 218, 300
Pyrimidine *see* Amino acids
Pyroxychlor, 159, 203, 213
Pythiaceae
    antheridium, 124–125, 130
    causing crop loss, 152–154
    causing feeder root infection and de-
      cline diseases, 136–139
    causing foliage, stem, crown and fruit
      rots, 140–142
    chlamydospores, 125
    control, biological, 156–157
      chemical, 158–159, 202–208
      cultural, 155–156
    criteria for subdividing, 125–131
    differentiated from Peronosporaceae,
      74, 129
    distribution and redistribution, 144–
      148
    ecology, 149–152
    environmental influence on diseases
      caused by, 143–144
    heterothallism, 126–127, 130–131
    homothallism, 126–127
    infecting algae and fungi, 134–135
      seeds and seedlings, 135–136
    interaction with other pathogens, 144
    isolation of, 131–133
    mycology, 124–131
    non-parasitic pathogens, 142–143
    oogonium, 124–130
    oospore, 124–129
    parasitised by other zoosporic fungi,
      251
    production of growth regulators by,
      142–143
    toxins produced by, 142
    zoospore, 123–125, 128–129, 131, 136,
      139
Pythiopsis, 222
Pythium, 9–10, 14, 20, 22, 28, 32–33, 35,
    38–41, 45–46, 56–57, 60–64, 66–68,
    72, 74, 87–90, 110–111, 124–130,

Pythium—cont.
132–145, 147–160, 202–203, 206, 216–217, 220–224, 228, 242, 250, 253–255, 257–258, 278, 280–283, 288, 290, 297, 300
acanthicum, 135, 280
afertile, 35
aphanidermatum, 14, 22, 33, 35, 39, 57, 128, 132, 140–141, 144, 152, 159, 280
aristosporum, 67, 280
arrhenomanes, 134, 139, 141, 280
butleri, 128
catenulatum, 128, 141, 152
coloratum, 128
compectens, 140
debaryanum, 39, 90, 126–128, 140–141, 152, 203, 280
deliense, 140
dissotocum, 128, 280
echinulatum, 90
graminicola, 67, 141, 144
heterothallicum, 127
intermedium, 127
irregulare, 126–128, 136, 139, 142–143, 152, 155, 159, 280
iwayamai, 141, 280
mamillatum, 152
middletoni, 152
monospermum, 134
multisporum, 90
myriotylum, 136, 140, 142, 150
okanoganense, 280
paraoecandrum, 142
periplocum, 152
perniciosum, 140
proliferum, 14, 280
pulchrum, 280
spinosum, 152, 280
splendens, 140, 158
sulcatum, 140
sylvaticum, 110–111, 126–128, 136, 139, 142, 152, 159
torulosum, 90, 141, 280
ultimum, 90, 128, 135–136, 141–144, 150, 152, 155–157, 159, 228, 280
vanterpoolii, 141
vexans, 60, 143, 152

Q

Quintozene, 133, 163, 203, 209–212, 216, 221

R

RE26745, 198, 222
RE26940, 222–223
Recognition, between zoospore and plant, 25
Recombination, somatic, in Peronosporales, 105–107
Replant disease, 139
Resistance
to drugs, in Peronosporales, 96–99, 104, 107, 119
to fungicides, general, 226–230
in Peronosporaceae, 199–200
in Pythiaceae, 159
genetics of, in Peronosporales, 92–94
by host to infection/disease, 33–34, 111–118, 156–158, 162, 197, 201, 212, 231
Respiration, inhibited by fungicide, 220
Resting spore
Blastocladiales, 58
Chytridiomycetes, 58
culture, 58
epibiotic, possibly following rhizoid fusion, 58
Plasmodiophoromycetes
absence of centrioles, 177
development, 169, 173–178, 277–278
exit pore, 177
germination, 178
internal structure, 177–178
numbers, methods of counting, 188
in soil, longevity, 214
variability, 281, 282
viability, methods of assessing, 188
Rhizidiomyces, 45, 54–55, 62, 65
apophysanus, 55
Rhizoctonia, 144, 217, 220
solani, 144
Rhizoids, 32
fusion, 58
and sporangium development, 52–53
types, 250
Rhizomania, 248
Rhizophlyctis, 12, 21, 64, 68
rosea, 12, 21, 64, 68
Rhizophydium, 14, 52, 55, 58, 78, 300
graminis, 52, 58, 78, 300
sphaerothecae, 55
Rhizoplast, 12, 20–21

Rhodochytrium, 80
Rhysotheca, 299–300
Ribosomes, 268
    arrangement of, 13–15
    cytoplasmic RNA, 65
    and action of streptomycin, 219–220
Ridomil, 27
RNA
    enzymes involved in synthesis, 194
    hybridisation with DNA, 120
    messenger, 31
    ribosomal, in taxonomic studies, 65,
        166–167
Root
    extracts, 131
    exudates, 135–136
    feeder, infection of by Pythiaceae,
        136–139
Root rot, 208
Rootlets, 20–21, 47
    in Chytridiomycete zoospores, 77
    in Hyphochytriomycete zoospores, 75
    in Oomycete zoospores, 71
    in Plasmodiophoromycete zoospores,
        79
    phyletic significance of, 50–52
Rose bengal, in selective media, 133
Rots, caused by Pythiaceae, 134, 140–
    142, 205–207
Rozella, 13, 21, 252, 254, 256
    achlyae, 252
    allomycis, 13, 21, 252
Rumposome
    and kinetosome, 17
    and light sensitivity, 17–18
Rusts, spores as virus vectors, 234
Rusty root disease (carrot), 242

S

Saprolegnia, 10, 18, 20, 65–66, 68, 71–72,
    87, 90, 175, 222, 252, 280
    ferax, 18, 20, 252
    terrestris, 90
Saprolegniaceae
    cytoplasmic microtubules, 18
    DNA analysis in, 65
    evolved into Peronosporales and
        Leptomitales, 73
    meiosis, 57

parasitised by other zoosporic fungi,
    251–253
relationship   to   Plasmodiophoro-
    mycetes, 164
    Vaucheriaceae, 46
sexual reproductive morphology, 57
Saprolegniaceae,   relationship   to
    Vaucheriaceae, 46
Saprolegniales, 44, 46
    dehydrogenases, 69
    hyphae, significance of, 72–73
    mastigonemes, 48
    nuclear division, 89
    parasitic on other fungi, 250
    requirement for sulphur, 60
Saprophytism
    action of thiram on, 216
    and ancestry of Chytridiomycetes and
        Plasmodiophoromycetes, 168, 171
    loss of in Oomycetes, 73
Satellite virus, 236, 239–240
Schizochytrium, 72
Sclerophthora, 9, 27–28, 40, 74, 116, 200,
    301
    macrospora, 28, 116, 301
Sclerospora, 27, 30, 32, 40, 61, 74, 87–88,
    108, 206, 289, 296, 298, 301
    graminicola, 30, 32, 108, 301
Seed and seedling diseases, 135–136, 140,
    149, 153, 202–203, 205–206, 212,
    220
Selective media
    to isolate Pythiaceae, 132–133, 145
    to culture zoosporic organisms, 41–42
Selfing, in Peronosporales, 101–103
Serology, 59, for establishing species
    limitations, 67, 128
Sexual reproduction
    Chytridiomycetes, 58
    Hyphochytriomycetes, lack of con-
        clusive evidence for, 57–58
    Oomycetes, 56–58
    Plasmodiophoromycetes, 59
    zoosporic fungi, 56–59
Shikimate *see* Tryptophan
Sirenin *see* Hormones
Siphonales, 46
Sleeping sickness, 81
Snow rot, 141
Sodium, salts as fungicides, 195–196
Sodium dimethyl dithiocarbamate, 214
Sodium ethyl phosphonate, 206, 222

Soil amendments
  to control Plasmodiophoromycetes, 212
  Pythiaceae, 208
Soil stabilisers, to control Pythiaceae, 207
Soil sterilisation, 203, 206, 210–212, 214 *also see* individual chemical names
Sorghum downy mildew, 200–202
Sorodiscus, 165, 280, 301
  cokeri, 280
Sorosphaera, 27, 49, 51, 54, 59, 166, 175, 177, 186, 189, 266–268, 276, 279, 282, 297, 301
  radicalis, 301
  veronicae, 49, 51, 54, 59, 175, 186, 266–268, 276, 279, 282, 301
Spermatozoa, of primitive plants, 50
Spizellomycetaceae, arrangement of ribosomes, 13
Spizellomycetales
  chitin degradation, 61
  mitosis, 55
  nitrogen requirement, 61
  ploidy, 58
  rootlets, 51
  zoospore types, 22–23
Spongospora, 27, 59, 80, 166, 187, 209, 211, 234–236, 242–243, 246–247, 257, 282, 301
  subterranea, 59, 80, 234–236, 242–243, 246–247, 282, 301
    var. *nasturtii*, 211, 243, 257
    var. *subterranea*, 171, 187, 211, 257
Sporangiophore, in Oomycetes, 73
Sporangium *also see* Zoosporangium
  development, 259, 266
  endogenous and exogenous, 52–53
Stachel, 39, 79, 168, 178
Steam, for soil sterilisation, 203
Sterols
  loss of ability to synthesise, in Oomycetes, 73, 194
  presence of in different taxonomic groups, 64
  synthesis and fungicide activity, 218, 229
Stomata, zoospores attracted to 35
Strawberry leathery rot, 206
Strawberry red stele (red core), 206
Strawberry wilt, 206

Streptomyces, 158–159, 218
  griseus, 218
Streptomycin, 199–200, 202, 218–220, 227 *also see* Antibiotics
Sugar cane downy mildew, 196
Sulphate, reduction of, 60
Sulphur
  as fungicide, 210–211, 246
  requirement for in different groups, 60
  in Oomycetes, 73
  from sulphate, 60
Sulphuric acid
  to control Pythiaceae, 155–156
Sunflower downy mildew, 202
Suppressive soils, 244
Surfactants, effect on zoospores, 213, 247
Synaptonemal complexes
  in Lagenidiales, 57
  in Plasmodiophoromycetes, 186
Synchytriaceae, colonial polycentric development in, 53
Synchytrium, 3, 9, 13, 17, 26, 28–30, 32, 34, 37, 39–40, 58, 77–78, 80, 212–213, 234, 288–289, 302
  anemones, 28, 40
  borreriae, 80
  endobioticum, 9, 13, 17, 26, 29–30, 32, 37, 40, 78, 213, 234, 288–289, 302
  fulgens, 34
  macrosporum, 13, 26, 78
  psophocarpi, 302
Systemic fungicides *see* Fungicides and individual fungicide names

T

Taxonomy *also see* Classification
  historical development of, 45–47
  in relation to ultrastructure, 26
Tea red rust, 80
Temperature
  influence on fungicides, 222
  response to, 62, 143, 245–246
Tetracycline, 227 *also see* Antibiotics
Tetramyxa, 302
Thiabendazole, 210–211, 214
Thiamine *see* Vitamins
Thiophanate-methyl, 210, 214, 218
Thiram, 195, 203, 216–217

Thraustochytriaceae, relationship to Oomycetes, 72
Thraustochytriales, rootlet system, 51
Thraustochytrium, 72, 80
Thraustotheca, 61
Tip rot (sugar beet), 208
Tobacco blue mould, 199–200
Tobacco mosaic virus, 234, 238
Tobacco necrosis virus, 236, 239–240, 242, 244
Tobacco stunt virus, 238–239, 247
Tomato blight, 204
Toxin, produced by Pythiaceae, 142
Trachysphaera, 302
  fructigena, 302
Transition zone (flagellar), 47, 50
  in algae and fungi, similarities, 50
  in Chytridiomycete zoospores, 77
  in Hyphochytriomycete zoospores, 75
  in Oomycete zoospores, 71
Triadimefon, 213
2,4,5-Trichlorophenol, 211
Trichocysts, 79, 168
Trichoderma, 102, 109, 157, 159
  viride, 157
Trichomycetes, relationships of, based on rRNA, 166–167
Trifluralin *see* Dinitroaniline herbicides
Triforine, 214, 229
Triphenyltin fungicides, 201
Trypanosomatidae, 81
Tryptophan, biosynthesis of, evolutionary significance, 68
Tubulin, affinity of fungicides for, 218

U

Ubiquinone, 220
Ultrastructure *also see* Zoospore and individual cell components
  studies and use of, 25, 27
  in studies of ploidy, 90–91
Ultraviolet radiation, induction of resistance by, 228
Urea, 212
Uric acid, 212
Uridine, 212
Urophlyctis, 299, 302
Ustilago maydis, 221

V

Vacuoles, 21–22
  contractile, 22
  in plasmodia of Plasmodiophoromycetes, 174–175, 258, 267
Valine *see* Amino acids
Variation, genetic, in Peronosporales, 91–94
Vaucheria, 46, 78, 257
Vaucheriaceae, relationship to Saprolegniaceae, 46
Vesicle, 21–22, 268, 277, 279
  adhesion, 25
  in exit lobes, 268
  glycoprotein and encystment, 22
  ingestion and cleft, in plasmodia, 174–175
  'peripheral', 22
  pleomorphic, 268
  water expulsion, 22
Vinclozolin, 221
Viruses *also see* individual virus names
  chemical control of diseases caused by, 246–247
  entry into host plant, 240
  filamentous, 234, 236, 240
  isometric, 234, 236, 240
  non-persistent, 234, 238, 240
  persistent, 234, 239
    stability of, 241
  straight tubular, 234, 236, 240
  types, transmitted by zoosporic fungi, 234
  -like particles in Peronosporales, 114
  zoospores as vectors of, 37, 41, 211, 213, 233–248
Vitamins
  in Oomycetes, 73
  synthesis, 60

W

Water
  hydrophilicity of fungicides, 225
  requirement for, 28–30, 244
  soil moisture, affecting disease incidence caused by Pythiaceae, 143–144
  as source of inoculum, 149–150
  survival in, 25

Water moulds
    definition, 44
    evolutionary trends, 149
Wheat mosaic virus, 238, 244, 246
Wheat spindle streak mosaic virus, 211
Wheat streak mosaic virus, 238
White blister (rust) disease, 209, 293
Woronina, 49, 79, 165, 171, 186, 250, 253–258, 266–268, 276, 278–283
    glomerata, 257
    polycystis, 79, 257, 280
    pythii, 49, 250, 253–258, 266–268, 276, 278–283

X

Xanthophyceae, 81

Z

Zinc, salts as fungicides, 195, 203, 210–211, 247
Zineb, 195–197, 199, 207, 209–210, 216
Ziram, 195
Zoosporangium, 3
    definition, 3
    discharge, 276
    meiosis in, 57
    types, 250
Zoospore
    biflagellate and multinucleate, 58
    in classification of aquatic fungi, 164
    definition, 3
    and disease control, 27, 293–294
    dispersal of, 30–31
        in hydroponic culture, 203
    encystment, 14, 33
        glycoprotein vesicles and, 22
        microtubules and, 20
        in Plasmodiophoromycetes, 178–179, 182
        and root infection by Pythiaceae, 139
        time of, 58
    type, in relation to classification, 38–40
    and virus transmission, 240
    essential features, 2, 3
    excystment, microtubules and, 20
    and disease control, 27, 293–294
    incomplete differentiation, 58
    infection by, compared with conidia, 200
    longevity of, 31, 33, 76
    morphology, 5, 9
    movement, 2, 49–50
        amoeboid, 76
        and fungicides, 223, 245
    maximising production of, 32
    numbers of, related to infection of host, 136
    release, 56
    shape, 5, 20
        in Oomycetes and Chrysophyceae, 71
    size, 9
    types, 3
        ultrastructural, 22, 26, 268
    virus transmission by, 37, 41, 80
    water requirement, 28–29, 149, 244
Zoosporic organisms (general)
    asexual reproduction, 55–56
    chemotaxonomy, 59–70
    culture, 41–42, 145, 171, 251–256, 287–290
    ecological habitats, 194
    flagellation in relation to classification and phylogeny, 167–168
    inter-relationships (synopsis), 81–82
    life cycle, 52–59
    meiosis, 56–59
    methods for collecting, 251, 287–290
        electron microscopy, 291–292
        isolation, 252–253, 287–290
        light microscopy, 290–291
        obtaining infection, 254–256, 290
        obtaining zoospores, 255–256
        two-membered culture, 252–256, 290
    mitosis, 54–55
    obligate plant parasites compared with facultative and saprophytic, 3–4, 27
    phylogeny, 25–26
        in relation to mitosis, 55
    saprophytic, 4
    sexual reproduction, 56–59
    spread
        (epidemiology), 30–31, 145–148
        within host tissues, 40
    technical difficulties of working with, 3–4, 43, 145, 250–251

Zoosporic organisms—*cont.*
  thallus, evolution of, 52–54
  as virus vectors, 233–248
Zoosporogenesis
  centriole alignment in, 49
  inhibited by herbicides, 209
  microtubules and, 20
  as model systems, 25
  in Oomycetes, loss of, 73
  in Peronosporaceae, conditions for,
    196
  in Plasmodiophoromycetes, 259, 266
  prevented in water-logged soil, 152
  rapidity of, 31
Zygomycetes *also see* Zygomycotina
  insensitivity to fungicides, 222

possibly derived from Conjugatae, 46
relationship to Chytridiales, 47
relationships of, based on rRNA, 166–
  167
Zygomycotina, 44 *also see* Zygomycetes
  dehydrogenases, 68–69
  evolution, based on fatty acid syn-
    thesis, 64
  lysine synthesis, 67
  origin, based on cell wall analysis, 62
  rRNA, 65
  sterols, 64
Zygotes, 3, 93
  definition, 3
  meiosis in, 87–90